THE DINGO DEBATE

For all dingoes
Past, present, and future

THE DINGO DEBATE

Origins, Behaviour and Conservation

Editor

BRADLEY SMITH

With contributions by Rob Appleby, Chris Johnson, Damian Morrant, Peter Savolainen and Lyn Watson

CSIRO PUBLISHING

National Library of Australia Cataloguing-in-Publication entry

The dingo debate : origins, behaviour and conservation /
editor, Bradley Smith.

9781486300297 (paperback)
9781486300303 (epdf)
9781486300310 (epub)

Includes bibliographical references and index.

Dingo – Australia.
Dingo – Origin.
Dingo – Behavior.
Dingo – Effect of human beings on.
Dingo – Folklore.
Aboriginal Australians – Folklore.
Dingoes as pets.

Smith, Bradley Philip, editor.

599.772

Published by
CSIRO Publishing
36 Gardiner Road, Clayton VIC 3168
Private Bag 10, Clayton South VIC 3169
Australia
Telephone: [+613] 9545 8555
Email: csiropublishing@csiro.au
Website: www.publishing.csiro.au

Cover photographs by Bradley Smith

Edited by Adrienne de Kretser, Righting Writing
Cover design by James Kelly
Typeset by Thomson Digital
Printed by Ingram Lightning Source

Feb26_RP_ILS

Contents

List of contributors

Bradley Smith Research Fellow in human and animal psychology, Central Queensland University, Appleton Institute.
Bradley is an animal behaviourist with a background in psychology. He specialises in the cognition and behaviour of canids, as well as the relationship between humans and animals. Bradley has spent the last decade conducting experimental and observational studies of dingoes in both captive and wild settings, and investigating the dingo's controversial past. Bradley's research and photographs have been featured in all forms of media, including scientific publications. He currently serves as director of the Australian Dingo Foundation and scientific advisor for one of Australia's largest dingo sanctuaries.

Peter Savolainen Associate Professor in Evolutionary Genetics, KTH – Royal Institute of Technology, Science for Life Laboratory.
Peter is a Royal Swedish Academy of Sciences Research Fellow, with a specialisation in population genetic studies of the origin and history of the domestic dog. Based on a unique sample collection from more than 3000 dogs worldwide, and collaborations with international research groups in countries such as China and Australia, Peter has been instrumental in uncovering detailed information about the origin of the domestic dog and the Australian dingo.

Rob Appleby PhD candidate (Ecology), Griffith University, Environmental Futures Research Institute.
Rob is an ecologist, animal behaviourist and technology developer with a keen interest in combining these disciplines to further dingo conservation efforts. He has been studying Fraser Island dingoes since 2003, and is a member of the Fraser Island Dingo Working Group. Rob is currently completing a PhD thesis centring on human–dingo interaction dynamics.

Chris Johnson Professor of Wildlife Conservation, University of Tasmania.
Chris is an ecologist with special interests in the conservation of mammals, environmental history, and the complex effects of the extinction of native species and arrival of new species on ecosystem processes. He has studied the ecological effects of dingoes (and other predators) throughout Australia.

Damian Morrant PhD candidate (Zoology and Tropical Ecology), James Cook University, Centre for Tropical Environmental and Sustainability Sciences.
Damian is a ecologist with a particular interest in the movement patterns of terrestrial mammals, and the behavioural responses of predators to olfactory stimuli. Damian's PhD research in the lowland Wet Tropics of Australia employed a range of investigative approaches,

spanning ecology and social science, to investigate the perceived and real threats posed to native fauna by dingoes and other wild dogs.

Lyn Watson Director, Dingo Discovery and Research Centre, Toolern Vale, Victoria.
Lyn is the co-founder and managing director of one of the largest dingo sanctuaries in the world, and has over 25 years experience in breeding and managing captive dingoes. Lyn is passionate about preserving the pure dingo, educating the public about the plight of the dingo, and achieving conservation and protection for the dingo. She is also an experienced and successful breeder of various domestic dog breeds and is a licensed international all-breeds dog judge.

Preface

> 'They're a magnificent-looking animal, they really are … Just as long as they stay north of the dog fence!'
>
> Audrey Sheehan (farmer from South Australia).

The dingo is Australia's largest wild mammalian predator, and it is the king of the Australian bush. Though dingoes are migrants, their arrival was distant enough that they have become quintessentially Australian and, in this sense, unique to this part of the world. Dingoes possess many traits that make them the ultimate survivors; they are extremely adaptable, resilient and intelligent. Together, these traits have allowed them to carve out a living in almost all parts of the country. Their toughness is legendary, and is the basis of many a tale. Yet they also have a softer side that we have slowly come to know and appreciate; they are graceful, playful, loving, loyal and protective – a true testament to the beauty of nature.

A book such as this could not have been written about any other Australian animal. None elicits such strong emotions of fear and hatred, nor can they avoid controversy on virtually every aspect of their existence, be it time and place of origin, identity or belongingness. Sadly, despite its uniqueness, the dingo is not as popular as other Australian native animals and is not recognised, unlike many of our iconic species, on a coin or a coat of arms. Rather, dingoes are often only associated with negative connotations, their very name commonly used as a slur and their presence evoking a most entrenched form of vilification. They remain a source of significant contradiction: they are considered both a native animal and an invasive pest; a tourist drawcard and a dangerous threat; devastating to the agricultural industry yet crucial for a healthy ecosystem; a baby-killer and a beloved companion. They are arguably the most maligned and misunderstood of our native fauna.

Ultimately, like so many of their predator cousins, when the economic interests of humans conflict with the activities of wild animals, the wild animals are invariably the losers. Such is the paradox that surrounds the dingo. We can admire them, so long as they stay out of our way. Lionel Hudson neatly summed up the quandary in his 1974 book *Dingoes Don't Bark*, when he said: 'the dingo has had nothing but a raw deal from its countrymen … worst of all we show no respect for him as a wild animal' (p. 72). The way the dingo is perceived and treated in Australia, as I hope you will discover after reading this book, is completely undeserved.

Konrad Lorenz, one of the fathers of ethology, once argued that 'if you wish to really study an animal, you must first love it'. By this he means that you need to form a connection with, and develop empathy for, a particular animal before you can really begin to learn from it, and comprehend exactly what is going on. This statement is considered somewhat controversial, because scientists are supposed to remain objective observers. But there is much truth in Lorenz's observation, and it is one that has always resonated with me. I find that the more I engage with dingoes (and those associated with them), the more I am able to learn from them. The more I discover, the more intrigued I become and the more I realise how much there is still left to know. The dingo has now become a large part of my career and my life. One of the reasons why I dedicated this book to dingoes is to thank them for all that they have given me. I see the same changes in people who meet a dingo for the first time – they are immediately captivated by its energy and beauty, and their lives and connection with nature change from that moment.

A dingo may look similar to your dog at home, but if you ever get a chance to meet one you'll likely realise that they are worlds apart. When you first encounter a dingo, whether in captivity or in the wild, it stares back at you, always at a distance, and appears to sum you up methodically. It positions itself downwind, so as to take in your scent. It rapidly expels air from its mouth, a silent snuff bark designed to discreetly warn others of danger. It prefers to remain hidden and isn't quick to fight, unless compelled to. Dingoes are secretive animals whose trust is difficult to earn. Throughout this book, I intend to reveal some of their secrets and offer insight into their fascinating and little-known story.

I must point out that this book is not intended to be a biased or romanticised tale of the dingo. Rather, it provides an objective account, with all opinions and conclusions based on the best available evidence. By providing a scientific and scholarly account of the dingo, rather than hearsay, the animal beneath the popular myth and stereotypes is revealed. This book is deliberately focused away from the ecological aspects of the dingo, which have been covered in previous publications. Instead it is focused on how truly unique they are, and how our interactions, behaviours and attitudes towards them have shaped their past, present and future. This book offers a much-needed comprehensive and up-to-date background on the behaviour, origin and history of the dingo, describes the role that humans have had in shaping the dingo, provides a detailed review of human–dingo conflicts and, importantly, discusses dingoes in a global context, with comparisons to other old and new world canids including domestic dogs. A concerted effort is made to show all sides of the debates, as not all scientists seem to agree. To conclude, a discussion about the future of the dingo in Australia is presented, which draws together all the available evidence in support of a new direction for managing Australia's native canid.

Writing this book has been a long but rewarding process. It was made easier by the help of many people, whom I would like to acknowledge. I am particularly indebted to Rob Appleby, Damian Morrant, Chris Johnson, Peter Savolainen and Lyn Watson for contributing their expertise and knowledge to various chapters. I believe this has greatly improved the book's overall quality and impact. I would also like to thank the many individuals who generously allowed the use of their photographs, Eloise Deaux for providing assistance and spectrographs of dingo vocalisations, James Serpell for statistics relating to pet dingo behaviour, as well as Danielle Every, Brad Simpson, Damian Morrant, Rob Appleby and Stacey Smith for reading and commenting on various chapter drafts. I thank Briana Melideo and the team at CSIRO Publishing for their helpful advice and support, and for making the writing of this book such a smooth and enjoyable experience.

Working with dingoes has brought me in touch with many wonderful people with whom I have developed life-long friendships. I owe a great deal to Carla Litchfield from the

University of South Australia for introducing me to the field of animal behaviour, inspiring me with her passion and love of animals and providing the foundation of my scientific career. Rob Appleby has given me many experiences with wild dingoes on Fraser Island, and I thank him for many years of friendship, support and great philosophical discussion. Thanks also to Lyn and Peter Watson from the Dingo Discovery and Research Centre, Toolern Vale, Victoria. My research could never have been possible without your love, enthusiasm, knowledge and support.

Last but not least, I want to sincerely thank my work colleagues at the Appleton Institute (Central Queensland University), my friends and family, Zara and Shadow and, most of all, Stacey, for their ongoing love, support and encouragement.

Dr Bradley Smith

1

Characteristics of the Australian dingo (*Canis dingo* Meyer, 1793)

Bradley Smith

Dingoes are not dogs. Although they may look the same, they are distinct in many ways. Some of the differences aren't immediately obvious or well-known, but they quickly reveal themselves when you look a little closer. The dingo is actually classified as a medium-sized free-ranging canid indigenous to Australia. Even though it is a member of the genus *Canis*, one of the most widespread mammal genera, the dingo is unique for several reasons. It is the only representative of the genus in Australia, and was one of only a handful of terrestrial eutherian mammal species found in Australia at European arrival. From its physical characteristics, senses, personality, intelligence, social life, communication and hunting methods, the dingo has developed numerous specialisations that have helped it become a highly successful predator in a landscape that is both harsh and ever-changing. It is these characteristics that will be examined in more detail in this chapter, and in various sections throughout the book.

Because most people know a thing or two about dogs, I have found that one of the best ways to explain what a dingo is, is by describing how they differ. It is important that we distinguish dingoes and dogs for a variety of reasons. These will become apparent as the book progresses, but perhaps the most important reason is that dingoes are a unique and significant Australian species both historically and ecologically, and their conservation will not be taken seriously as long as people think of them as *just dogs*. A secondary aim of this chapter, therefore, will be to highlight some of the differences between dogs and dingoes.

Much of the information on dingo behaviour presented in Chapters 1 and 2 has been pieced together from various ecological studies and scientific observations of dingoes conducted in the wild and supplemented by captive studies. Due to the dingo's inherent wildness and shy nature, combined with the remoteness of the landscapes in which it often lives, much of the dingo's life takes place in private. But myth and legend regarding dingo behaviour are slowly being replaced by truths based on scientific research. There are still many gaps in our knowledge because ecological studies have focused on specific regions or groups, the scientific methodology in some cases is outdated or obsolete, and anecdotal accounts of behaviour have yet to be scientifically verified. Added to that, much of the early research on dingoes was potentially biased, as it was focused more towards their extermination (working under the philosophy of 'know thine enemy' according to the main CSIRO researchers focusing on the dingo),[1] than on gaining a comprehensive understanding of the species. Fortunately, over the last few decades our knowledge of the dingo has greatly improved as modern surveying methods and technology become available, as the research questions diversify and as interest in the dingo and its conservation grows.

Classification: *Canis lupus dingo*, *Canis familiaris dingo* or *Canis dingo*?

The order of mammals known as the Carnivorans (meat eaters) includes families of bears, cats, weasels, raccoons, civets, hyenas and the dogs and foxes (or Canidae). The Canidae are grouped together because they have a similar basic form that is adapted to chase prey, eat meat, have long pointed canine teeth (like fangs), sharp claws, a simple digestive system and a highly developed brain. All members have 42 teeth (with the exception of the dhole *Cuon alpinus*), all walk on their toes, possess long tails and have four toes on each hind foot. The dingo belongs as a member of the family Canidae and genus *Canis*. The genus *Canis* comprises seven species, including the side-striped jackal, golden jackal, coyote, black-backed jackal, Ethiopian wolf, red wolf and grey wolf (*Canis lupus*), in which the domestic dog is generally recognised as a subspecies[2] (Table 1.1).

Much debate surrounds the accurate scientific classification of dingoes and, like many of the other canids, its scientific name has changed over time.[3] Some proposed names have included, for example: *Canis antarcticus*, Kerr, 1792; *Canis dingo*, Meyer, 1793; *Canis familiaris australasiae*, Desmarest, 1820; *Canis australiae*, Gray, 1826; and *Canis familiaris novaehollandiae*, Voigt, 1831.[4] The official scientific name of the dingo, however, as listed in the Official List of Specific Names in Zoology, is *Canis dingo* after German naturalist Friedrich Meyer, 1793.[5] This name was officially ratified by the International Commission of Zoological Nomenclature (ICZN) in 1957.[6] Part of the ratification process involved suppressing the name *Canis antarcticus* (Kerr, 1792). Even though Kerr's reference came earlier than Meyer's, the Committee believed that *antarcticus* had never been in common usage, and to change the name of the dingo some 150 years later would cause unnecessary confusion.

Whether the dingo should be *Canis dingo* or *Canis lupus dingo* has been hotly debated, and even though the official taxonomic name is *Canis dingo* there are still inconsistencies in the way the dingo is referred to, particularly within the scientific community.[7] The use of

Table 1.1. Scientific hierarchy of the dingo according to the Integrated Taxonomic Information System (Taxonomic Serial No. 726820)

Taxonomic classification		
Kingdom	Animalia	
Phylum	Chordata	
Subphylum	Vertebrata	
Class	Mammalia	Linnaeus, 1758
Subclass	Theria	Parker & Haswell, 1897
Infraclass	Eutheria	Gill, 1872
Order	Carnivora	Bodwich, 1821
Suborder	Caniformia	Kretzoi, 1938
Family	Canidae	Fischer, 1817
Genus	Canis	Linnaeus, 1758
Species	*Canis dingo*	Meyer, 1793
Common names	Dingo	
Other names	Warrigal; Australian native dog	

Canis lupus dingo follows the arguments led by Corbett[2, 8] and the convention that wolves are thought to be the ancestor of dogs and dingoes. However, the relationship between wolves and dogs may not be that simple. We now know, for example, that the history and relationships between wolves, dingoes and dogs are complex and not yet definitive.[9, 10] In 1982, *Canis lupus dingo* was recommended to the ICZN,[11, 12] but the application was rejected.

The earliest scientific descriptions of the dingo provided by Kerr in 1792, and by Meyer a year later, were based upon a painting and a brief description in the journal of Australia's first colonial Governor, Arthur Phillip.[13] This means that there is no original or reference dingo specimen. This has led to some confusion about what a pure dingo is, and how to distinguish dingoes from dog or hybrids. Most morphological and genetic descriptions of dingoes use modern specimens, which is not ideal given that cross-breeding with dogs (hybridisation) might have influenced the dingo genome. A research team led by Mathew Crowther recently sampled pre-20th-century dingo specimens (69 skulls), so there is now a morphological description and benchmark of what can be considered a dingo.[14]

In 2003, the ICZN made an effort to rectify the confusion in the naming of domestic derivatives that have separate names from their wild ancestors. Misunderstandings particularly occur where the domestic name predates or is contemporary with that of the wild ancestor.[15] Such a problem particularly applies to *Canis lupus* (as the wild progenitor) and domestic dogs, and dingoes (the domestic derivatives). The Commission ruled in favour of keeping the names separate, and that the first available specific name that was based on a wild population should be used. Gentry, Clutton-Brock and Groves,[16] the team that first proposed the change in 1995, argued that in practice, 'since wild species and their derivatives are recognisable entities, it is desirable to separate them nomenclaturally when distinct names exist' (p. 649). Thus, given that the dingo was first described in 1793 as a distinctive wild canid (subsequently upheld by the ICZN), that the dingoes are reproductively isolated in undisturbed natural environments, that dingoes show many behavioural, morphological and molecular characteristics that differ from those of dogs and wolves (discussed in later chapters) and that the latest convention is to keep the names for wild and domestic animals separate, it appears clear that the dingo continue to be referred to as *Canis dingo*.

First documented observations of the dingo

The first report of a dingo in Australia was made by Dutch navigator Jan Cartensz, who in 1623 observed dingo tracks while in the Gulf of Carpentaria on an expedition to find people with whom to trade spices. In 1688, William Dampier observed dingoes on the north-west coast, and described those he saw as being 'little hungry wolves – lean like so many skeletons'. The earliest reference or record of the term dingo came from the observations of early explorers of the Indigenous people and the dogs that lived alongside them. For example, in 1770 Captain James Cook saw tame dogs with the Indigenous people when exploring along the Queensland coast,[4] and British explorer Watkin Tench, who was part of the first fleet of ships sent to settle Australia in 1789, wrote that 'the only domestic animal they have is the dog, which in their language is called dingo, and a good deal resembles the fox of England'. According to Iredale,[4] this was the first printed use of the word dingo. The first graphical

Figure 1.1: John Gould image of the dingo. This is one of the better early depictions, as dingoes in other illustrations were very fox-like.

Source: John Gould, *Mammals of Australia*, Vol. III, Plate 51.

representation was the aforementioned publication concerning the events of Captain Arthur Phillip's voyage to Botany Bay in 1879 and settlement of Australia. He entitled the drawing *Dog of N.S.W*; it depicted a female dingo that Phillip ended up sending to a friend in England. Most of the early drawings or paintings of the dingo are based upon animal skins or taxidermy dingoes, and are not entirely good representations. Perhaps the most accurate and attractive depiction was painted by John Gould. It appeared in one of his three volumes of works titled *Mammals of Australia* published between 1849 and 1861 (Fig. 1.1).

General description

The dingo is a medium-sized canid. As a generalist hunter, it is built for pursuing, attacking, killing and eating prey that varies from small (e.g. insects and rodents) to much larger than itself. Although very dog-like in appearance, dingoes show departures from the characteristics of the ordinary dogs towards those of wolves (see later in this chapter for a comparison between dogs, wolves and dingoes). The dingo is extremely elegant in appearance and comes across as alert, intelligent, very hardy and lean. Their short coat comes in three main coat colours including ginger, black and tan, and a creamy white, often with small white markings on the chest, feet/legs and tail tip. Their body is designed for speed, agility and stamina (see Plate 1). For example, the chest is narrow and the fore limbs are pressed into the chest, with elbows turned inward and paws turned outward to allow both fore and hind legs on the same side to swing in the same line. Dingoes possess strong jaws and a flexible neck that is suited for both small and large prey. The shape and position of the eyes and ears allow for an excellent awareness of their surroundings. Dingoes are extremely social animals that live, hunt and raise young in family groups and often mate for life. They possess strict social hierarchies and a complex system of communication, particularly howling for long-range communication. They are driven by instincts that ensure their survival, including hunting, procreation and the protection of territory and family. They are generally cautious, independent and rarely obedient or affectionate towards people. In sum, the dingo is an example of a dog more removed from all the influences of domestication than any other, and is by far the most ancient of any of the living species of dogs. See Figs 1.2 and 1.3 for examples of the typical dingo phenotype.

Physical description

This section lists features that are considered typical (hallmarks) of the pure dingo (assisted by the use of terminology that has been developed to describe the anatomy of horses and domestic dogs, as well as dingoes).[17] Figure 1.4 will assist with identifying the location of the relevant anatomy.

- **Build.** The dingo's body is longer than it is high, and appears long-legged. Dingoes are of light build and do not carry excess layers of fat or flesh (raw-boned). When looking at the dingo from above, the head is the widest part of the body, and the shoulders tight knit to the ribcage. A slight waist appears at the loin area. The dingo's appearance (type), its proportion of body parts and the way it moves and stands (balance) is more important than its size, as this can vary. The dingo looks ever-alert and responsive to its surrounds.
- **Gait and movement.** The dingo appears light on its feet and moves in an extremely efficient way without any unessential movement (flexion) of muscles or joints. They are capable of suspended gallop, canter, brisk trot and a loping walk. When the dingo is moving, the fore and hind legs on the same side swing in the same line, assisted by a narrow chest and fore limbs pressed onto its chest. When walking, their hind foot steps in line with their front foot. Dingoes mostly travel by walking and trotting, and usually in fairly straight lines (see Fig. 1.6).
- **Head and skull.** The skull is broad, and longer than it is wide. The head is wedge-shaped and appears large in proportion to its body. The skull is high behind the ears (i.e. the ears sit forward of the skull) with a well-developed occiput and very broad cheekbones (zygoma). It is flat between the ears. When viewing the dingo in profile, it is slightly rounded from the top part of the skull (crown) to the stop (the indentation between the eyes where the nasal bones and skull meet), with a distinct median furrow (a slight indentation of median line down the centre of the skull to the stop). The skull narrows in front of the eyes to the muzzle, which tapers only slightly to the well-opened nostrils. The muzzle has a well-developed underjaw. In the mouth, the incisors are even, well-developed and arranged in a scissor bite. The lips are tight-fitting and cover all teeth. The neck is thick and well-developed, and relatively long.

Figure 1.2: (a) Side and (b) front profile of a Fraser Island dingo. The dingo's ribcage extends almost along its entire body to protect all the organs. The dingo's chest is narrow so the legs appear to be close together, causing the large feet to splay to the side when standing still.

Source: Bradley Smith.

Figure 1.3: (a) Profile of the head of a Fraser Island dingo. The ears sit forward of the skull and have a clean outside edge and do not flare towards the base. (b) The eyes are slanted and almond-shaped with heavy dark eye lining. A line drawn through the corners of the eye should dissect the base of the outer ear, where head proportions and expression are correct.

Source: Bradley Smith.

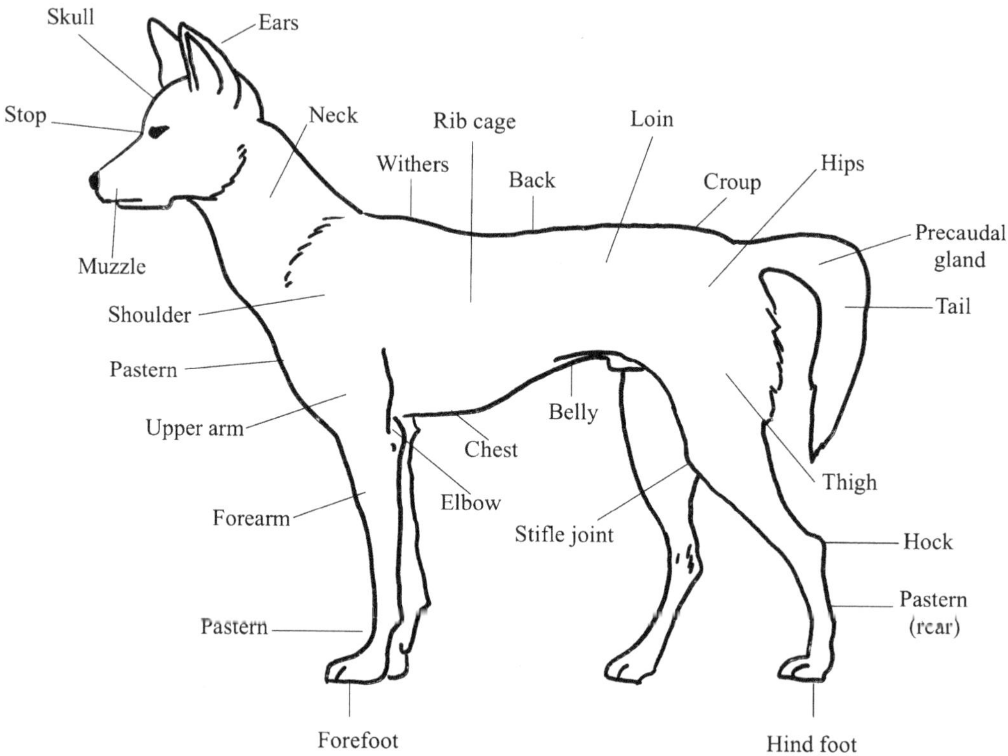

Figure 1.4: The parts of the anatomy used to describe dogs and dingoes. Outline adapted from the assessment sheets developed by Lyn Watson and Barry Oakman for the Australian Dingo Club Inc. (now defunct) to assist determination of dingo purity for registration as a dingo with the Australian National Kennel Council.

- **Eyes.** The eyes are triangular or almond in shape and obliquely placed. They are medium-sized and usually hazel to dark, and have dark rims. In pure dingoes, a line drawn through the corners of the eye should dissect the base of the outer ear, where head proportions and expression are correct.
- **Ears.** The ears are upright, hooded and placed high on the skull. They are small to medium in size, never large or thick, and slightly rounded at the tip. They are positioned forward of the occiput (the bony bump at the back of the top of the skull). The ear lobe is small, giving the ear an exceptionally neat and vertical clean outer edge.
- **Forequarters.** The chest is narrow in width, and the depth of the chest does not extend below the elbow. The sternum (brisket) is long and carried well back beyond the elbows. The shoulder is high and flat at the highest point of the dingo's shoulder (wither). The humerus is long, and the elbow to pastern very long. The front feet are medium-sized, oval shaped, thickly padded and slightly turned out. Nails are strong and short.
- **Body.** The back section of the dingo is strong and straight. The ribcage is long and extends to the rear. The back (wither to loin) is straight. The loin (the part of the

body that runs on either side of the spine from the ribs to the hipbone) is arched and long, the croup (region around the pelvis) is long. The loin shows a rise and, along with the structure of the muscle, indicates speed and agility.

- **Hindquarters.** The croup is broad and long, and angled to suit diverse terrain (e.g. flatter for open country, and steeper for mountainous country). There is great length from hip to hock (the dingo's true heel – the joining on the hind leg between the second thigh and the metatarsus), the stifle (knee) angle is moderate, the hock angle is moderate and the rear pasterns are parallel. There is enough length of rear pastern to act both as a spring for jumping and as an efficient lever for speed and endurance. The entire hindquarter is sound and powerfully muscled. The feet are oval shaped and of medium size.
- **Tail.** The tail is set well behind the hipbones, and is of medium length. The tail is flattish, broadening from one third behind the base to mid length then tapering to the end (i.e. shaped like a bottle). The tail is typically carried low, and does not curl over the back. A scent gland is positioned on the tail, identified by a dark spot.
- **Coat.** Dingoes come in one of the three colours (ginger, black and tan, or white/cream), and there is no sable, ticking or brindle. Often, white tips are present at the extremities. The coat has a dry/hard outer, and an undercoat should be present. There is no doggy odour or large patches of white, and the fur should not be oily. Dingo coats will be discussed in more detail in relation to genetics later in this chapter.

Body measurements

The average adult male dingo weighs ~15 kg (ranging to 24 kg), stands 55 cm at the shoulder and measures 123 cm from nose to tail tip. Males are generally larger than females (Table 1.2). Dingoes rarely carry excessive amounts of fat, and seeing them with exposed ribs is a common sight in the wild. A few individuals may weigh more, but these are averages and depend greatly on the environment and amount of prey available. Dingoes from northern and north-western Australia are generally larger than those found in central and southern regions. Variations in weight may be dependent on the environment, availability of prey and the ease with which it can be caught.[2, 18] Captive dingoes are often larger and heavier, due to constant access to food, and medical care.

Coat colours and genetics

The dingo's beginnings and relationship to Asiatic wolves, combined with a limited number of founding animals (see Chapter 3) have resulted in a clearly defined and unchanging colour palette. Pure dingoes have only three basic coat colours, and the majority are light ginger or sandy colour. Most dingoes also have white points – white toes, feet or socks, a white tail tip and a white chest patch,[7, 19, 20] although the amount of white present varies considerably between individuals.[21] The absence of white tips does not reflect impurity. The dingo's coat varies considerably from a wiry single coat in the tropical far north, to a thick double coat in the colder southern mountain regions (see Plate 2). The coat is seasonal and in general has no body (or 'doggy') odour.[22] Although there are nine different genes that control colourations in

Table 1.2. Average body measurements for 97 wild dingoes (unknown ages) Australia-wide and 22 captive adult dingoes (average age 6 years)

Dingo body measurements	Wild		Captive	
	Male	Female	Male	Female
Body weight	15.8 ± 4.1	14.1 ± 3.9	18.91 ± 2.1	16.18 ± 1.7
Shoulder height*	586.9 ± 22.5	562.3 ± 17.3	564.55 ± 21.1	533.64 ± 24.6
Total length	1246.86 ± 136	1219 ± 39.1	1359.0 ± 36.3	1280.91 ± 57.2
Chest/girth	–	–	624.55 ± 24.2	594.55 ± 13.7
Head length	222.4 ± 5.8	214.5 ± 6.5	–	–
Skull circumference	–	–	420.0 ± 15.5	389.09 ± 13.8
Ear length	103.9 ± 3.2	98.5 ± 3.5	–	–
Tail length	315.4 ± 32.5	304.6 ± 24.8	–	–
No. of dingoes	49	48	11	11

Values are means and standard deviations expressed in mm, except weight, which is kg.

* Measurements reported by Corbett were taken from deceased or anesthetised animals, which may be different from those taken from living animals (as in this sample). This is particularly the case with shoulder height, where Corbett's measurements are likely to be inflated. When a living dingo is standing, the legs are not completely straight or extended as they would be when it is lying down and the limbs are relaxed and stretched.

Sources: Wild dingo information adapted from Corbett.[2] Captive dingo information based on animals housed at the Dingo Discovery Centre (courtesy of Lyn Watson).

the dog, there are only three genes likely to affect coat colour in the dingo. These genes interact to control the pattern and type of pigmentation present in the coat. Table 1.3 provides information relating to the known and likely genotypes of possible dingo coat colour variations.[23] In some ways, determining the genotypes for dingo coat colour is difficult, as there is such variation in the shading and patterns of the colours due to several modifier genes.

According to Cairns *et al.*[23] there are five coat colour patterns (phenotypes) found in dingoes. A description of these colours and their occurrence across Australia (based on 3129 dingoes collected across Australia)[2] includes the following.

- **Ginger** (74%). A dingo with a red coat varying in shade from deep rust to pale cream. Ginger dingoes are more likely to have dark muzzles[18] with the dorsal surface usually darker than the ventral surface[2, 21] – this is not to be mistaken for sable. There is often white colouration on muzzle, neck, chest, feet and tail tip. In some areas (e.g. Fraser Island), dingo pups are born with black guard hairs but this fades and disappears as they age, occasionally leaving a few black hairs on the scent gland at the top of the tail. Ginger or red colouring may provide camouflage in central Australia where the sand is a reddish colour (see Plates 3 and 4).
- **Black and tan** (12%). A black dingo with tan points, brow pips, muzzle, chest, belly, feet and legs. Tan may vary from deep tan to cream. Many people do not realise that dingoes come in black and tan. One of the reasons for this is that they look remarkably like the Australian kelpie. So much so, that I know of multiple individuals who own dingoes (and perhaps are not supposed to) and have registered them as kelpies with local councils and veterinarians. This colouration provides useful camouflage in heavily wooded and forest surroundings (see Plate 5).

Table 1.3. Genetic determinants of dingo coat colour

Coat phenotype	**Likely genotypes and variations**			**Allele descriptions**
	Agouti*	**MC1R (*E*)#**	**CBD103 (*K*)^**	
Ginger	a^ya^y, a^ya^t	*EE*, *Ee* or *ee*	k^yk^y	• Homozygote a^ya^y and heterozygote a^ya^t (black and tan carrier) causes the common ginger dingo colouration pattern • *EE*, *Ee* or *ee* genotype would produce coat colouration patterns as determined by the more dominant Agouti gene
Black and tan	a^ta^t	*EE* or *Ee*	k^yk^y	• Homozygote a^ta^t genotype only • a^y is the dominant allele over a^t; heterozygote a^ya^t causes a ginger colouration in dingoes • These animals are carriers for both a^y and a^t, thus have the potential to produce ginger or black and tan progeny • *EE* or *Ee* genotype would produce coat colouration patterns as determined by the more dominant Agouti gene
White	*aa*	*ee*	k^yk^y	• A homozygous *ee* genotype causes the dingo to be white regardless of the Agouti genotype
	a^ya^y	*ee*	k^yk^y	
	a^ya^y or a^ya^t	*ee*	K^BK^B	
Solid black	*aa*	*EE* or *Ee*	k^yk^y	• A homozygous *aa* genotype causes solid black coat colouration • *E* is required to allow dingoes to produce black pigment
	a^ya^y or a^ya^t	*EE* or *Ee*	K^BK^B, K^Bk^{br} or K^Bk^y	• It is the most dominant allele on CBD103 • The K^B allele causes a dog to have a solid black coat in the presence of *E* at MC1R regardless of the Agouti genotype

Note: Homozygote refers to identical pairs of genes for a specific trait. If both parents carry the same form of the gene for a specific trait, the organism is said to be homozygous for that trait. In heterozygous organisms, the genes for a specific trait are different (i.e. the phenotype will reflect the dominant trait).

* Agouti controls the distribution of black or red pigment in a dingo's coat. There are four Agouti alleles, classified as a^y, a^w, a^t and *a* in order of dominance. In dingoes it is likely that the a^y, a^t and *a* (although recessive and extremely rare) alleles exist in current wild and captive populations. a^y is the dominant allele for coat colouration in dingoes, causing red/yellow hair colour. The a^t allele describes the black and tan point colour pattern and the recessive black allele *a* causes solid black coat colour. The a^w allele is the wild type. It is often called wolf-sable or agouti, and is the natural colouration of wild grey wolves. There is no defined role for the a^w allele in dingo coat colouration, and it is even plausible that a^w is not present in dingoes.

\# *Melanocortin* 1 receptor (MC1R), more commonly known as the extension gene *E*, controls whether an animal is capable of producing both red and black pigment or red pigment only. There are four known *E* alleles: (in order of decreasing dominance) *E*, E^M and *e*, while E^G is dominant to *E* and *e* but recessive to E^M. Dingoes code for *E* and *e* alleles only. *e* is the wild type allele, and dogs with this allele (as either homozygote *EE* or heterozygote *Ee*) can produce both red and black pigment. The recessive allele of the extension locus is *e*, which prevents dogs from producing black pigment, thus a homozygote *ee* dog has a red or white coat (depending on the Agouti genotype).

^ *Beta-defensin* 103 (CBD103) gene causes solid black colouration in dogs. It is also known as the *K* locus and has three known alleles, K^B, k^{br} (brindle) and k^y, in order of decreasing dominance. Dingoes are most likely fixed at the *K* locus for k^yk^y, but there is speculation that K^B is ancestral in dingoes and causes solid black coat colouration.

Source: Cairns *et al.* (2011).[23]

- **White** (2%). A dingo with a solid white uniform coat. May have lighter nose, paw pad and eye rim pigmentation. It can still exhibit true white spotting on the extremities of toes and tail tip. Quite often the tips of the ears and top of muzzle can be an apricot colour. The right combination of genes (two ee's – refer to Table 1.3) inhibits all black pigment and can act on nose leather and eye rims, making them liver coloured. This is not to be confused with an albino (complete absence of colour pigments or melanin). This colouration may be helpful in areas where there are white sands and dry salt lakes (see Plate 6).

- **Solid black** (1%). A dingo with a solid black coat. May have white markings on muzzle, neck, chest, feet and tail tip.
- **Hybrid coats – patchy, sable, brindle** (12%). The remaining 12% of dingoes Corbett found were patchy (i.e. either ginger or black with large and/or small white patches on dorsal, ventral and leg areas), sable (i.e. black or dark dorsum extending to about halfway down the sides of the body and legs) and brindle (i.e. dark narrow vertical stripes, often broken on the dorsum and sides of the body and legs). Generally, coat colours other than the three primary colours indicate hybrids and/or domestic dogs.[11, 19] Cross-breeding experiments show that first-generation dingo–dog hybrids look very similar to dingoes in terms of shape and colour. If a dingo is mated with dingo-like dogs such as kelpies, blue heelers and collies, first-generation hybrids are extremely difficult to distinguish from pure dingoes; however, some may have ticking or spotting (particularly in the dingo/heeler crosses). Second-generation hybrids and back-crosses to the domestic parent show obvious signs that it is not a dingo (a mongrel). When cross-bred with other dogs such as labradors and dobermanns, the resulting animals are easily recognisable as hybrids/mongrels based on the shape of the head, ears and tail.[24]

The outer coat and the soft crimped undercoat, if present, each have their own genetic colour-regulating mechanisms. Most ginger dingoes south of the Tropic of Capricorn (the lower half of Australia) retain a fine crimped wolf grey undercoat in varying degrees. The colder the winter climate, the denser the undercoat will be. For example, in extreme tropical conditions the undercoat is practically absent, but when moved to colder climates dingoes begin to develop an undercoat. Some ginger dingoes have an agouti undercoat, while others have a white fleece colour. The undercoat on some black and tan dingoes is sometimes a light sandy colour. This can be easily detected by spreading the hair against its natural growth grain.

Ginger is a dominant colour, and carries the genetic material for all three of the primary colours. Black and tan, and white both breed true, and are both recessive. That is, two black and tan dingoes will produce only black and tan offspring, and two white/apricot parents will produce only white coloured puppies. When white meets black and tan, a sandy gold is the likely result. Figure 1.5 provides the coat colour inheritance (genotype) of the dingo according to parents of different coat colours.

The role of the dingo in developing Australian dog breeds

Although the dingo was considered a threat to the agricultural industry, the early European settlers could see the inherent genetic value of dingoes as a potential contributor to domestic

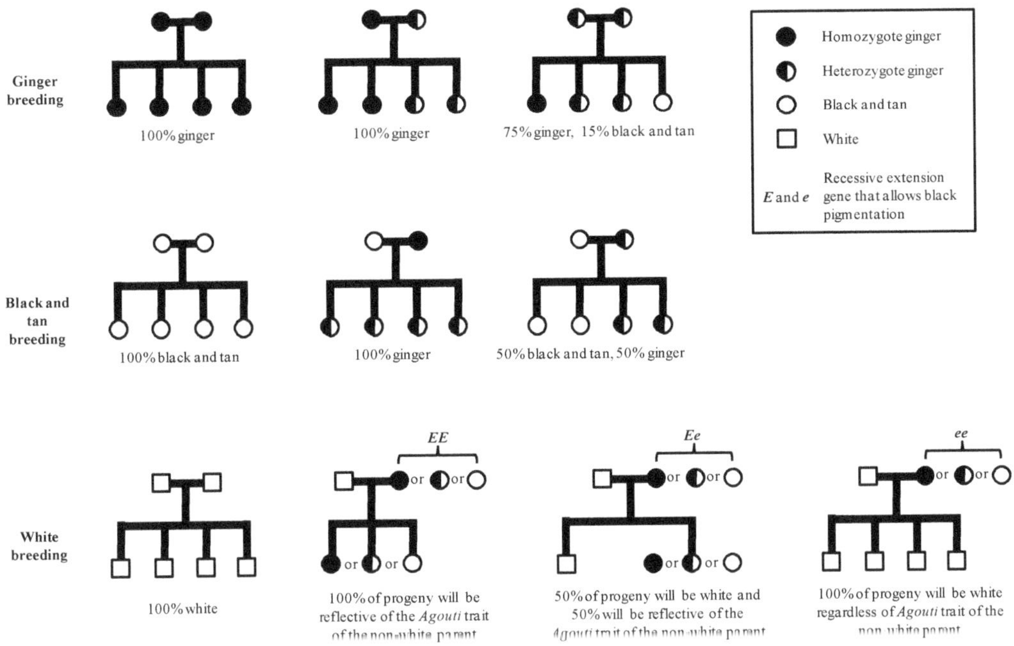

Figure 1.5: The inheritance of coat colour in the dingo according to different genotypic combinations of breeding pairs. The top half of each diagram represents the phenotype and genotype of the breeding pair, and the bottom half represents the first-generation progeny. Note there will be variation in coat colour depending on the influence of modifier genes.

dog breeds. Thus, dingoes have been utilised in the creation of some Australian breeds of domestic dog. Early dog-breeders in Australia experimented by crossing the dingo with European breeds (e.g. collies from northern Britain and Scotland). For example, the Australian cattle dog was created by interbreeding the dingo with working cattle dogs such as smooth-haired blue merle collies, the dalmatian and the bull terrier as early as the 1840s.[25] These dogs were admired for their courage and stamina when herding over long distances in the harsh Australian conditions, and for their well-muscled bodies and tough feet and legs.[25, 26] There is a possibility that the dingo was involved in the development of the Australian kelpie, a sheep dog.[25] It is believed that breeders of such breeds occasionally back-cross with pure dingoes to maintain the dingo's characteristics; however, this was done secretly and it is difficult to determine how much it still occurs.

Differences between dingoes and domestic dogs and their usefulness in determining dingo purity

Dingoes show biological, phenotypic, behavioural and ecological characteristics that more closely resemble wild rather than domesticated canids. These will be outlined in the following section and are also summarised in Table 1.4. Many of these differences are subtle, and it remains somewhat challenging to distinguish dogs and dingoes and/or to determine the level of purity in hybrids.

Table 1.4. Summary of the main differences between dingoes, wolves and dogs

	Wolves	Dingoes	Dogs

Classification	*Canis lupus*	*Canis dingo*	*Canis familiaris*
Weight (kg)	M: 43–45, F: 36–38	M: 15–19, F: 14–16. For a size comparison between dingoes and wolves see Fig. 1.7 and Plate 7	Highly variable (1–79)
Lifespan	In wild up to 9 years. In captivity up to 17 years	In the wild up to 10 years, most around 5–7 years. In captivity up to 20 years	Highly variable among breeds, median age ranges from 10–13. Oldest reached 26 years
Length (cm)	104–165	130	Highly variable, ranging from 9.5–250
Shoulder height (cm)	81–86	55	Highly variable, average 15–84, range 6.3–106
Coat	A very dense and fluffy winter coat, with short undercoat and long coarse guard hairs. Coat colour ranges from white through various shades of blond, cream and ochre to grays, browns and blacks	Double-coated (in colder climates). Three main colours (ginger, black and tan, white). White on the chest, feet, tail. Hair covers all body, including belly and inner thighs. Lack body odour	Can have a double or single coat. Colour and pattern highly varied. Includes sable, ticking, brindle and patchy coats. Doggy odour
Skull/head	Wolf skulls differ from domestic dogs: the width, height and length are larger. The sagittal crest (raised area of bone at the back of the skull) is more pronounced. Very large canine teeth, which are curved and thick. Smaller, rounder and thicker ears, with hair within ear	Dingoes have a longer muzzle, larger canine and carnassial teeth, with canine teeth slightly longer and more slender, larger auditory bullae and a flatter cranium with larger nuchal and sagittal crests. Extremely flexible neck – can rotate cranium laterally through 200° and rotate vertically by more than 120°. Cheeks are the widest part of the body	Highly variable. Compared to wolves, dogs have a 20% smaller skull and 30% smaller brains, as well as proportionally smaller teeth. Usually have pointed ears and no hair within ear

Ears	Well-furred erect ears, rounded at tip	As per wolf	Variety of shapes (e.g. erect, floppy) and sizes, lack fur in ears
Eye	Slanted almond-shaped eyes, with dark/black eye lining. Wide orbital angle of 45°	Almond-shaped eyes, obliquely placed and dark eye lining	Orbital angle of 53°
Track size (front paw) (cm)	Front paw length = 9–13 Front paw width = 7.5–12.5 Wolves have large feet with two very large protruding front toes. Hind paw smaller	Adult front paw length = 8 Adult front paw width = 7. Appear standard size regardless of height, development or build	Highly variable, although a large dog is around 7. Often have smaller and rounded feet
Gait	When moving, the fore and hind legs on the same side swing in the same line, assisted by a narrow chest and fore limbs pressed onto chest. When they walk, hind foot steps in line with front foot. Mostly travel by walking and trotting, and usually in fairly straight lines	As per wolf (also see Fig. 1.6)	Most dogs do not place each hind foot in the track made by its front foot, rather, they put each hind foot beside its front foot track
Body	Chests are narrow, with legs appearing to be side by side causing large feet to splay to the side when standing still. Legs are long. Absence of rear dew claws	As per wolf. Has narrow chest and a long sternum that protects all vital organs	Chest is proportionally wider. Rarely does the rib cage cover all internal organs. Many have rear dew claws
Tail	Carry tail low or straight. Don't have defined tail tips. Have a precaudal gland (dark spot positioned approx. 7–10 cm from base of tail)	Thick and brushy, bottle shaped. Tail carriage low. Precaudal gland present. White tail tip common	Tail carriage is often up and curled upwards over their backs. Precaudal gland absent or rare
Social organisation	Live in and maintain a defined territory. Strict hierarchies. Only dominant male and female breed. Long-term pair bond formed	As per wolf	No strict hierarchies formed, even in free-ranging dogs
Vocalisation	Wide range of vocalisations in different contexts. Howling most common	As per wolf, with snuff barking a common vocalisation	Dogs bark in most situations (including fear, aggression and joy). Capable of howling but do so rarely
Seasonality	One breeding season per year. Sexually mature around second year. Males and females fertile only during breeding	One breeding season per year. Sexual maturity after 12–24 months. Females fertile only during season. Males fertile all year but show a cycle	Multiple seasons per year. Male fertile all year. Reach sexual maturity at 6–12 months. Capable of multiple seasons per year
Parental behaviour	Alloparental care of young. Parents regurgitate, and food provision pups	As per wolf	Rarely regurgitate, or show alloparental care or care provided by the father

(*Continued*)

Table 1.4. (*Continued*)

	Wolves	Dingoes	Dogs

Classification	*Canis lupus*	*Canis dingo*	*Canis familiaris*
Hunting ability/ prey size/diet	Medium to large prey – ungulates and fish. Cooperatively hunt to kill prey larger than themselves (e.g. bison). Do not easily digest starch	Small and medium prey. Most small to medium-sized mammals (72%). Do not easily digest starch or high fat content. Cooperatively hunt to kill larger prey (e.g. kangaroos)	Feral dogs are mostly scavengers. Although carnivores they are often omnivores, and can digest food such as grains, fruits, plants and meat. Can digest starch
Intelligence	Highly intelligent and adaptable. Independent problem-solvers. Capable of higher-order behaviours	Wolf-like in their problem-solving ability. Exhibit higher-order behaviours (Chapter 8)	Have lost some independent problem-solving abilities compared to wild canids. Reliant on humans for assistance when faced with problems/ challenges
Domestication, and reliance on humans	Do not make suitable pets. Difficult to train and socialise. Suited to living in the wild	Suited to living in the wild, but can make successful pets under the right conditions. Highly aloof, difficult to train (Chapter 9)	Can be domesticated and great companions. Suited to the human niche. Highly responsive to training. Always reside around humans, even if ownerless (e.g. villages, human waste sites)

Genetic analysis

Dingoes are genetically distinct from dogs as they display different microsatellite markers (mitochondrial DNA),[27] mtDNA halotypes (nuclear DNA)[28] and single nucleotide polymorphisms (SNPs).[29] These differences can be substantial enough to determine whether a dingo is pure or hybrid. In the late 1990s and early 2000s, Alan Wilton from the University of New South Wales developed a dingo purity test.[28, 30] He compared microsatellite variation (short sequences of DNA in captive dingoes that were of the believed dingo phenotype) with previously described microsatellites from domestic dogs, and found distinct differences. The outcome was a pay-for-service dingo purity test that provides a probability (relative likelihood) as to whether the genetic sample was more likely to be from a domestic dog, a hybrid dingo–dog or a pure dingo.

Figure 1.6: Unlike most dogs, dingoes and wolves generally make single tracks when they walk. (a) The hind foot steps in the place of the front foot. (b) Dingoes mostly travel by walking and trotting, usually in fairly straight lines. Dog travel is more variable, using a lot of gallops and bounds and often weaving about. (c) A front view of a dingo, showing the placement of the paws when walking.

Source: Bradley Smith.

Figure 1.7: The differences in size between the dingo and wolf. Also see Plate 7.
Source: John Williams.

The use of genetic testing is ideal for ensuring the purity of captive dingoes involved in captive breeding programs, and for gaining a level of purity within wild canid populations.[31] Such methods would be useful for the selective removal of hybrids from the wild. Unfortunately, because blood or tissue samples are required for laboratory analysis, genetic analysis or examination in the field is not currently practical. Genetic analysis is a very time-consuming practice and would require, for instance, a tested animal to be maintained in captivity or identified and recaptured after analysis is complete and it is classed a hybrid. Technology is advancing rapidly and the development of a genetic test that can be implemented in the field is expected within the next 10 years.[32]

Skull morphology

Several studies have used skull morphology to discriminate dingoes from dogs, primarily led by Macintosh[33] and a team that included CSIRO scientists Newsome and Corbett.[2, 34] This was the only method of assessing purity before the advent of a DNA test. After identifying characteristics that distinguished between dingoes and dogs, the researchers reduced the number of measurements in the final analysis to those with the greatest statistical significance and that were more easy to accurately measure.[32] The result was a formula or composite scores for determining purity. Over time the method and calculations and definition of purity according to skull morphology have changed,[2] rendering some of findings from early studies questionable.

There are many skull characteristics that can be used to distinguish dingoes from dogs and dingo-like canids from Asia.[2, 18, 34, 35] Dingoes have a longer muzzle,[34, 35] larger canine and carnassial teeth, with canine teeth slightly longer and more slender; larger auditory bullae,[2, 22, 34, 35] a flatter cranium with larger nuchal[34, 35] and sagittal crests.[21, 36] It is important to note that none of the dingo skulls examined by Jones[21] displayed any of the more extreme modifications or changes caused by domestication (e.g. more rounded and forward-pointing eyes, shortening of muzzle and subsequent overcrowding of teeth, reduction in size of the sagittal and nuchal crests, tympanic bulla). Actually, the dingo has changed little over the last 3500–5000 years, according to a comparison between the oldest complete skeleton from Fromm's Landing (3500 years old) and modern dingoes.[33]

Like genetic analysis, skull morphology is of limited use in the field for assessing the purity of a dingo because the animal must be dead before any reliable measurements can be taken. CAT scans and X-rays could be taken of live animals, but many of the internal measurements are still not possible. Also, measurements of juvenile dingoes are often not accurate because the animals are not fully developed. Even when the skull can be measured, it is difficult to do so accurately due to differences in equipment and level of expertise.

Annual breeding season

Pure dingoes have a single breeding cycle per year, whereas domestic dogs and hybrid dog–dingoes may have two or more;[11, 37] see Chapter 2). It is believed that first-generation hybrid bitches lose this annual breeding cycle. Thus, the absence of breeding season can be used as an indicator of purity. For example, if a bitch is pregnant outside the typical breeding season, it is likely she is not pure. Purcell argues that the presence of seasonality should be the primary factor in assessing purity.[38] Breeding seasons, however, differ depending on location (longitude and latitude), which makes determining exact dingo seasons difficult, particularly in areas near the equator. The marker would be effective only if a bitch were on heat outside this extended window, which only permits identification for just over half the year. This method would be relevant only to females, as males are capable of breeding throughout the year. The presence or absence of an annual breeding season should thus be used in conjunction with the other indicators, as there are many ways to determine purity.

Physical characteristics

Phenotypic (observable) characteristics are useful indicators of genetic status, although they are highly subjective and therefore not always accurate. For example, second-generation hybrids may still look like a typical dingo and have no obvious phenotypic changes.[21,39] Coat colour (e.g. a uniform coat colour, white markings) may be used in some circumstances[2,19,21] and would appear to be the most practical method in the field. However, this is not considered reliable because coat colour varies (e.g. Corbett[18] observed that only 5% of wild dingoes had a large tail tip and only 23% had white feet), and hybrids may display 'typical' dingo colouration.[32,40] Elledge and others[40] conducted a study that evaluated methods of assessing dingo purity. They found that, for experienced wild dog managers, the use of visual characteristics was quite accurate in determining hybrids and pure dingoes. The most common characters used in assessment included coat colour, presence/absence of white points, tail form and brushiness, the presence of sable and the presence of ticking or spotting in the coat. Head and body conformation and the presence of floppy ears were also used. Visual characters remain useful, but unfortunately their use is highly subjective and dependent upon prior experience. To be effective at making such assessments, wild dog management personnel would need to receive standardised training.

Beyond coat colour, there are other observable characteristics of dingoes that can be used to distinguish them from dogs. For example, dingoes have extremely flexible joints, a scent gland located on their tail, no body odour, the head/skull is the widest part of their body, and no hind dew claws. They differ from dogs regarding the placement of fore and hind feet when walking (see Fig. 1.6).

Behavioural differences

Dingoes are distinguishable from domestic dogs in many aspects of behaviour (see Chapter 2). Whereas dingoes have been influenced only by natural selection for at least the last 5000 years, dogs are the result of intensive selection by humans and thus are highly variable depending on breed (and the purpose of that breed). In general, dingoes exhibit the same behaviours that have been described for other wild canids, the majority of which are not apparent in domestic or free-ranging dogs. For example, they live in highly structured packs led by a breeding pair, perform pack/social hunting in certain areas, exhibit mutual defence of territories and paternal male and alloparental care of young.[2,41–44] Dingoes howl as the main form of communication, and rarely bark.[11] One behaviour unique to the dingo and the New Guinea singing dog[45] (and the Norwegian kundehund) is the head toss, i.e. the rotation of head through a 90° arc, considered useful for spotting objects or prey located above, such as in trees. Lastly, dingoes develop physically and mentally a lot earlier than dogs (e.g. are able to run and play at three weeks), and are fully mature at 12 months whereas domestic dogs are much slower to develop.

Temperament, personality, intelligence and sociability

The dingo can accept humans as social companions, but only if raised from a pup (around five to seven weeks) and provided with a suitable environment. Dingoes are far more difficult than dogs to keep as pets as they have not been selected for the traits that humans seek in a companion animal[46] (see Chapter 10 for further detail about the dingo's temperament and suitability as a pet). Part of the reason why dingoes are difficult to keep as a household pet, are difficult to trap and have been so successful in the Australian bush, is because they are extremely intelligent and adaptable. There are many cognitive differences between domestic and wild canids, particularly in regard to problem-solving abilities and the ability to follow human communicative gestures.[47–49] Chapter 9 provides a detailed comparison between the cognitive abilities of dingoes and domestic dogs.

Ecological differences

The dingo is critical for healthy biodiversity and essential for ecological integrity, as it is a top-order predator and performs a role in suppressing mesopredators such as feral cats and foxes (see Chapter 8 for an in-depth discussion of the ecological role of the dingo). It is currently unknown whether dogs (domestic or feral) become an ecologically integrated predator like the dingo, but it is likely that they do not. Wild dogs are more likely to stay around areas of high human activity and to prey on domestic animals rather than native species.

Conclusion

Unfortunately for the dingo, it looks remarkably like the domestic dog. But as I hope this chapter (and indeed the remainder of the book) clarifies, the dingo is clearly identifiable, and distinguishable from domestic dogs. Chapter 2 moves beyond observable characteristics of the dingo, and highlights the many biological and behavioural differences.

References

1. Newsome AE, Corbett LK, Best LW, Green B (1972) The dingo. *Australian Meat Research Committee Review* **14**, 1–11.
2. Corbett LK (2001) *The Dingo in Australia and Asia.* 2nd edn. JB Books, Adelaide.
3. Corbett LK (1974) Contributions to the biology of dingoes (*Carnivora: Canidae*) in Victoria. MSc thesis. Monash University, Sydney.
4. Iredale T (1947) The scientific name of the dingo. *Proceedings of the Royal Society of New South Wales*, 35–36.
5. Meyer FA (1793) *Systematisch summarische uebersicht der neuesten zoologischen entdeckungen.* Dykische Buchhandlung, Leipzig.
6. Nomenclature, ICoZ (1957) Opinion 451. Use of the plenary powers to secure that the specific name 'dingo' Meyer, 1793, as published in the combination '*Canis dingo*', shall be the oldest available name of the dingo of Australia (class Mammalia). In *Opinions and Declarations Rendered by the International Commission on Zoological Nomenclature.* (Ed. F Hemming) pp. 329–338. International Trust for Zoological Nomenclature, London.
7. Fleming P, Corbett L, Harden R, Thomson P (2001) *Managing the Impacts of Dingoes and Other Wild Dogs.* Bureau of Rural Sciences, Canberra.
8. Corbett LK (1995) *The Dingo in Australia and Asia.* University of New South Wales Press, Sydney.
9. Freedman AH, Gronau I, Schweizer RM, Ortega-Del Vecchyo D, Han E, Silva PM, Galaverni M, Fan Z, Marx P, Lorente-Galdos B, Beale H, Ramirez O, Hormozdiari F, Alkan C, Vila C, Squire K, Geffen E, Kusak J, Boyko AR, Parker HG, Lee C, Tadigotla V, Siepel A, Bustamante CD, Harkins TT, Nelson SF, Ostrander EA, Marques-Bonet T, Wayne RK, Novembre J (2014) Genome sequencing highlights the dynamic early history of dogs. *PLOS Genetics* **10**, e1004016.
10. vonHoldt BM, Pollinger JP, Earl DA, Knowles JC, Boyko AR, Parker H, Geffen E, Pilot M, Jedrzejewski W, Jedrzejewska B, Sidorovich V, Greco C, Randi E, Musiani M, Kays R, Bustamante CD, Ostrander EA, Novembre J, Wayne RK (2011) A genome-wide perspective on the evolutionary history of enigmatic wolf-like canids. *Genome Research* **21**, 1294–1305.

11. Corbett LK (1995) Dingoes: expatriate wolves or native dogs? *Nature Australia* **25**, 46–55.
12. Honacki JH, Kinman KE, Koeppl JW (1982) *Mammal Species of the World: A Taxonomic and Geographic Reference*. Allen Press/Association of Systematic Collections, Kansas.
13. Phillip A (1789) *The Voyage of Governor Phillip to Botany Bay: With an Account of the Establishment of the Colonies of Port Jackson and Norfolk Island. Compiled from the Several Departments*. London.
14. Crowther M, Fillios M, Colman N, Letnic M (2014) An updated description of the Australian dingo (*Canis dingo* Meyer, 1793). *Journal of Zoology* **293**, 192–203.
15. Nomenclature, ICoZ (2003) Opinion 2027 (case 3010). Usage of 17 specific names based on wild species which are pre-dated by or contemporary with those based on domestic animals (Lepidoptera, Osteichthyes, Mammalia): Conserved. *Bulletin of Zoological Nomenclature* **60**, 81–84.
16. Gentry A, Clutton-Brock J, Groves CP (2004) The naming of wild animal species and their domestic derivatives. *Journal of Archaeological Science* **31**, 645–651.
17. Australian National Kennel Council (2009) *Australian Dingo* [24 April 2014]. http://www.ankc.org.au/Breed_Details.aspx?bid=103.
18. Corbett LK (1985) Morphological comparisons of Australian and Thai dingoes: a reappraisal of dingo status, distribution and ancestry. *Proceedings of the Ecological Society of Australia* **13**, 277–291.
19. Newsome AE, Corbett LK (1985) The identity of the dingo. III. The incidence of dingoes, dogs and hybrids and their coat colours in remote and settled regions of Australia. *Australian Journal of Zoology* **33**, 363–375.
20. Thomson PC (1992) The behavioural ecology of dingoes in north-western Australia. I. The Fortescue River study area and details of captured dingoes *Wildlife Research* **19**, 509–518.
21. Jones E (1990) Physical characteristics and taxonomic status of wild canids, *Canis familiaris*, from the eastern highlands of Victoria. *Australian Wildlife Research* **17**, 69–81.
22. Walters BJ (1995) *The Company of Dingoes: Two Decades with our Native Dog*. Australian Native Dog Conservation Society, Bargo.
23. Cairns KM, Wilton AN, Ballard JWO (2011) The identification of dingoes in a background of hybrids. In *Advances in Genetics Research*. (Ed. KV Urbano) pp. 309–327. Nova Science Publishers, New York.
24. Corbett LK (2001) The conservation status of dingoes (*Canis lupus dingo*) in Australia, with particular reference to New South Wales: threats to pure dingoes and possible solutions. In *A Symposium on the Dingo*. (Eds CR Dickman and D Lunney) pp. 10–19. Royal Zoological Society of New South Wales, Sydney.
25. Sanderson A (1981) *Complete Book of Australian Dogs*. New Holland Publishers, Sydney.
26. Hudson L (1974) *Dingoes Don't Bark*. Rigby, Adelaide.
27. Savolainen P, Leitner T, Wilton A, Matisoo-Smith E, Lundeberg J (2004) A detailed picture of the origin of the Australian dingo, obtained from the study of mitochondrial DNA. *Proceedings of the National Academy of Sciences of the United States of America* **101**, 12387–12390.
28. Wilton AN (2001) DNA methods of assessing dingo purity. In *A Symposium on the Dingo*. (Eds CR Dickman and D Lunney) pp. 49–56. Royal Zoological Society of New South Wales, Sydney.
29. vonHoldt BM, Pollinger JP, Lohmueller KE, Han E, Parker HG, Quignon P, Degenhardt JD, Boyko AR, Earl DA, Auton A, Reynolds A, Bryc K, Brisbin A, Knowles JC, Mosher DS, Spady TC, Elkahloun A, Geffen E, Pilot M, Jedrzejewski W, Greco C, Randi E, Bannasch D, Wilton A, Shearman J, Musiani M, Cargill M, Jones PG, Qian Z, Huang W, Ding ZL, Zhang YP, Bustamante CD, Ostrander EA, Novembre J, Wayne RK (2010) Genome-wide snp and haplotype analyses reveal a rich history underlying dog domestication. *Nature* **464**, 898–902.
30. Wilton AN, Steward DJ, Zafiris K (1999) Microsatellite variation in the Australian dingo. *Journal of Heredity* **90**, 108–111.

31. Stephens D (2011) The molecular ecology of Australian wild dogs: hybridisation, gene flow and genetic structure at multiple geographic scales. PhD thesis. University of Western Australia, Perth.
32. Elledge AE, Leung LK, Allen LR, Firestone K, Wilton AN (2006) Asessing the taxanomic status of dingoes *Canis familiaris dingo* for conservation. *Mammal Review* **36**, 142–156.
33. Macintosh NWG (1975) The origin of the dingo: an enigma. In *The Wild Canids: Their Systematics, Behavioural Ecology and Evolution.* (Ed. MW Fox) pp. 87–106. Van Nostrand Reinhold, New York.
34. Newsome AE, Corbett LK, Carpenter SM (1980) The identity of the dingo. I. Morphological discriminants of dingo and dog skulls. *Australian Journal of Zoology* **28**, 615–625.
35. Newsome AE, Corbett L (1982) The identity of the dingo. II. Hybridization with domestic dogs in captivity and in the wild. *Australian Journal of Zoology* **30**, 365–374.
36. Gollan K (1984) The Australian dingo: in the shadow of man. In *Vertebrate Zoogeography and Evolution in Australasia: Animals in Space and Time.* (Eds M Archer and G Clayton) pp. 921–927. Hesperian Press, Perth.
37. Catling PC, Corbett LK, Newsome AE (1992) Reproduction in captive and wild dingoes (*Canis familiaris dingo*) in temperate and arid environments of Australia. *Wildlife Research* **19**, 195–209.
38. Purcell B (2010) *Dingo.* CSIRO Publishing, Melbourne.
39. Corbett LK (2008) Dingo: *Canis lupus.* In *The Mammals of Australia.* (Eds S Van Dyck and R Strachan) pp. 737–739. Reed New Holland, Sydney.
40. Elledge AE, Allen LR, Carlsson B-L, Wilton AN, Leung LKP (2008) An evaluation of genetic analyses, skull morphology and visual appearance for assessing dingo purity: implications for dingo conservation. *Wildlife Research* **35**, 812–820.
41. Corbett LK (1988) Social dynamics of a captive dingo pack: population regulation by dominant female infanticide. *Ethology* **78**, 177–198.
42. Thomson PC (1992) The behavioural ecology of dingoes in north-western Australia. III. Hunting and feeding behaviour, and diet. *Wildlife Research* **19**, 531–541.
43. Thomson PC (1992) The behavioural ecology of dingoes in north-western Australia. II. Activity patterns, breeding season and pup rearing. *Wildlife Research* **19**, 519–530.
44. Thomson PC (1992) The behavioural ecology of dingoes in north-western Australia. IV. Social and spatial organisation, and movements. *Wildlife Research* **19**, 543–563.
45. Koler-Matznick J, Brisbin IL, Feinstein M, Bulmer S (2003) An updated description of the New Guinea singing dog (*Canis hallstromi*, Troughton 1957). *Journal of Zoology* **261**, 109–118.
46. Smith B (2014) Living with wild dogs: personality dimensions in captive dingoes (*Canis dingo*) and implications for ownership. *Anthrozoos* **27**, 423–433.
47. Smith BP, Litchfield CA (2010) How well do dingoes, *Canis dingo*, perform on the detour task? *Animal Behaviour* **80**, 155–162.
48. Smith BP, Litchfield CA (2010) Dingoes (*Canis dingo*) can use human social cues to locate hidden food. *Animal Cognition* **13**, 367–376.
49. Smith BP, Litchfield CA (2013) Looking back at 'looking back': operationalising referential gaze for dingoes in an unsolvable task. *Animal Cognition* **16**, 961–971.

2
Biology and behaviour of the dingo

Bradley Smith

Habitat, distribution and hybridisation

The dingo's present distribution covers the majority of mainland habitat types, including alpine moorlands, temperate eastern Australia, tropical wetlands and forests in northern Australia and arid hot deserts in central Australia.[1, 2] Since European settlement in Australia, there is a noticeable absence of dingoes from densely settled areas and prime grazing country on the southern side of the dingo exclusion fence, a 5400 km fence which runs across the south-eastern part of Australia.[3, 4] Unfortunately, there have been few, if any, thorough surveys of dingo populations or distribution. What is known comes from studies investigating hybridisation between dingoes and dogs. Thus, the focus of this section relates to such studies.

The population and distribution of dingoes in Australia is influenced not only by human control measures, it is also greatly influenced by hybridisation with domestic dogs. Today, hybridisation is the primary threat to the survival of the Australian dingo outside of captivity. It is problematic because it can lead to the breakdown of adaptive gene combinations, the loss of genetic diversity, the extinction of wild populations and changes to ecosystem function.[5] Stephens[5] conducted an Australia-wide study looking at the extent and spatial pattern of dingo–dog hybridisation for her doctoral studies. She obtained genotypes from 3637 wild dogs using microsatellite DNA markers, and looked at the proportion of dingo ancestry in those populations. Stephens found that all the regions sampled had some degree of dog ancestry, but the extent varied substantially. Overall, 46% of the free-roaming dogs that were sampled were classified as pure dingoes. The south-east of Australia had an extremely high proportion of hybrids, with only 1–4% of tested animals considered pure dingoes. This is not surprising given the density of human settlement and the location of the best land for agriculture in the south-east – factors which have been flagged as a strong predictor of dingo hybridisation levels.[6, 7] The incidence of hybridisation has changed little over the past few decades. For example, Newsome and Corbett,[6] using skull morphology, found that the incidence of pure dingoes have been high for quite some time (e.g. only 23% pure dingoes on the south coast of New South Wales and 65% dingoes in the Victorian highlands). The findings from the Stephens' study were not all bad, because high populations of pure wild dingoes were found in other parts of Australia (particularly central Australia, where only 13% sampled were hybrids, and 41% in Western Australia). The proportion of dingo genes in a population will decline at a slower rate if dingoes breed with hybrids rather than dogs. For example, a first-generation hybrid (F_1) resulting from a dingo and a dog will have 50% dog genes, and the gene pool will be diluted by half every generation that breeds with dogs. However, a hybrid born as a result of a cross between a dingo and a dingo–dog hybrid (with 50% dog genes) will have only 25% dog genes.[8]

Almost all the wild dogs (99%) showed some dingo ancestry, which Stephens suggested indicates that either domestic dogs find it difficult to survive in remote Australia, or that the majority of hybridisation is with roaming dogs that then return to their owners. Indications from mtDNA suggest that all dingo–dog matings that lead to successful rearing of progeny in the wild had the dingo as a mother,[1, 8, 9] basically because of the difficulty in raising a litter outside of captivity (discussed below, 'Reproductive activity and parental behaviour'). The hybrid offspring are more likely to survive and integrate into dingo societies than are domestic dogs raised in captivity, because the latter have little experience hunting or socialising with dingoes.[10, 11]

However, there is the possibility that the number of dingo–dog hybrid numbers are overstated in many studies, for two reasons. First, most studies of purity in wild dingo populations have been conducted to determine the distribution of dingo–dog hybrids rather than that of pure dingoes. Samples used for genetic studies, such as those by Stephens, are often heavily weighted towards animals collected (killed) from areas around high human activity and artificially abundant food resources. They are not always reflective of dingoes that actively avoid feral dogs, or live remotely and do not encounter them. Second, behavioural differences between dingoes and dogs appear to assist in preventing hybridisation. That is, it is difficult for a feral domestic dog to successfully integrate and compete with a large dingo population that has a stable pack structure.[1, 12] Like wolves,[11] the behavioural and physiological differences are so great that mating is unlikely to occur (although it clearly does in many instances), and hybrid offspring rarely survive to breed in the wild. In fact, for these reasons, it is argued by some[11] that hybridisation is not a real management issue, and that nature will in essence weed out those that don't belong in the wild. Natural selection ensures that the behavioural sequences necessary for survival and reproduction are maintained and passed onto subsequent generations.[13] Ecologically, dingoes perform a role as a top-order predator as compared with domestic or feral dogs[14] (see Chapter 8) but it has not been reported whether dogs (domestic or feral) become an ecologically integrated predator.

Home range

A home range is the 'area traversed by the individual in its normal activities of food gathering, mating, and caring for young'.[15] Its territory is the portion of the home range it defends. For example, the dominant male may have a much larger home range than the pack's territory.[15] A family or social group of dingoes maintains a distinct territory or territories within a home range, and is usually loyal to this site from year to year.[16, 17] The sizes of home ranges may vary and overlap, and appear to be influenced by the availability of resources.[18, 19] In radio-tracking studies, dingoes in areas with high rainfall, a temperate climate and large and diverse potential prey species may have smaller home ranges of ~21.5 km^2 (e.g. in the forest areas of south-eastern Australia)[20] and 27 km^2 (in the northern tablelands of New South Wales.[21] Ranges are usually larger in arid environments, which are less productive (e.g. 95.8 km^2 in north-west Western Australia[22] and 67 km^2 in Central Australia)[16]. In areas where prey populations fluctuate, home ranges may range from 90–300 km^2.[22] According to tracking studies by Harden[21] and Thomson,[17, 23] dingoes may travel around 10–15 km per day. While they may be active at any time of the day (e.g. in far north Queensland, dingoes are most active during the day: Damian Morrant, pers. comm), they are least active during the hottest parts of the day and most active at dawn and dusk.[24] Estimates of home ranges are subject to error and depend greatly on the method used to calculate them (see Chapter 7).

The flexibility in their social organisation and territorial boundaries helps dingoes to survive and adapt to many different situations. This plasticity is particularly evident in the population on Fraser Island, which is a large sand island ~1800 km^2 that is home to ~150 dingoes. Although living in family groups, the island's dingoes do not appear to have formed defined territories; dingoes are regularly found all over the island both alone and with other families (Rob Appleby, pers. comm.). The high human presence on the island may be a contributing factor to their behaviour. Fluid or mobile territories have been reported in other areas such as in Kakadu National Park, where the floodplains and wet seasons determine where dingoes hunt (forest or floodplains), what they hunt and how they hunt (group or solitary).[16]

Social organisation and population regulation

Wild dingoes encounter a wide variety of habitats, prey species, climatic conditions and levels of human exploitation,[25] all of which influence their social organisation and pack size. In most cases, dingoes are social animals. Where conditions are favourable, they form stable family groups and maintain distinct territories that overlap little with those of neighbouring packs.[23] Early reports by anthropologists and scientists using aerial radio-tracking studies (before GPS technology) suggested that dingoes were essentially solitary animals because the studies often witnessed wild dingoes travelling alone. However, such anecdotal observations are not reflective of the social dynamics of the dingo. Corbett,[16] for instance, argued that because dingoes are mostly inactive during the day people mistakenly conclude that they are solitary, but often there is an unseen companion nearby. Various observational studies of dingo packs have also discredited the idea that dingoes are solitary. In the Simpson Desert, large stable packs of up to 10 individuals have been reported; in Kakadu National Park dingoes may be solitary, in pairs or in groups of three to eight; and in the Fortescue River region stable packs consist of around three to eight dingoes.

Groups contain a dominant male and female pair (most often the parents) and their offspring of various ages, with pack sizes ranging between three and 12 individuals; unrelated animals are rarely allowed to join a pack.[16, 23, 26] Pack members cooperate to hunt and in communal activities such as eating, resting and raising pups.[27] Each pack typically raises only one litter per year.[16] The complex use of behaviours such as dominant, submissive, affiliative and aggressive (discussed in the following section) helps to foster cohesion among pack members.[16, 26]

Social organisation, however, is highly variable and appears most dependent on the type and availability of prey. As a very efficient solitary hunter, the dingo may spend much of its time alone outside breeding season, but will engage in cooperative hunting to hunt larger prey. Other factors which influence dingo population numbers include immigration and emigration, natural mortality (e.g. disease, starvation), the direct impact of humans (e.g. poison baits, shooting, trapping, motor vehicles, culling by management), interactions with domestic dogs, inhibition of breeding in subordinates, mating preferences by dominants and possibly dominant female infanticide.[25, 28] It should be noted that infanticide has been observed in only a single captive study conducted by Corbett[28] and has not been reported in any wild populations. It is more than likely that the conditions of the study site (i.e., size and lack of opportunity for individuals to disperse) resulted in this abnormal behaviour.

Reproductive activity and parental behaviour

A recent review of the reproductive and parental behaviour of the genus *Canis* by Kathryn Lord and colleagues[29] revealed that dingoes are true wild canids in relation to their reproductive and parental behaviours. In contrast to wild canids, domestic dogs are relatively poor parents and in general do not exhibit the complex parental behaviours of wild canids. This section gives summaries of these reproductive and caregiving categories in the dingo.

Seasonality

Like wolves[30] and other wild *Canis* species,[29] male and female dingoes reproduce seasonally.[31–33] That is, sexual activity, mating and parturition (giving birth) occur during the same annual period. The timing of the season is likely triggered by changes in daylight/sunlight (photoperiodism). This is the developmental or physiological responses of plants and animals to the relative lengths of the light and dark periods, and can provide information directly relating to seasonal changes. Births are thus perfectly timed to coincide with periods that are optimal for raising young (e.g. at the beginning of seasonal blooms of food).[29] The amount of parental investment required to raise a litter to adulthood is extremely high, so the ability to have multiple litters within a year has been lost.

To understand the triggers of the breeding season, it helps to plot the length of daylight patterns across the year. The duration of daylight (daylength) for low and mid-latitude locations is defined as the interval of time between sunrise and sunset. The breeding season for dingoes (in the southern hemisphere) begins when the daylight hours are decreasing (i.e. when the days are getting shorter). Lord *et al.*[29] conducted a comparison of seasonality between domestic dogs, captive dingoes and captive wolves. They illustrated the differences between dogs and wild canids under controlled conditions by plotting conception dates against the length of daylight across a calendar year (Fig. 2.1). Almost all dingo bitches conceived before the winter solstice (shortest day of the year) and after the summer solstice (longest day of the year) whereas dogs conceived throughout the year. Wolf females primarily conceived between late January and early March. From this we can conclude that canids follow a seasonal pattern related to (but not completely determined by) the length of daylight.

For dingoes, the breeding season typically occurs between February and June, with most births in the southern hemisphere winter months (June to August).[24, 31–33] Figure 2.2 provides mating and breeding times for the captive dingo population used in the daylength analysis. When moved to the northern hemisphere, their season adjusts accordingly, with the majority of births occurring in late winter/early spring (February–March).[34] Breeding season is determined by the females' annual oestrus cycle, as males are continuously fertile in most

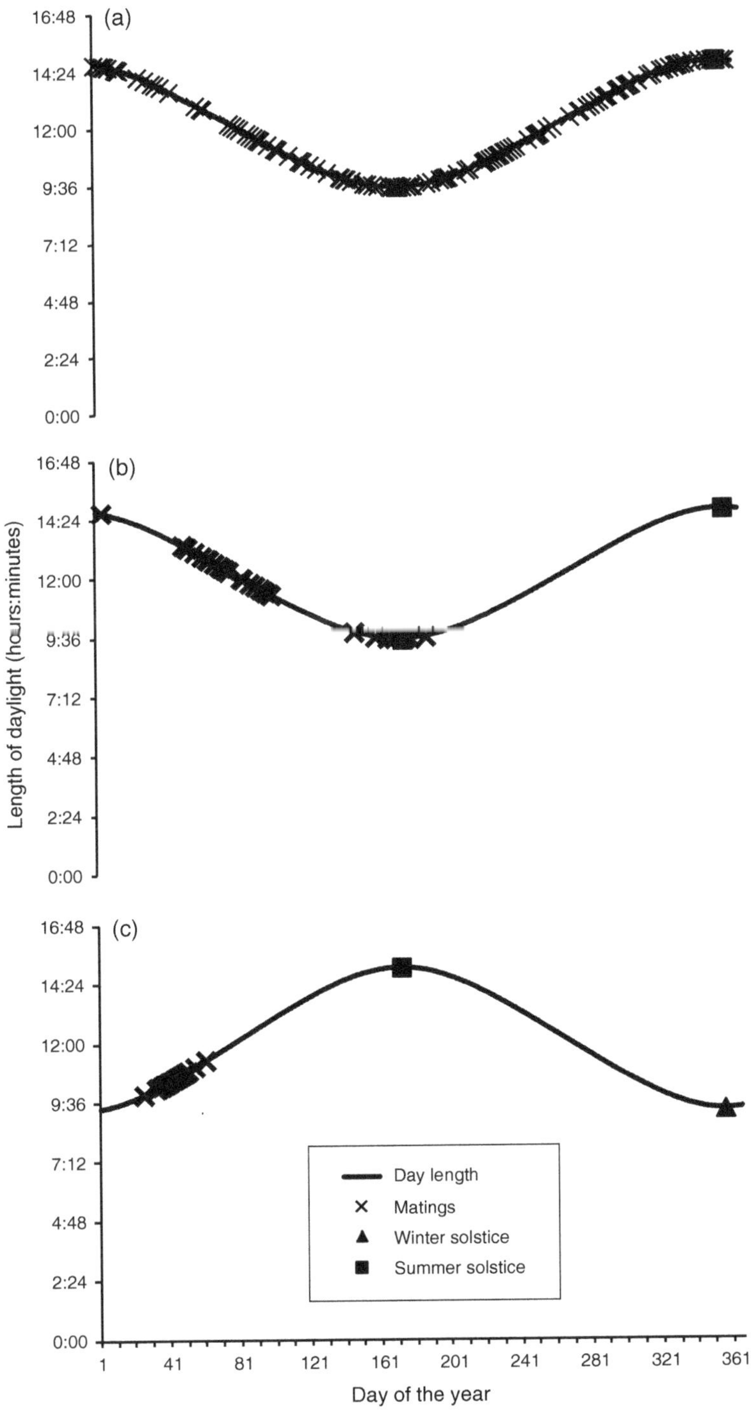

Figure 2.1: Number of conceptions (a) of domestic dogs (golden retriever and labrador retrievers, N = 141), (b) captive dingoes (N = 36) and (c) wolves (N = 37) plotted against the length of daylight over the 2008 calendar year.

Source: Lord *et al.* (2013).[29]

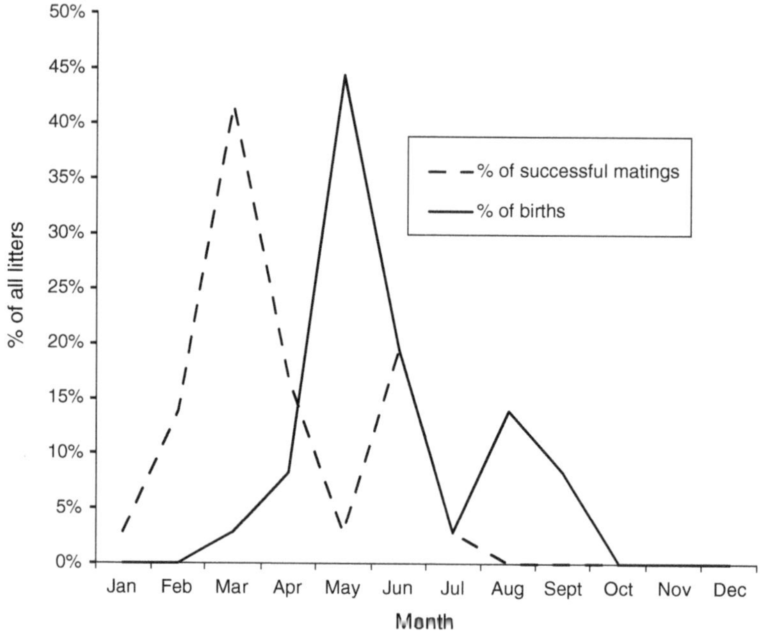

Figure 2.2: Month of conceptions and births at the Dingo Discovery Centre in Melbourne, 1999–2008 (total of 36 litters).
Source: Bradley Smith, unpublished data.

regions.[33] Thomson[24] found that over a nine-year period on the lower Fortescue River in north-western Australia (21° 22′S), pups were born from mid-May to mid-August, with more pups born in July than any other month. He estimated the breeding season to occur from mid-March to early June. Catling and colleagues[32] reported that both wild and captive dingoes in Central Australia (25°S) bred from March to July. They also reported that captive dingoes in Canberra (35°S) bred from April to July. Jones and Stevens[31] reported in a four-year study that female dingoes in Victoria (37°S) were either in oestrus or conceived between January and April.

There are clear physiological patterns that occur in male and female dingoes during the onset of mating season. Females have an annual oestrus cycle, where they exhibit tumescence (vulval swelling) between April and July and uterine weights increase in April, peaking in May/June.[32] Male cycles may be influenced by temperature.[16] For instance, males in hot and arid environments exhibit a true annual cycle, where there is an increase in testes and prostate weight, semen volume and associated changes in testes histology from January to March, with a peak in April/May.[31, 32] Males in other habitats are capable of breeding all year, but there is a sharp peak between February and July in prostate and testes weight, as well as in sperm count.[16, 32] Like all canids, during mating the male mounts the female from behind (Fig. 2.3a). The pair then become coupled (a copulatory tie) and cannot break apart for anywhere between 5 and 30 min (Fig. 2.3b). The dingoes become locked together during intercourse due to a bulbous structure at the base of the penis that increases inside the vagina before the male ejaculates, thus preventing withdrawal until it reduces in size. Due to vulnerability (e.g. to

Figure 2.3: (a) A pair of dingoes on Fraser Island, mating. (b) In a copulatory tie.
Source: Bradley Smith.

rivals), they often stand facing opposite directions during the tie. Although the copulatory tie is not necessary, it is believed to increase the chances of conception.

The onset of the breeding season varies according to the following factors.

- **Age.** Dingoes can reproduce in their first year in flush conditions, but most do not bear pups until they are two years of age.[16, 31, 32, 35] Jones and Stevens[31] reported that oestrus was variable, occurring between one and four years of age, with only 36% of females and 63% of males breeding before two years, while 97% of males over three years old reproduced.
- **Social status.** Most dominant females (usually the oldest) in captive environments tend to come into oestrus before subordinate females.[28]
- **Geographic latitude.** In northern Australia, dingoes often whelp two months earlier than those 480 km south.[36] Corbett[16] provided a summary of births across Australia, that vary according to latitude. Northern Australia (13°S 130′E), predominantly June and July; Western Australia (21°S 116′E), June–August; Central Australia (24°S 133′E), May–August; south-east Australia (37°S 139′E), May–August. Corbett[16] also reported on dingo-like canids in the northern hemisphere, with those at the equator (0°) giving birth all year and those in Thailand (14°N) primarily during October–November.
- **Seasonal conditions.** Dingo seasonality appears to be quite responsive to environmental conditions. For example, males in hot and arid environments exhibit a true annual cycle while in other habitats males are capable of breeding all year.[16, 31, 32] In periods of drought, testes responded several months later than during flush periods, uterine weight can be delayed several months, and the average litter size decreases.[32, 36]

In general, seasonality in dingoes can be broken down into periods when the focus of dingo society is on reproduction and then on rearing puppies. Table 2.1 provides a summary of the dingo's activity across months and seasons.

Dingoes, and their close relative the New Guinea singing dog (*C. hallstromi*), have a relatively unique oestrus cycle. That is, if the bitch is not impregnated during her first cycle, she can have a second cycle closely following the first.[38] Jones and Stevens[31] reported that some dingoes displayed a relatively long breeding season and, although the peak of births occurred in winter, the duration would have allowed time for some early breeders to have had a second litter in the one year (however, they could not confirm this). It is uncertain whether such behaviours occur in the wild.

Domestication has severe consequences for seasonality in canids. Domestic dogs are not seasonal, regardless of whether they are free-living or under human control. Females are capable of coming into season every seven months on average, and males are constantly capable of reproducing. Thus, pups can be born at any time of year.[39] It is suspected that first-generation female dingo–dog hybrids of similar size to dingoes (as with coyote and wolf–dog hybrids) immediately lose any reproductive seasonality, although this has yet to be shown scientifically.[31]

Table 2.1. Summary of the reproductive activities of the dingo calendar. Timings may vary with location and circumstances (e.g. wild and captivity)

Month	Season	Activity
March April May	Autumn	• Breeding season • The frequency of howling, scent-marking, ground-scratching and aggression increases and signifies pre-breeding • Before mating, dingoes spend this time either finding a mate and/or strengthening existing pair bonds • Pre-oestrus lasts 10–12 days: oestrus lasts 7–16 days in captivity and 30–60 days in the wild[16] • Gestation period is around 63 days[16]
June July August	Winter	• This time of year is focused upon whelping and weaning pups • Female selects suitable den for whelping • Nursing occurs in two stages. (1) Intensive maternal care of pups, which are confined to the den. (2) Pups are weaned and moved between locations (total food requirements of the pups is much greater)
September October November	Spring	• Post-nursing • Pups start to become independent around 3–4 months • Parents teach and train pups
December January February	Summer	• Non-breeding period • Pups either remain with the family and share pack activities, are abandoned or leave territory

Sources: Observations by the author, Thomson[23] and Corbett and Newsome.[37]

Pair bonding and rearing of young

In almost all cases, wolves exhibit long-term pair bonding, maintain a shared territory and both parents raise the young.[26, 40–42] Dingoes too form long-term pair bonds, defend shared territories and cooperatively care for young.[16, 23, 24, 37] Overall, dingoes appear to have a lower chance of success raising litters when acting alone.[24] The establishment of a pair bond is critical for successful conception, as well as the successful raising of a litter. In fact, the pre-oestrus period of dingoes is longer than that of dogs (30–60 days). According to Corbett,[16] a protracted period of pre-oestrus may allow wild canids sufficient time to strengthen pair bonds and is important for the induction of full sexual potency in males. The formation and maintenance of pair bonds is in stark contrast to the behaviour of domestic dogs, as they do not form pair bonds but remain generally promiscuous.[43, 44] Male dogs also do not guard females for exclusive reproductive access, and they do not assist females with the rearing of young.[39]

Parental feeding behaviour

Dingo pups are nursed for seven to 10 weeks.[24, 35] Alloparental behaviour (care of pups by parents and relatives) has been identified in both wild[16, 24, 37] and captive dingoes[28] beginning at 15 days.[35] In a wild population in central Australia, two females had dens within 20 m of each other; one was observed feeding four of her own pups and two of her neighbour's.[45] In a captive *C. dingo* colony,[28] a subordinate female continued to nurse another female's pups after her pups were killed. Likewise, grey wolf pups are reported to be weaned between five and 10 weeks of age.[29, 46] There are rare cases of communal denning (including allonursing) by wolves both in the wild[40, 46] and in captivity[47] when more than one female has a litter. In comparison, dog mothers nurse their young for five to 10 weeks[48–51] and, although uncommon, allonursing has been occasionally observed in dogs. Pal[49] reported two breeding females with adjacent dens nursing each other's pups. Daniels and Bekoff[52] reported an observation of two females giving birth in the same den.

Regurgitation and provisioning

Wild canids regularly regurgitate partly digested food to their offspring when a pup gives the appropriate begging signal. Pups are also provisioned with chunks of meat from large prey or whole small prey items and, when more developed, will follow parents to food.[29] In wolves, for instance, adults (including breeding male and female as well as siblings from previous years) regurgitate to pups when pups are around four weeks of age.[40] Wolf pups receive approximately two regurgitations a day.[46] Regurgitation by adult domestic dogs to pups is reported, but it is rare to see pups being regularly fed in this manner.[29] Both dingo parents regurgitate food to their young, but rates of regurgitation have not been reported. Thomson[24] observed that, in addition to the mother, other pack members brought food to pups. In one instance, Thomson[24] observed a mother and an adult male feeding on a kill near the den. They returned separately to the den, where both regurgitated food for the pups. On other occasions, when individual dingoes could be recognised, pups were seen to solicit food from dingoes other than their mother. Corbett and Newsome[37] briefly described some observations of four wild dens. Five pups in one den were visited by both parents as well as one pair of non-breeding yearlings. The adult male was greeted with much affection, but neither he nor the yearling males were observed retrieving food for the pups. In another instance, after periods away from the den, the mother and possibly a second female from the group were observed consistently bringing back rabbits for the pups (rabbits were abundant at the time). On one occasion the mother was seen returning from the direction of a nearby watering point (2.5 km away), and was seen regurgitating water to her pups when they greeted her and licked her mouth. There is one reported instance where a bitch that had recently whelped was seen to eat some of the food regurgitated by a male, suggesting that pack members may provide for bitches confined to the den with very young pups.[24]

Birth, growth and development of puppies

Gestation lasts between 61 and 69 days (average 63 days).[16, 24] Females become visibly pregnant (e.g. larger belly, prominent teats) around 10 days before birth. Around this time, the female selects a suitable den in which to whelp and raise her young. Den sites have been found in hollow logs, enlarged rabbit warrens and rock shelters, under large tussocks of spinifex and at the bases of large trees[16] (see Plate 8). The female spends time cleaning and enlarging the area before the birth.[35] In captivity, the female refuses food on the day/night before the birth.

Dens may also be selected for their ability to provide young with hunting opportunities. Corbett and Newsome,[37] during observations of a wild dingo pack, noticed that the rabbits near the chosen den were not preyed upon by the adults. However, as the pups got older, one of the yearling bitches was observed apparently coaching the pups into hunting. The researchers concluded that the rabbits around the den were left for the exclusive use of the pups while the adults hunted further away. Harden[53] noted that bitches rotated their litters around different den sites in the same order each year, which he believed may help ensure pups have enough game to hunt in the immediate area.

The size of the litter varies according to seasonal conditions as well as the condition and past experience of the mother. The average litter size is four to five pups, with a possible range of one to 10.[16, 31, 32, 37] This may be the same in captivity. In a captive setting in Victoria, the average litter size is 5.8, with an average sex ratio of 2.2 males to 3.4 females (Table 2.2).

The growth and development of dingo pups, particularly during the first 30 days of life, is not well understood. Perhaps the most detailed account, though not comprehensive or definitive, was given by Breckwoldt. Breckwoldt[35] kept a pack of captive dingoes in a semi-natural compound, which he used for filming a documentary on the dingo. In the book that accompanied his film, he described the behaviour of a litter of 11 puppies in Candelo, New South Wales. The dingo pups began to open their eyes around day 12 and all were open at day 14, around the time the mother began regurgitating food for the pups and bringing food into the den. The mother spent most of the day and all night with the pups until they were four days old. Then she began spending less and less time with them. Seven days after birth, she spent only the night with them. At three weeks the pups first ventured outside the den. At around five weeks, the litter were confidently and freely moving outside the den for most of the day.[35] Pups usually become independent at three to six months or, in a pack, at 12 months when the next breeding season begins.[16, 37] Thomson[24] observed adults taking pups aged nine to 14 weeks to feed on nearby kills. Pups aged between nine and 24 weeks progressively accompanied the adults on longer journeys. By ~16 weeks, the pups were less reliant upon intensive parental care for their immediate survival.[24]

Jackson[54] provided some details of bodyweight growth from dingoes born at Healesville Sanctuary in Victoria. Catling *et al.*[32] also documented growth and development in dingoes, however, the data represented a combination of dingoes and cross-breds, so is of limited value. However, they did determine that head length (from the back of the sagittal crest to the tip of the nose) was the most reliable measurement of age until 120 days. After that, eye-lens weight

Table 2.2. Average size and sex of litters born at the Dingo Discovery Centre, Victoria

Year	No. of litters	Litter size (average)	Sex ratio (average)	
			M	F
2008	8	6.5	2.5	4.0
2007	4	6.2	3.0	3.2
2006	5	6.0	2.4	3.6
2005	5	7.6	4.5	3.0
2004	5	5.2	1.7	3.2
2003	2	5.5	1.0	4.0
2001	3	4.3	1.3	3.0
2000	3	4.0	1.0	3.0
1999	1	5.0	–	–
Mean	–	5.8	2.2	3.4

of dead dingoes can be accurate up to 500 days of age. The best way to age adult dingoes is by the canine teeth. Pulp cavity ratios from radiographs have been found to be a reliable measure of the approximate age of dingoes.[55]

In an attempt to gain more information about the developmental stages of dingoes, I conducted a small study of the pups born at the Dingo Discovery and Research Centre during the 2013 season. There were six litters, and a total of 29 puppies. Notes on the behaviour and physiological changes of the pups were recorded on an *ad hoc* basis, as well as their measurements and weights. When the puppies were being weighed, their gums were checked to determine if their upper canine teeth had erupted, and their eyes inspected to determine to what degree they were open. Figure 2.4 outlines the beginning of various stages of the puppies' physiological and behavioural development (by no means comprehensive), and Figs 2.5 and 2.6 shows growth according to weight, skull size, chest circumference and body length. Dingo pups were around 300 g at birth, and grew quickly. Evidence of the size differences between males and females (sexual dimorphism) was evident from birth. Weight changes were evident when weaning off mother's milk to semi-solid food began at about day 20. It must be pointed out that these figures are based on the growth and development of well-fed captive pups, and thus may not be entirely reflective of what occurs in the wild. They do, however, provide a useful guide to the developmental stages of dingo pups in the first few weeks of life.

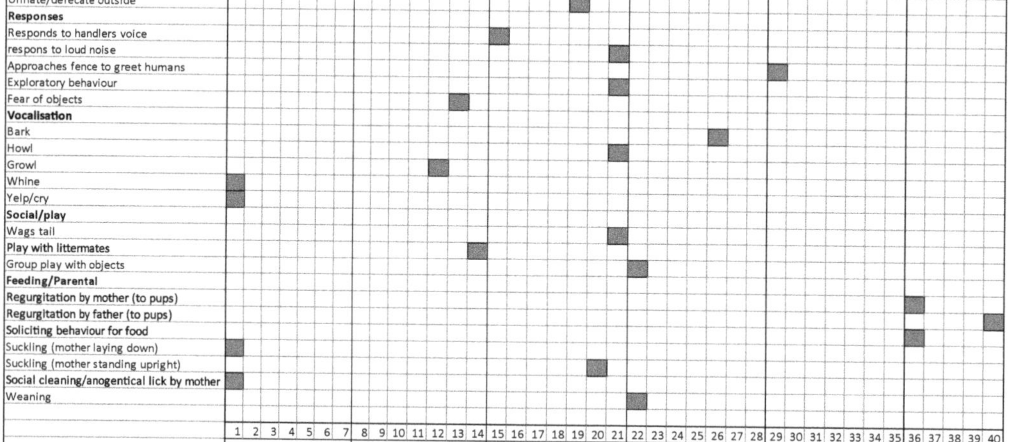

Figure 2.4: A summary of dingo development stages according to the earliest day that they were observed on an *ad hoc* basis by sanctuary staff. Data represents 29 dingoes from six litters born in 2013 at the Dingo Discovery and Research Centre (Victoria).

Source: Bradley Smith, unpublished data.

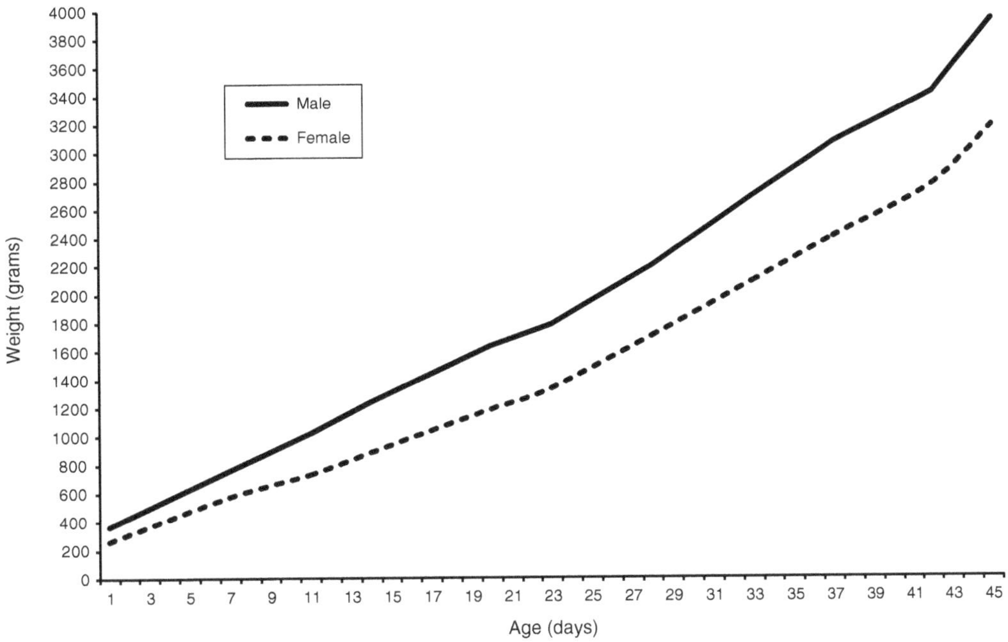

Figure 2.5: Changes in bodyweight of a litter of captive dingo pups (three males, two females) during the first 50 days.

Source: Bradley Smith, unpublished data.

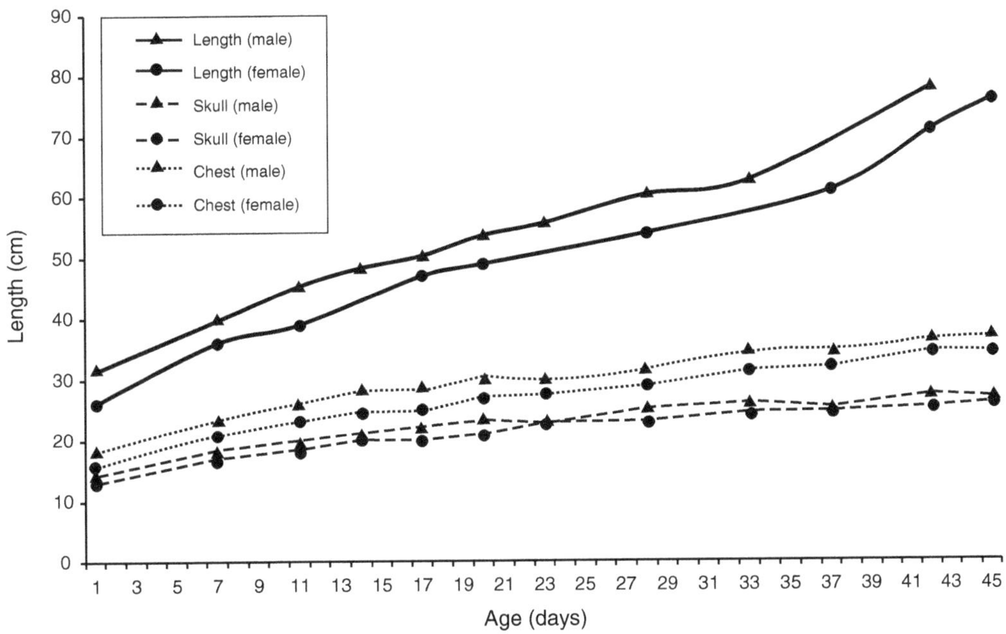

Figure 2.6: Growth in body length (from nose to tail tip), chest size (taken just behind elbows) and skull size (taken just behind the eyes) of a litter of captive dingo pups (three males, two females) during the first 45 days.

Source: Bradley Smith, unpublished data.

Communication among dingoes

Dingoes are a highly social species that require a social order, expression and communication systems. It is clear to any observer that social displays and interaction among all members of the dog family form a large part of a canid's daily life, and are most important for maintaining cohesion within packs and living areas so that resources can be used efficiently. The type of communication that the dingo and its wolf ancestor are most famous for is howling, but there are many ways that dingoes communicate beyond vocalisations, including their posture and use of odour. We know quite a lot about the communication of dogs and wolves, but very little about that of dingoes. It has, however, been shown that dingoes are very similar, so we are able to infer some information from studies of dogs and wolves.

Postural communication and its relationship to social behaviour

Dingoes display a wide array of observable patterns of behaviour such as postures and gestures that express their inner states or feelings, as well as intensity of those feelings. These specialisations in behaviour and postures have evolved to hold the social group together and to reduce aggression among its members.[56] Social behaviour patterns result from different combinations of basic motor capacities that are shared by both dogs and wolves. Kleiman[56] found that social behaviour was similar throughout the entire dog family, differing according to degree rather than kind. Animal behaviourist John Scott[57] described visible aspects of 50 common behaviour patterns in domestic dogs which, for the most part, he and others have also observed in wolves.[58] These social behaviour patterns involve almost every aspect of the animal's body. That is, they consist of the coordinated activity of the face, head, back, tail and anal area, in addition to vocalisations such as whimpering or growling. To date there has not been a systematic description or breakdown of the postural communication in dingoes, although Corbett[16] mentioned them briefly. Corbett and Newsome[37] reported that social postures of the dingo are very similar to those of the domestic dog, but many patterns are more strongly developed and stereotyped than in the domestic dog (particularly the rituals around dominance and submission). Given the similarities with wolves, much of the descriptions given here are those relating to wolves summarised by Mech[40] based on his observations in the wild, and Schenkel's[59] studies of various canid species in captivity.

In the 1930s and 1940s animal behaviourist Rudulph Schenkel[59] conducted what still remains the most extensive study of social behaviour in canids. He began with a study of wolves housed in the Basle Zoological Garden in Switzerland, and later extended his observations to include a pair of dingoes held in the Zurich Zoological Gardens, jackals, foxes and domestic dogs. Unfortunately, his studies were conducted with only captive animals. However, with the exception of his now debunked notion of α or lead wolves being at the centre of the social structure and a large motivator of social interactions (as opposed to living as family groups),[26] most of his findings in relation to dominance–subordinate or leader–follower relationships have been supported by later studies of wild wolf populations (e.g. by wildlife biologist David Mech).[40]

Schenkel[59] (1947) described many canid interactions (particularly agonistic ones), and found that the posture of the animal indicates social status in the group. Social status can be broken down into one of two extremes – dominant/superior status and inferior/subordinate status (and any degree between these two extremes). Social status is most frequently expressed by the movement of facial muscles, movement/activity of the eyes and position of the ears. For example, bared teeth, open mouth, mouth corners pulled forward, a wrinkled and swollen forehead and ears erect and pointed forward all indicate a full threat by a dominant dog. Complete insecurity and subordination are expressed by a closed mouth with corners pulled far back, a smoothed forehead with squinted eyes, and ears drawn way back and held close to the head. The tail, which varies according to its position, shape and movement, is also a helpful indicator of mood and status. During times where there is no social tension, the tail hangs loosely. A dingo threatening or showing dominance raises its tail above the level of its back; a submissive dingo holds its tail very low, often tucked between its legs or curved forward alongside its legs (see Schenkel's[59] *Expression Studies of Wolves* for diagrams of expressive positions of the tail and of facial expressions in the wolf). Tail movements are also indicative of various mood states. For instance a loose, free tail-wagging indicates general friendliness. The position of the tail in relation to the anal genital area is also important, as behaviours relating to inspecting (and therefore presenting or withdrawing) the anal area have various implications.[40, 59]

Social control asserted by higher-ranking individuals is mostly achieved by displaying dominance rather than by constantly displaying threatening behaviour. This can be achieved by the dominant animal doing things as simple as staring intently at others. Lower-ranking individuals will respond by cringing, and will restrict their own glances. This is very different from the interactions we have with dogs, whereby we encourage dogs to look us in the eyes to establish communication and interaction. The use of eye contact has implications for many of our interpretations of dog and wolf behaviour during cognitive experiments that involve human interactions (see Chapter 10). Other ways of achieving social control include standing over, where the dominant animal stands stiffly across the forequarters of a lying individual, or riding up, where the dominant individual places its fore legs across the shoulders of a subordinate either from the side or behind.

Under usual social stability, displays by high-ranking dingoes reaffirm their dominance and subordinate dingoes assume protective patterns of active or passive submission. Active submission is a general reflection of friendliness and tolerance towards the dominant members, and is often seen in group greetings where the inferior pushes the muzzle of the superior with its nose and licks it as well as tail wagging (see Plate 9). Dingoes (and domestic dogs) socialised to humans will greet them in the same way as they greet their conspecifics. Passive submission, however, is often characteristic of timidity and helplessness and is normally displayed by subordinates under direct threat and self-assertion by a higher-ranking animal. Schenkel described this as 'the inferior lies half on its side and half on its back to expose the ventral side of its chest and sometimes the abdomen … The ears are directed backwards and lie close to the head. The tail is more or less bent ventrally so that it passes between the thighs. The often enthusiastic and friendly activity of active submission is reduced … replaced by a passive posture full of trust, devotion and demonstrated helplessness'[60] (p. 323; see Plate 10).

These behavioural patterns form the foundations of dingo society. Like any society there are rules and customs, many of which are learned from parents and older relatives during development and are well and truly entrenched by adulthood. Essentially dingoes need to learn *how* to be a dingo. Corbett[16] described an isolation experiment he conducted (which would now be considered unethical), to test the importance of dominant behaviour with dingo packs. He raised very young pups in isolation then introduced them at varying ages to captive dingoes that had been raised in normal society. He also released a three-month-old pup raised in isolation into a litter of five wild dingoes of similar age. Neither the captive nor the pup released in the wild displayed submissive behaviour. The captive-reared isolated pups were 'mauled' by normal aggressive behaviour from the socialised pups; the pup in the wild became the leader of the group. Corbett concluded that dominant behaviour was innate and that submissive behaviours were learned. Thus, like all social animals, it is important for both wild and captive dingoes to maintain natural behaviours by spending as long as possible with their parents and siblings. Researchers conducting observations of dingoes in captivity must account for the animals' histories when interpreting behaviours.

Communication through odour

Dingoes have a highly developed sense of smell that is employed in social communication. Olfactory behaviour patterns with interacting dingoes involve either the head and neck or the anal genital area. For instance, licking and sniffing the muzzle or mouth may indicate whether that individual has eaten recently, and/or if performed by pups may stimulate an adult to regurgitate food[40] (see Plate 11). Anal genital sniffing (presentation and withdrawal) in relation to dominance was briefly mentioned in relation to visual communication, but it is also thought to have an olfactory aspect. It is likely to play an important role during courtship and mating.[40]

Scent-marking, or leaving chemical signals (such as pheromones) originating from urine, faeces and scent glands[16] on objects at particular sites, is used as a method of communication with members of their own and rival packs (Fig. 2.7). The true details of the information being transmitted remains a bit of a mystery. One purpose may be to synchronise reproduction. For example, it has been shown in other canids that urine can indicate gender, endocrine information can be conveyed in urine, and testosterone and oestrogen both increase urination and thus become attraction to the opposite sex. It may also provide information about an individual and when they were last in the area, and assist with navigation (e.g. marks along roads and near junctions[16] and maintenance of territories. This is a regular behaviour in the life of wild[16, 17] and most certainly in the captive dingoes that I have observed. Thus it is likely that urinating on an object provides enough information that individuals can be identified, at least in terms of age, gender and perhaps rank. It is possible that dingoes will not scent-mark and instead urinate in water or waterways when they are in an unfamiliar territory to avoid detection, or when attempting to avoid communicating. For example, in captivity pregnant dingoes often urinate in their water buckets, presumably to prevent other females from detecting their pregnancy.

Scent-marking is oriented towards conspicuous landmarks (e.g. grass tussocks, fence posts, rocks) presumably so it is more effective or likely to be noticed on novel objects or odours (e.g. animal carcasses, other species' faeces) which may warn other conspecifics of dangerous items

Figure 2.7: Dingo scent-marking alongside a road on Fraser Island.
Source: Bradley Smith.

such as poison baits. These scent marks are often located at shared resources such as water in arid areas, hunting grounds, trails and roads, and particularly intersections.[16] Corbett was curious as to the role of dingo scent-marking, so conducted several observations of the scent-marking behaviour of wild dingoes from arid and tropical Australia, and of a captive colony he kept. He found that urination varied throughout the year but that it was linked to breeding seasons (two to three months before mating period). Males urinated more than females, raked the ground more frequently, urinated with raised leg more often and were more likely to urinate on conspicuous objects. Where there were no conspicuous objects, they used faeces. Corbett believed that at one site at Kapalga in the wet tropics and in the Tanami Desert, Northern Territory, dingoes were using scent posts to indicate which hunting grounds they were using and sharing with other groups. Unfortunately, due to the complexity of the chemicals that make up such olfactory cues and scents and our inability to smell or experience what dingoes can, a lot of what we know remains speculative and is based upon what we can observe.

Vocal communication

The dingo has quite a broad vocal repertoire which is used in several situations such as during social interactions, in food-associated contexts, when changes in the environment are noticed, and for long-range communication. Corbett and Newsome[37] found that dingo vocalisations served six purposes: (1) to locate mating partners, (2) to locate other dingoes for regrouping,

(3) to locate and avoid strangers, (4) to signal intention to come in to water, (5) to warn pups and probably others of danger and (6) to express alarm. The vocalisations of dingoes also closely resemble those of other canids, particularly wolves.[61]

Most of our knowledge of dingo vocalisations comes from two sources, on which the information in this section is based. Ecologist Laurie Corbett[16] is probably the only person to extensively study the vocalisations of dingoes in the wild. He observed a population of wild dingoes in the Simpson Desert, and recorded over 5000 howls and the interactions of two packs around a large waterhole they shared (results of his studies will be discussed throughout this section). In captivity, Deaux and Clarke[61] studied the acoustic repertoire of dingoes and produced a comprehensive summary of dingo vocalisations by video-recording a large captive dingo colony of ~35 individuals living as pairs or in small family groups. They also described the context in which the vocalisations occurred. From this, Deaux and Clarke[61] identified 11 categories of sounds, including nine discrete vocalisations (laryngeal sounds) and two types of nasal sounds. Table 2.3 provides a detailed description of these vocalisations according to the primary contexts in which they were produced. Details of the spectrographic measurements and associated primary contexts are summarised in Table 2.4. Contexts and descriptions of vocalisations are aided by those previously reported by Schassburger[62] in relation to wolves.

It is widely believed by the public and many dingo owners that dingoes are rather quiet and incapable of barking. Although it is rare to hear a dingo bark, it is incorrect to say that they do not, or cannot. Both wolves and dingoes can and do bark, but they do so infrequently and in a limited range of contexts.[16, 64, 65] The vocal repertoire of domestic dogs, on the other hand, largely consists of barking and variations of barking that are used for numerous contexts.[66] For example, dogs might bark when they are happy and excited, but also use a similar bark to warn of an intruder.

A dingo bark is a lot sharper and more abrupt than that of domestic dogs (Fig. 2.8). The bark-howl is an agitated cry, started with one or several barks usually followed by a plateau howl. In the wild, Corbett[16] witnessed this in situations of extreme alarm. The bark-howl is usually directed towards a threat, but also to warn pups or companions of immediate danger. In one instance, Corbett recalled a female that remained near her male companion who had just been shot, repeatedly bark-howling. When the shooter repositioned for a better shot at the female, a previously unsighted pup ran off. The female only stopped bark-howling and moved away when the pup was well away from the humans.

An interesting vocalisation is a snuffing or snuff barking, which generally occurs when dingoes are startled or unsure whether something may represent a threat. The sound is produced by the rapid and repeated explosion of air through the nose, almost like a sneeze (see Fig. 2.9). Corbett[16] believed they did this to 'increase the intake of fresh air containing the scent of the intruder so they can quickly identify and decide whether to fight or flee' (p. 75). The accuracy of this interpretation is uncertain, but the sound most certainly acts as a subtle warning to others of impending danger. Snuffing is common among populations of captive dingoes that I have observed, as well as in the captive dingoes kept by Corbett, and is directed at humans (particularly strangers) when approaching the enclosures. I have seen this behaviour in wild dingoes on Fraser Island, where researchers have closely approached dingoes in

Table 2.3. Summary of dingo vocalisations based on quantitative acoustic parameters and the contexts in which they occur

Dingo vocalisation	Description and context
Nasal sounds	These are classified as nasal sounds, produced by expulsion of air through the nasal passages. Snuffs and snorts are both short in duration, snuffs high in peak frequency whereas snorts are low in peak frequency. Both are produced in the context of frustration and anxiousness and in individuals exhibiting an anxious posture. They are associated with pacing or retreating behaviours. Snorting individuals display an enhanced state of nervousness. These sounds are produced in response to a disturbance in the environment (threat) but their true function remains unclear.
Woof and bark	Woofs and barks are short explosive vocalisations, with barks relatively louder. Woofs and barks are produced across many contexts, but barking individuals display higher levels of arousal. When elicited by a threat, barks are associated with defensive or aggressive postures. Woof vocalisations appear to serve as both a warning and threat signals. Barks seem to warn conspecifics of a threat, with the intensity of the stimulation being conveyed through the gradual transformation of woofs into barks.
Growl and snarl	Growls are uttered with a closed mouth and are highly variable in duration, whereas snarls are produced with an open mouth and are of much shorter duration. Growls show the lowest peak frequency of all dingo sounds, and little variation. Snarls are similarly low in peak frequency. Both are produced in agonistic short-range social and threat contexts. Growls can be directed at dingoes or other species, snarls are specific to fight situations and indicate a high level of aggressiveness. Growling individuals display aggressive postures, indicating that growling is related to threat signals.
Bark-howl	A successive and uninterrupted emission of an extended bark and a howl. Bark-howls have a low peak frequency, are of medium to long duration and are produced when the signaller notices a disturbance in the environment. They seem to serve as a warning to conspecifics, e.g. the presence of an unfamiliar human or a venomous snake. Individuals typically display a defensive posture.
Howl	Typically tonal although some howls do show broadband components. They are long in duration and peak frequency varies widely due to individual variation. The spectral structure varies gradually from a sustained flat tone to a frequency-modulated vocalisation. Used for long-range social communication. Dingoes may be able to identify other individuals based on the auditory characteristics of their howl, as reported in wolves by Tooze *et al.*[63] There are several types of howls, e.g. flat howls, barking howls, combination howls and yip howls. Howls are associated with a distinct posture (lips drawn forward and muzzle pointed towards the sky). Eliciting stimuli are not fully understood, but howls tend to stimulate a reply from others. Responding individuals show strong alert behaviour and increased activity. Howling often escalates into a chorus, where several individuals howl simultaneously.
Whine and whimper	These are tonal vocalisations of short to medium duration. Whines are low and whimpers high in peak frequency. These sounds are used in short-range contexts (social, threat or feeding situations). During threat contexts, vocalising individuals simultaneously display anxious behaviours such as pacing, suggesting they are linked with nervousness, anxiety and/or high levels of arousal. During social interactions, these vocalisations are likely to serve as appeasement signals and are associated with submissive postures and displays. Whines and whimpers are most often associated with distress, anxiousness, appeasement and submission.
Yelp	These are short tonal vocalisations with a high peak frequency. Yelps are produced during agonistic short-range social interactions and are associated with submissive displays. Individuals may yelp as a consequence of receiving pain, encouraging the aggressor to stop its action.

Source: Deaux and Clarke.[61]

Table 2.4. Spectral structure of dingo vocalisations

Sound category	Spectral structure	Primary contexts of production	Duration (ms)			Peak frequency (kHz)		
			Median	Min	Max	Median	Min	Max
Snuff	Broadband	Frustration, anxiousness	12	8	24	4.1	2.7	6
Snort	Broadband	Frustration, anxiousness	12	8	20	1.3	0.2	4.7
Woof	Broadband	Warning, defence	13	8	24	0.5	0.2	2.9
Bark	Broadband	Warning, defence	21	11	30	0.6	0.5	1.5
Growl	Broadband	Threat, attack	128	18	489	0.2	0.1	0.8
Snarl	Broadband	Threat, attack	70	12	150	0.5	0.2	2.9
Bark-howl	Composite	Warning, defence	171	80	360	0.6	0.4	1.5
Howl	Tonal	Long-range social	243	91	615	0.6	0.4	1.5
Whine	Tonal	Frustration, anxiousness	32	6	172	0.6	0.3	3.9
Whimper	Tonal	Frustration, anxiousness	18	5	91	4.9	4.1	10.5
Yelp	Tonal	Fear, pain	13	4	177	1.7	0.9	3.1

Broadband = broad frequency bandwidth. Tonal = narrow frequency bandwidth. Composite = both broadband and tonal components are present.

Source: Recreated from Deaux and Clarke.[61]

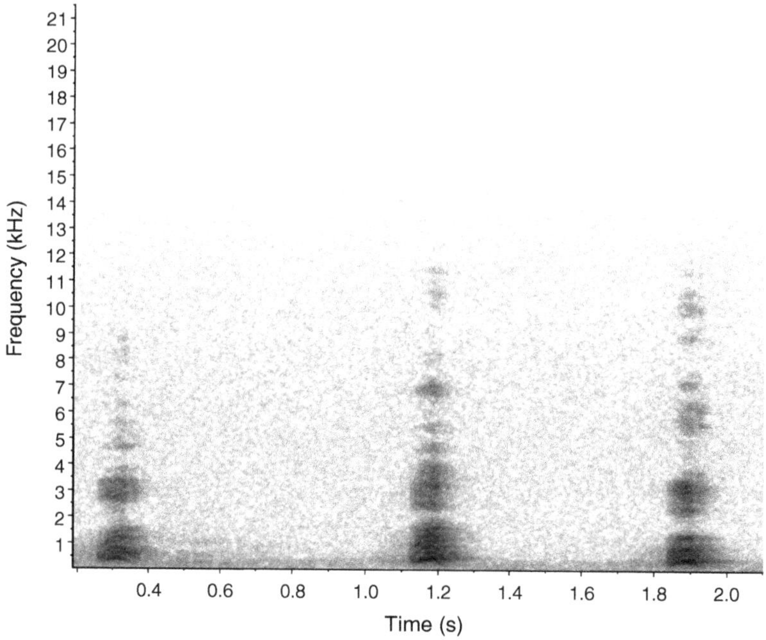

Figure 2.8: A visual representation of a sequence of three barks from a dingo. Time (sec) is displayed on the horizontal axis), frequency (kHz) on the vertical axis and relative power (loudness) on the grey-scale axis. The spectrogram shows that barks are short (10–20 ms) and explosive vocalisations. While the frequency of the vocalisation ranges from 500 Hz to 12 kHz, the pitch (the 'loudest' frequency) is about 500 Hz. Hann window with 800 samples/s, 75% overlap, DFT = 1024 samples.

Source and analysis courtesy of Eloise Deaux.

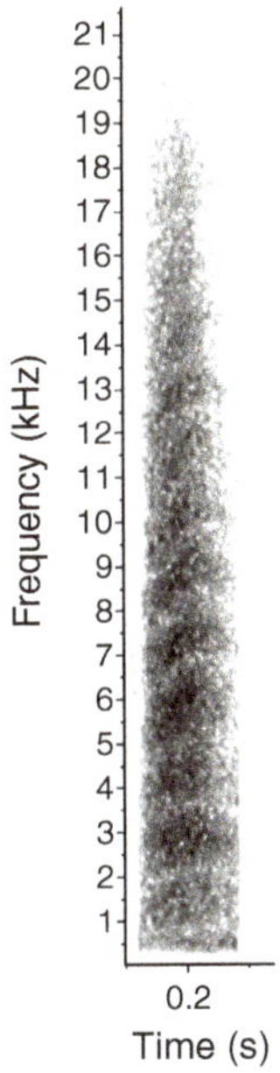

Figure 2.9: Spectrogram of a snuff vocalisation, produced by the expulsion of air through the nasal passages. Unlike howls, where clear frequencies are visible (i.e. dark, marked lines), snuffs are 'noisy' signals. In other words, they are made up of random energy distributed along the frequency range. Hann window with 600 samples/s, 75% overlap, DFT = 1024 samples.

Source and analysis courtesy of Eloise Deaux.

the forest areas. Although only seen directed towards humans, it would serve equally well for other threats that face a dingo group.

Like wolves, the vocalisation most expressed by dingoes is howling (Figs 2.10, 2.11). From his observations, Corbett determined that there were three basic types of howling: plateau (holding one note, with an even rise and fall of amplitude), inflected (yodel, an abrupt rise and protracted fall) and abrupt (yelp, a rise held as a brief high note). He suggested there are 10 variations of these three main types.[16, 37] He noticed that the frequency and duration of howling were highest at dawn (6 am) and dusk (7 pm). In Corbett's study area, howling was highly seasonal, occurring almost exclusively during the mating season (March–May).[16] Thus, the frequency of howling is affected by breeding season, dispersal, lactation and social stability.[16] Corbett noted another small peak in howling October/December, which coincides with the training and weaning of pups. On Fraser Island, I have witnessed puppies howl to greet parents that were out of sight but on their way back from a hunting excursion. Pup howling is quite distinct, with a much higher pitch. It is often of short duration or cuts off abruptly after the maximum pitch is reached.[16]

Howling is a crucial communication tool common to all wild canids because they are often spatially separated.[67] Because howls often travel over large distances, they help determine the location of others. From his observations, Corbett concluded that there were two main purposes of howling in dingoes – to attract pack members or to repel intruders from the pack home range. Plateau howls, which are the most commonly used howl, appear not to identify

Figure 2.10: Dingo howling at the Dingo Discovery Centre.
Source: Bradley Smith.

the individual but simply to provide information on their whereabouts. They also convey a message likened to 'I am here, where are you?' In captivity, such howls are regularly used when dingoes cannot see each other, to determine that others are still nearby. This often leads to chorus howling where multiple, if not all, members howl at the same time.

Like other wild canids, dingoes howl with distinct pitches during chorus howl. As the number of animals howling in a group increases, so does the variation in pitches.[16, 61] Thus, a chorus contains different howls, inflexions and changes in pitch and duration. These provide information about the individuals howling and the intention of the howl. The ability to do this suggests that canids can estimate the size of an unseen pack.[68] It is possible that these highly variable chorus-howls might generate a confounding effect by making pack size appear larger and therefore repel potential rivals from their territory and resources.[69]

Armed with the knowledge that howling is used for locating group mates, a popular method used by doggers (dingo hunters) when shooting dingoes is 'howling them up', i.e. imitating a dingo howl. For a time there were regular competitions in rural Australia to see which farmer/hunter had the best howl.[70–72]

Dingo diet and hunting strategy

Habitat type and prey availability determine the dingo's diet and hunting strategies. These range from opportunistic feeding and hunting, to targeted predation of large kangaroos and wallabies.[25] Thus, dingoes are considered opportunistic predators as they tend to prey upon

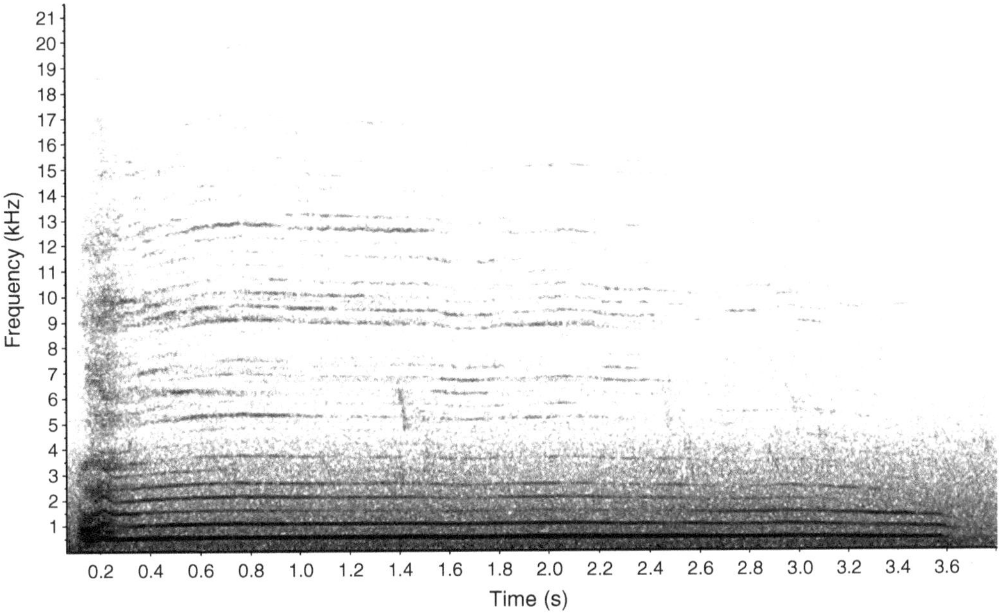

Figure 2.11: Spectrogram of a plateau howl vocalisation produced by an adult wild dingo. The howl starts with an abrupt on-glide (0.1–0.2 sec) and continues in a 'plateau' shape until 3.6 sec. The pitch is on the fundamental frequency (i.e. first frequency) at ~600 Hz. The following 'lines' are called harmonics and are integer multiples of the fundamental frequency. Harmonics up to 20 kHz are visible on this spectrogram. Hann window with 800 samples/s, 75% overlap, DFT = 1024 samples.

Source and analysis courtesy of Eloise Deaux.

species at a rate similar to the expected rate of encounter.[73] In most locations, however, dingoes focus on only a narrow range of species.[16] Corbett[16] provided a comprehensive analysis of the diet of dingoes determined from studies conducted by various research teams between 1966 and 1986. The summary covered many areas of Australia (there was little variation across regions) and included data from over 12 000 faecal and stomach samples. A total of 177 species were detected, the majority (72%) being mammals, with the remaining comprising birds (19%), vegetation (3%), reptiles (2%) and insects (1.3%). Corbett concluded that around 20% of the mammalian prey in the dingo diet were large mammals (those that weigh substantially more than a dingo, >15 kg) and require a pack of dingoes hunting cooperatively to kill it. The remaining mammalian prey species (small and medium, <15 kg) could be killed by a dingo hunting alone. More recent fieldwork around Australia supports those findings.[27, 74, 75] It appears that livestock animals (cattle and sheep) account for only a small amount of the dingo's diet, and that dingoes actually prefer relatively smaller prey.[27, 75] There have been observations of dingoes scavenging on the remains of cattle, red kangaroos and other dingo carcasses, suggesting that the contents of stomachs or scats are not necessarily reflective of predation.[76]

When foraging, the size of hunting groups is determined by the type and abundance of prey, with groups forming only to hunt large prey.[24, 27] Dingoes have well-developed sense organs, which are utilised when hunting. Individuals hunting alone rely more on scent than sight to trail the prey; they are often several hundred metres behind the prey animal during a chase that sometimes lasts hours.[25] Small prey (e.g. rabbits) is often hunted by sighting followed by chasing, or by scenting and pouncing in grassland. Dingoes are particularly successful at locating rabbit warrens, which are usually inspected on a dingo's regular hunting circuit.[16] Dingoes rely on hearing and smell to find rodents, grasshoppers and other prey moving in grass or vegetation, then pounce on them.[25] Caching (storing food for later use) of prey has also been observed, as has frequent returning to carcasses and the defence of such carcasses from other dingoes.[16]

Dingo diets throughout Australia consist mainly of animals from the Family Macropodidae ('large foot'),[27, 77–81] which includes kangaroos, wallabies, tree-kangaroos and pademelons. In a study by Jarman and Wright,[82] dingoes did not seem to select their prey based on size of kangaroo group but rather on whether groups contained large adult males and females with medium pouch young. A technique used by dingoes to hunt kangaroos is to bite and injure the hind leg as well as the neck and throat.[74, 83] Of the seven occasions when Jarman and Wright[82] observed dingoes chasing kangaroos, six involved solitary dingoes and one a pair. Of the 10 dingo–kangaroo interactions observed by Marsack and Campbell,[74] all involved a solitary dingo and only two were successful. Three methods of predation were observed. (1) The dingo stalked a stationary kangaroo, downwind, to within 20 m before running at it. (2) The dingo trotted directly towards the kangaroo (who was aware of the dingo), then ran fast when the kangaroo hopped away. (3) Both animals were unaware of each other, until the kangaroo was flushed from cover by a passing dingo. Chases lasted from 200 m to 4.4 km. Of the hunts observed, 15.6% involved three or more dingoes; however, these chases never involved large prey. The success rate was typically low, with only 4.5–10.8% ending in a kill. Low success rates are likely to be a result of herbivore defence systems, such as group members clustering around weaker members or fleeing to water,[82, 84, 85] and times of low prey densities (e.g. in drought).[27]

Dingoes in groups are far more successful at 'bailing up' large prey than are solitary dingoes.[24] Higher numbers of dingoes provide an advantage because the leading dingo can focus on changing the prey animal's direction of travel, driving it into the path of the other dingoes, which in turn can cut off the prey animal.[24] The hunting methods of dingoes are similar to the techniques employed by wolves.[24, 40] Pack hunting depends on the size of the prey and is more likely in areas of high prey density; it is less used in desert regions where smaller prey items are more common. For example, in the Nullarbor Plain Marsack and Campbell[74] witnessed only solitary hunting by dingoes. They hypothesised that 'since solitary dingoes were more successful at obtaining rabbits, solitary hunting of rabbits represents a better return for effort than hunting in a group' (p. 356). This might suggest that cooperative hunting of larger prey is performed only during periods of drought where smaller and medium-sized prey (more easily caught) are less abundant.[86]

Robertshaw and Harden[78] found evidence suggesting that dingoes may practise 'farming'. A common reaction of kangaroo and wallaby mothers (*Wallabia bicolor*) to harassment by dingoes is to abandon young-at-foot or 'eject' pouch young during flight. The latter was observed by those authors on three occasions. At this particular study site (compared to those nearby), the wallaby breeding season had been extended and there was greater turnover of pregnancies to account for the constant loss of young. It is possible that dingoes deliberately target mothers during predatory interactions; however, the authors stated that the selection of *W. bicolor* individuals was not biased towards one sex.

Webb[87] reported a case where three dingoes successfully killed a large lace monitor (*Varanus varius*). Lace monitors are potentially dangerous (given their large sharp claws and formidable teeth) and difficult prey (easily able to escape by climbing a tree or entering burrows). Two dingoes simultaneously attacked the lace monitor, one biting it behind the head while the second grabbed the tail. Both dingoes held on, pulling in opposite directions. About 30 sec into the two-minute event, a third dingo appeared; it bit into the monitor's exposed abdomen and disembowelled it. As soon as the monitor appeared dead, the two holding dingoes dropped it and the three started to consume the lizard. Webb[87] believed the attack was 'well coordinated' and 'greatly reduced the chances of the lizard being able to bite or scratch the dingoes' (p. 55).

Conclusion

In every practical sense, the dingo is a wild canid and one that is uniquely Australian. Perhaps the dingo could be considered the Australian canid ecotype or ecospecies. An ecotype is a genetically distinct geographic variety, population or race within (or among closely related) species. It is adapted to specific environmental conditions yet is still capable of interbreeding with other geographically adjacent ecotypes without loss of fertility or vigour. This is because the dingo is extremely abundant with only small variation across the Australian continent, and has presumably changed little over the last 3500–5000 years.[88] A limited founding population[9, 89] and isolation would have resulted in close breeding, with natural selection weeding out the least fit to leave only the fittest, hardiest and most opportunistic.

Yet, as Chapter 1 highlighted, the definition of a dingo remains a topic of much debate. In some ways it is difficult to determine where to draw a line between a dingo and its close relatives such as the dog, hybrid dog–dingoes, New Guinea singing dogs or the village dogs in Asia (see Chapter 3 for a more detailed discussion of genetic comparisons). Some argue that it doesn't really matter what a dingo is – that as long as it is able to survive in the wild (hunt its own food and raise young) and is a functional member of the ecosystem, then we should just consider it a dingo and leave it at that.[90] After all, any canid that's not like a dingo (e.g. in anatomy, intelligence, hunting skills, social and parenting behaviour) is surely not likely to survive. There is still benefit in trying to determine and identify what a pure dingo is, so that we know what we are attempting to conserve and what traits are most important.[91] However, if the dingo is constantly evolving to meet its ever-changing environment, then maybe our definition of dingo needs to change with it.

By identifying and distinguishing pure dingoes from hybrids according to levels of dog ancestry (instead of 'hybrid' or 'not hybrid'), or even purity based on visual methods (which has been shown to be a moderately reliable method),[92] theoretically, dog ancestry in wild populations could be removed, or at least the rate of hybridisation could be slowed. However, given the high proportion of hybrids in the wild, culling hybrids in some areas (e.g. eastern

Australia) is not a valid option because it could result in substantial demographic loss. When implementing control methods, it remains difficult to determine dingo purity (or the level of purity), particularly in the field and when using indiscriminate methods of control such as poison baits and traps. It is important that wildlife managers do their best to preserve what we understand to be pure dingo traits and to preserve populations with the highest known purity, and that we all learn to tolerate hybrids with a high percentage of dingo genes.[8]

References

1. Corbett LK (2001) The conservation status of dingoes (*Canis lupus dingo*) in Australia, with particular reference to New South Wales: threats to pure dingoes and possible solutions. In *A Symposium on the Dingo*. (Eds C Dickman and D Lunney) pp. 10–19. Transactions of the Royal Zoological Society of New South Wales, Sydney.
2. Corbett LK (2008) Dingo: *Canis lupus*. In *The Mammals of Australia*. (Eds S Van Dyck and R Strachan) pp. 737–739. Reed New Holland, Sydney.
3. Holden P (1991) *Along the Dingo Fence*. Hodder and Stoughton, Sydney.
4. Woodford J (2013) *The Dog Fence: A Journey Across the Heart of Australia*. Text Publishing, Melbourne.
5. Stephens D (2011) The molecular ecology of Australian wild dogs: hybridisation, gene flow and genetic structure at multiple geographic scales. PhD thesis. University of Western Australia, Perth.
6. Newsome AE, Corbett LK (1985) The identity of the dingo. III. The incidence of dingoes, dogs and hybrids and their coat colours in remote and settled regions of Australia. *Australian Journal of Zoology* **33**, 363–375.
7. Woodall PF, Pavlov P, Twyford KL (1996) Dingoes in Queensland, Australia: skull dimensions and the identity of wild canids. *Wildlife Research* **23**, 581–588.
8. Elledge AE, Leung LK, Allen LR, Firestone K, Wilton AN (2006) Assessing the taxonomic status of dingoes *Canis familiaris dingo* for conservation. *Mammal Review* **36**, 142–156.
9. Savolainen P, Leitner T, Wilton A, Matisoo-Smith E, Lundeberg J (2004) A detailed picture of the origin of the Australian dingo, obtained from the study of mitochondrial DNA. *Proceedings of the National Academy of Sciences of the United States of America* **101**, 12387–12390.
10. Daniels MJ, Corbett L (2003) Redefining introgressed protected mammals: when is a wildcat a wild cat and a dingo a wild dog? *Wildlife Research* **30**, 213–218.
11. Vilà C, Wayne RK (1999) Hybridization between wolves and dogs. *Conservation Biology* **13**, 195–198.
12. Walters BJ (1995) *The Company of Dingoes: Two Decades with our Native Dog*. Australian Native Dog Conservation Society, Bargo.
13. Jones E (1990) Physical characteristics and taxonomic status of wild canids, *Canis familiaris*, from the eastern highlands of Victoria. *Australian Wildlife Research* **17**, 69–81.
14. Glen AS, Dickman CR, Soulé ME, Mackey BG (2007) Evaluating the role of the dingo as a trophic regulator in Australian ecosystems. *Austral Ecology* **32**, 492–501.
15. Burt WH (1943) Territoriality and home range concepts as applied to mammals. *Journal of Mammalogy* **24**, 346–352.
16. Corbett LK (2001) *The Dingo in Australia and Asia*. JB Books, Adelaide.
17. Thomson PC (1992) The behavioural ecology of dingoes in north-western Australia. I. The Fortescue River study area and details of captured dingoes *Wildlife Research* **19**, 509–518.
18. Newsome TM, Ballard GA, Dickman CR, Fleming PJ, van de Ven R (2013) Home range, activity and sociality of a top predator, the dingo: a test of the resource dispersion hypothesis. *Ecography* **36**, 914–925.
19. Newsome TM, Ballard GA, Dickman CR, Fleming PJS, Howden C (2013) Anthropogenic resource subsidies determine space use by Australian arid zone dingoes: an improved resource selection modelling approach. *PLoS ONE* **8**(5), e63931.

20. McIlroy JC, Cooper RJ, Gifford EJ, Green BF, Newgrain KW (1986) The effect on wild dogs, *Canis familiaris*, of 1080-poisoning campaigns in Kosciusko national park, NSW. *Australian Wildlife Research* **13**, 535–544.
21. Harden RH (1985) The ecology of the dingo in north-eastern New South Wales. I. Movements and home range. *Australian Wildlife Research* **12**, 25–37.
22. Thomson PC, Marsack PR (1992) Aerial baiting of dingoes in arid pastoral areas with reference to rabies control. In *Wildlife Rabies Contingency Planning in Australia.* (Eds P O'Brien and G. Berry) pp. 125–134. Australian Government Publishing Service, Canberra.
23. Thomson PC (1992) The behavioural ecology of dingoes in north-western Australia. IV. Social and spatial organisation, and movements. *Wildlife Research* **19**, 543–563.
24. Thomson PC (1992) The behavioural ecology of dingoes in north-western Australia. II. Activity patterns, breeding season and pup rearing. *Wildlife Research* **19**, 519–530.
25. Fleming P, Corbett L, Harden R, Thomson P (2001) *Managing the Impacts of Dingoes and Other Wild Dogs*. Bureau of Rural Sciences, Canberra.
26. Mech LD (1999) Alpha status, dominance, and division of labour in wolf packs. *Canadian Journal of Zoology* **77**, 1196–1203.
27. Thomson PC (1992) The behavioural ecology of dingoes in north-western Australia. III. Hunting and feeding behaviour, and diet. *Wildlife Research* **19**, 531–541.
28. Corbett LK (1988) Social dynamics of a captive dingo pack: population regulation by dominant female infanticide. *Ethology* **78**, 177–198.
29. Lord K, Feinstein M, Smith B, Coppinger R (2013) Variation in reproductive traits of members of the genus *Canis* with special attention to the domestic dog (*Canis familiaris*). *Behavioural Processes* **92**, 131–142.
30. Mech LD (2002) Breeding season of wolves, *Canis lupus*, in relation to latitude. *Canadian Field Naturalist* **116**, 139–141.
31. Jones E, Stevens PL (1988) Reproduction in wild canids, *canis familiaris*, in the eastern highlands of Victoria. *Australian Wildlife Research* **15**, 385–394.
32. Catling PC, Corbett LK, Newsome AE (1992) Reproduction in captive and wild dingoes (*Canis familiaris dingo*) in temperate and arid environments of Australia. *Wildlife Research* **19**, 195–209.
33. Catling PC (1979) Seasonal variation in plasma testosterone and the testis in captive male dingoes, *Canis familiaris dingo. Australian Journal of Zoology* **27**, 939–944.
34. Zuckerman S (1953) The breeding seasons of mammals in captivity. *Proceedings of the Zoological Society of London* **122**, 827–950.
35. Breckwoldt R (1988) *A Very Elegant Animal: The Dingo*. Angus and Robertson, Sydney.
36. Newsome AE, Corbett LK, Best LW, Green B (1972) The dingo. *Australian Meat Research Committee Review* **14**, 1–11.
37. Corbett L, Newsome AE (1975) Dingo society and its maintenance: a preliminary analysis. In *The Wild Canids: Their Systematics, Behavioral Ecology and Evolution.* (Ed. MW Fox). Van Nostrand Reinhold, New York.
38. Koler-Matznick J, Yates BC, Bulmer S, Brisbin IL (2007) The New Guinea singing dog: its status and scientific importance. *Australian Mammalogy* **29**, 47–56.
39. Boitani L, Francisci F, Ciucci P, Andreoli G (1995) Population biology and ecology of feral dogs in central Italy. In *The Dog: Its Ecology, Behaviour and Evolution.* (Ed. J Serpell). Cambridge University Press, Cambridge.
40. Mech LD (1970) *The Wolf: Ecology and Behaviour of an Endangered Species.* Natural History Press, New York.
41. Harrington F, Mech L (1982) Patterns of homesite attendance in two Minnesota wolf packs. In *Wolves of the World: Perspectives of Behavior, Ecology, and Conservation.* (Eds FH Harrington and PC Paquet) pp. 81–105. Noyes Publications, New Jersey.
42. Ballard WB, Ayres LA, Gardner GL, Foster JW (1991) Den site activity patterns of gray wolves, *Canis lupus*, in south-central Alaska. *Canadian Field Naturalist* **105**, 497–504.

43. Daniels TJ (1983) The social organization of free-ranging urban dogs. I. Non-estrous social behavior. *Applied Animal Ethology* **10**, 341–363.
44. Ghosh B, Choudhuri D, Pal B (1984) Some aspects of the sexual behaviour of stray dogs, *Canis familiaris*. *Applied Animal Behaviour Science* **13**, 113–127.
45. Green BF, Catling PC (1977) The biology of the dingo. In *Australian Animals and their Environment*. (Eds H Messel and ST Butler) pp. 51–60. Shakespeare Head Press, Sydney.
46. Packard JM (2003) Wolf behaviour: reproductive, social, and intelligent. In *Wolves: Behaviour, Ecology, and Conservation*. (Eds LD Mech and L Boitani) pp. 35–65. University of Chicago Press, Chicago.
47. Paquet P, Bragdon S, McCusker S (1982) Cooperative rearing of simultaneous litters in captive wolves. In *Wolves of the World: Perspectives of Behavior, Ecology, and Conservation*. (Eds FH Harrington and PC Paquet) pp. 223–237. Noyes Publications, New Jersey.
48. Scott J, Fuller J (1965) *Genetics and the Social Behavior of the Dog*. University of Chicago Press, Chicago.
49. Pal SK (2005) Parental care in free-ranging dogs, *Canis familiaris*. *Applied Animal Behaviour Science* **90**, 31–47.
50. Pal SK (2008) Maturation and development of social behaviour during early ontogeny in free-ranging dog puppies in West Bengal, India. *Applied Animal Behaviour Science* **111**, 95–107.
51. Martins T (1949) Disgorging of food to the puppies by the lactating dog. *Physiological Zoology* **22**, 169–172.
52. Daniels TJ, Bekoff M (1989) Population and social biology of free-ranging dogs, *Canis familiaris*. *Journal of Mammalogy* **70**, 754–762.
53. Harden R (1981) A look at the dingo. *Australian Natural History* **20**, 191–194.
54. Jackson S (2007) Dingoes. In *Australian Mammals: Biology and Captive Management*. (Ed. S Jackson) pp. 381–407. CSIRO Publishing, Melbourne.
55. Thomson PC, Rose K (1992) Age determination of dingoes from characteristics of canine teeth. *Wildlife Research* **19**, 597–599.
56. Kleiman DG (1967) Some aspects of social behavior in the Canidae. *American Zoologist* **7**, 365–372.
57. Scott J (1950) The social behaviour of dogs and wolves: an illustration of sociobiological systematics. *Annals of the New York Academy of Sciences* **51**, 1009–1021.
58. Ginsburg B (1965) Coaction of genetical and nongenetical factors influencing sexual behaviour. In *Sex and Behaviour*. (Ed. F Beach) pp. 53–75. John Wiley and Sons, New York.
59. Schenkel R (1947) Expression studies of the wolf. *Behaviour* **1**, 81–129.
60. Schenkel R (1967) Submission: its features and function in the wolf and dog. *American Zoologist* **7**, 319–329.
61. Déaux EC, Clarke JA (2013) Dingo (*Canis lupus dingo*) acoustic repertoire: form and contexts. *Behaviour* **150**, 75–101.
62. Schassburger RM (1987) Wolf vocalization: an integrated model of structure, motivation and ontogeny. In *Perspectives in Vertebrate Science*. Vol. 4: *Man and Wolf: Advances, Issues and Problems in Captive Wolf Research*. (Ed. H Frank) pp. 313–347. Dr W. Junk Publishers, Dordrecht.
63. Tooze ZJ, Harrington FH, Fentress JC (1990) Individually distinct vocalizations in timber wolves, *Canis lupus*. *Animal Behaviour* **40**, 723–730.
64. Fox MW (1971) *Behaviour of Wolves, Dogs and Related Canids*. Jonathan Cape, London.
65. Harrington FH, Mech LD (1978) Wolf vocalization. In *Wolf and Man: Evolution in Parallel*. (Eds RL Hall and HS Sharp). Academic Press, New York.
66. Yin S, McCowan B (2004) Barking in domestic dogs: context specificity and individual identification. *Animal Behaviour* **68**, 343–355.
67. Theberge JB, Falls BJ (1967) Howling as a means of communication in timber wolves. *American Zoologist* **7**, 331–338.
68. Harrington FH, Mech LD (1979) Wolf howling and its role in territory maintenance. *Behaviour* **68**, 207–249.

69. Ortolani A (1990) Howling vocalizations of wild and domestic dogs: a comparative behavioural and anatomical study. BA dissertation. Hampshire College, Amherst, M.A.
70. Newsome C (1985) *Dingo Howlers*. Colin J. Newsome, Glen Innes, NSW.
71. Bacon JS (1960) *The Australian Dingo: King of the Bush*. Glasson and Oliver, Sydney.
72. Murray AG (1972) Trapping dingoes: a dying art. *Walkabout* **26**, 32–34.
73. Vernes K, Dennis A (2001) Mammalian diet and broad hunting strategy of the dingo (*Canis familiaris dingo*) in the wet tropical rainforests of north-eastern Australia. *Biotropica* **33**, 339–345.
74. Marsack P, Campbell G (1990) Feeding behaviour and diet of dingoes in the Nullarbor region, Western Australia. *Australian Wildlife Research* **17**, 349–357.
75. Brook LA, Kutt AS (2011) The diet of the dingo (*Canis lupus dingo*) in north-eastern Australia with comments on its conservation implications. *Rangeland Journal* **33**, 79–85.
76. Allen BL (2010) Skin and bone: observations of dingo scavenging during a chronic food shortage. *Australian Mammalogy* **32**, 207–208.
77. Newsome AE, Catling PC, Corbett LK (1983) The feeding ecology of the dingo. II. Dietary and numerical relationships with fluctuating prey populations in south-eastern Australia. *Australian Journal of Ecology* **8**, 345–366.
78. Robertshaw JD, Harden RH (1986) The ecology of the dingo in north-eastern New South Wales. IV. Prey selection by dingoes, and its effect on the major prey species, the swamp wallaby, *Wallabia bicolor* (Desmarest). *Australian Wildlife Research* **13**, 141–163.
79. Robertshaw JD, Harden RH (1985) The ecology of the dingo in north-eastern New South Wales. II. Diet. *Australian Wildlife Research* **12**, 39–50.
80. Robertshaw JD, Harden RH (1985) The ecology of the dingo in north-eastern New South Wales. III. Analysis of macropod bone fragments found in dingo scats. *Australian Wildlife Research* **12**, 163–171.
81. Whitehouse SJO (1977) The diet of the dingo in Western Australia. *Australian Wildlife Research* **4**, 145–150.
82. Jarman PJ, Wright SM (1993) Macropod studies at Wallaby Creek. IX. Exposure and responses of eastern grey kangaroos to dingoes. *Wildlife Research* **20**, 833–843.
83. Shepherd NC (1981) Predation of red kangaroos, *Macropus rufus*, by the dingo, *Canis familiaris dingo* (Blumenbach), in north-western New South Wales. *Australian Wildlife Research* **8**, 255–262.
84. Purcell BV (2010) A novel observation of dingoes (*Canis lupus dingo*) attacking a swimming eastern grey kangaroo (*Macropus giganteus*). *Australian Mammalogy* **32**, 201–204.
85. Wright SM (1993) Observations of the behaviour of male eastern grey kangaroos when attacked by dingoes. *Wildlife Research* **20**, 845–849.
86. Corbett LK, Newsome AE (1987) The feeding ecology of the dingo. III. Dietary relationships with widely fluctuating prey populations in arid Australia: an hypothesis of alternation of predation. *Oecologia* **74**, 215–227.
87. Webb JK (1996) Observation of three dingoes killing a large lace monitor (*Varanus varius*). *Australian Mammalogy* **19**, 55–56.
88. Macintosh NWG (1975) The origin of the dingo: an enigma. In *The Wild Canids: Their Systematics, Behavioural Ecology and Evolution*. (Ed. MW Fox) pp. 87–106. Van Nostrand Reinhold, New York.
89. Savolainen P, Zhang Y, Luo J, Lundeberg J, Leitner T (2002) Genetic evidence for an east Asian origin of domestic dogs. *Science* **298**, 1610–1613.
90. Purcell B (2010) *Dingo*. CSIRO Publishing, Melbourne.
91. Crowther M, Fillios M, Colman N, Letnic M (2014) An updated description of the Australian dingo (*Canis dingo* Meyer, 1793). *Journal of Zoology* **293**, 192–203.
92. Elledge AE, Allen LR, Carlsson B-L, Wilton AN, Leung LKP (2008) An evaluation of genetic analyses, skull morphology and visual appearance for assessing dingo purity: implications for dingo conservation. *Wildlife Research* **35**, 812–820.

3
The origin and ancestry of the dingo

Bradley Smith and Peter Savolainen

The dingo is a relatively recent arrival to Australia, yet much of its origin and ancestry remain unclear. It is a topic that has generated much interest, and troubled many scientists and natural historians since a dog-like creature was first observed on Australian shores. Several theories have been offered. Some have suggested that the dingo is indigenous to Australia that either arrived as a wild animal with the first people to Australia some 50 000 years ago, or came to Australia unaided by humans. Others suggest the dingo is a domestic dog brought to Australia by seafaring people (for a summary of these early theories see).[1] Macintosh, an anatomist and anthropologist researching the origins of Indigenous Australians and dingoes, summarised the situation quite well after failing to find any definitive answer after decades of study. At the end of his quest, he concluded 'the greatest problem in trying to write about the dingo is that one has no proof of the animal's identification, ancestry, affinity, place of origin, or precise time of arrival in Australia; even the questions whether it was on arrival a partly-domesticated animal, subsequently becoming feral, or whether it is truly a wild dog, is disputed by many writers, and on almost completely polemical grounds' (p. 87).[2] Some 40 years have passed since Macintosh made this statement. Fortunately, the advent of molecular biology, in combination with an improved understanding of archaeological and anthropological evidence, has greatly assisted in answering many questions. In all, there are many pieces that help us understand the history and origin of the dingo, including fossil records, morphological comparisons, human and physical geography, Indigenous culture, and genetics. Each of these will be discussed to help determine when dingoes first arrived in Australia, whether they are indigenous to Australia or introduced, and who their ancestor may be.

When did the dingo arrive in Australia?

Fossil evidence

The archaeological evidence indicates that dingoes have been in Australia for at least 3500 years (or since the mid Holocene epoch, according to the geological timeline that extends from present day to 11 700 years ago). This date is based on the discovery of dingo remains located in natural pit-traps, in open preserving environments and in deposits with archaeological components throughout Australia. Most of these fossils, whether full skeletons or elements such as teeth and bones, cluster around a similar period, the oldest dated to 3500 years old found at Madura Cave on the Nullarbor Plain (see Table 3.1). There are no reports or archaeological evidence of dingoes in the Pleistocene epoch (11 700 years ago – the period which included the last ice age). For a time the dingoes' arrival was thought to be 7500–8000 years ago based upon fragments of bone found in the lower levels of Mt Burr rockshelter in South Australia. These fragments were dated at around 8000 years old because they were

Table 3.1. Locations and radiocarbon dates (years before present ± measurement error) for dingo specimens dated older than 1000 years

Region	Lat.	Long.	Site name	Site type	Laboratory code	Material for dating	Type of remains	Depth (cm)	14c age	14c error	Source
Northern Territory											
Northern Kimberley	125.78	–14.90	Ngurini	Rockshelter	Wk-1615	Charcoal	Skeletal elements	50	2820	90	12, 13
Western Australia											
Nullarbor	127.04	–31.98	Madura (Mereguda N62)	Rockshelter	ANU 807	Charcoal	Skeletal elements	75–90	3450	95	14, 15
Nullarbor	127.44	–31.42	Thylacine Hole (N63)	Rockshelter	NSW 30	Tissue	Desiccated animal	–	2200	96	16, 17
South Australia											
Murray–Darling Basin	139.54	–34.79	Fromm's Landing no. 6	Rockshelter	NPL-63	Charcoal	Articulated skeleton	23	3450	90	6, 18
Murray–Darling Basin	139.61	–34.69	Devon Downs	Rockshelter	GaK-1021	Charcoal	Articulated skeleton	350	2980	90	4, 18
Naracoorte Coastal Plain	140.50	–37.62	Mt Burr	Rockshelter	Gak 426	Charcoal	Skeletal elements	61–121	1020	40	4, 19, 20
New South Wales											
South Western Slopes	150.34	–33.10	Capertee (Site 3)	Rockshelter	V-33	Charcoal	Skeletal elements	20–25	2865	60	4, 21
Sydney Basin	150.42	–35.37	Burrill Lake	Rockshelter	ANU 139	Scattered charcoal	Unknown	27	1660	70	22
Sydney Basin	151.25	–33.83	Balmoral Beach (site 2)	Rockshelter	Beta-63204	Charcoal	Skeletal elements	117	3310	50	23, 24
North Coast	153.17	–29.22	Wombah 1	Midden	Gak 568	Charcoal	Skeletal elements	55–66	3230	100	25–29

found between layers dated before and after that age. However, this date was shown to be incorrect, with later excavations indicating that the layers had been disturbed and a few pieces of bone probably moved into an earlier level.[3,4] Macintosh[5] believed that a tell-tale sign that dingoes have not been here much longer than this time, is the absence of dingo remains among older fossil sites where there are 'thousands of specimens of extinct marsupials'. Of course this does not directly tell us anything about the time of dingo arrival, but the absence of remains in areas where older specimens have been found could suggest a relatively recent arrival. The methods and reliability of dating may influence these times, and there is nothing to suggest that the archaeological record is complete. That is, there is a chance that there are older fossils that have yet to be uncovered. It does, however, provide the most tangible evidence available indicating the timing of their existence in Australia, and thus must provide the foundation for any theory of arrival.

Many of the fossils are fragments, with the exception of the near-complete skeleton of an 18-week-old dingo found by John Mulvaney at Fromm's Landing along the River Murray.[6] This skeleton, rebuilt by Macintosh and dated at 3000 years old, provides a useful standard comparison with present-day skeletons. To achieve these comparisons, Macintosh[1] held a large private collection of dingo samples collected from around Australia and kept captive dingoes on the grounds of the University of Sydney. From dog–dingo skeletal comparisons he was able to conclude, first, that the dingo is separated from the domestic dogs and, second, that the dingo was skeletally homogenous across the continent. He also compared the Fromm's Landing fossil and suggested that since arriving on the Australian mainland, the dingo has remained unchanged in morphology.[7] Comparisons made by Laurie Corbett[8] between modern dingo and early Asian *Canis* spp. fossils similarly suggest that the dingo's general morphology has remained unchanged for the last 5500 years, and therefore not subjected to further artificial selection pressures by humans.

At this point, it is important to emphasise the dingo's isolation in Australia. Prior to European arrival there had been no artificial selection by humans or interbreeding (admixture) between dingoes and domestic dogs or wolves. This is unlike wolves and domestic dogs around the world that have undergone intense artificial selection, and had multiple incidents of interbreeding over the course of their history.[9–11] This isolation distinguishes the dingo from other canids, and therefore can be used to help define exactly what a dingo is. Establishing a definition is useful when trying to interpret and understand the dingo's origins. For the purpose of this chapter, and indeed the book, the dingo is considered a free-ranging member of the genus *Canis* that is endemic to Australia and has no modern dog ancestry. The definition excludes free-living domestic dogs with modern domestic ancestry (e.g. an escaped pet German shepherd), and dingo–dog hybrids (i.e. mixed ancestry). The dingo is not to be confused with the occasionally dingo-like dogs found throughout southern Asia. The dogs found in southern Asia are considered true domestic dogs that live with or alongside humans.

Indigenous Australian perspective

Unfortunately the relationship with Indigenous Australians provides little help in determining a period of dingo arrival. Dreaming narratives (animist creation narratives) have been documented that describe the arrival of dingoes as being quite recent in contrast to other animals or creatures.[30, 31] For example, Trezise[32] reported that in Indigenous mythology, the dingo was introduced to Cape York when humans already populated the area. Macintosh[5] noted that there are clear Indigenous references to the arrival in the north of recent mythological beings (in comparison to other mythological beings), which were accompanied by warriors who performed rituals associated with the dingo totem. The dingo appears to be unique in the way it is represented in Indigenous Australian mythology. That is, from the first time the dingo appears or is mentioned, it is represented in the form of a dog-like animal. This is in stark contrast to the many characters or ancestral beings described in other narratives; at some point all went through some sort of transformation, e.g. from or into human beings, natural species or natural features such as rocks.[33] This could indicate the dingo's arrival is relatively recent in terms of Indigenous mythology, and was added onto the older strata of myths. Unfortunately, oral histories alone are inconclusive and do not allow us to determine an age or timeline.

A more concrete account of Indigenous mythology may be obtained from rock art (e.g. cave painting using charcoal and ochre, rock engravings or petrography), which were common in many areas and have been dated as old as 30 000 years using radiocarbon dating. Many archaeologists such as Pardoe[34] and Flood[4] have noted that thylacines are widely represented in rock art, but dogs or dingoes rarely appear. At a particular site in Arnhem Land (far north of Australia) a team of archaeologists led by Gunn found it unusual that dingo images were absent, considering that many other animals, including other quadrupeds, were well represented.[35] Pardoe surmised that 'it could look as if dingoes were here only 500 years or so. They have not been translated into rock art, with any accumulation of figures over the few thousand years they have been in the country'.[34] There are a few Indigenous sites where representations of dingo-like animals are featured. In some cases, the dingo appears quite heavily in the artwork (e.g. large in size, central placement and polychrome colouring), suggesting that the dingo in these areas was of considerable importance. One of the most comprehensively documented examples was recorded by Macintosh.[36] In a rockshelter at Mount Manning (north of Sydney), Macintosh described the occurrence of a male and a female dingo (along with stylised male and female human figures and an echidna, *Tachyglossus aculeatus*) painted with red ochre, estimated to have been painted 581 ± 120 years ago (the charcoal associated with matching ochre in the floor deposit was radiocarbon dated).

According to the artwork style and carbon dating, it appears that images of dingoes (such as Fig. 3.1) have been made only in the last few thousand years. Representations of dingoes may appear in sites that have other images dated quite early, but this does not necessarily imply that the dingo images were made at the same time. Also, according to rock art expert Bednarik from the International Federation of Rock Art Organisations, placing too much emphasis on rock art may be problematic. It is extremely difficult to accurately date rock art, and very few artworks have credible age estimates.[37] Furthermore, it is difficult to know whether it is actually a dingo, and not another species, that is represented in rock art (Robert Bednarik, pers. comm.).

Figure 3.1: Contemporary Indigenous rock art from a site in Jowalbinna, Cape York Peninsula. The image depicts two dingo-like creatures chasing a wallaroo.
Source: Mary Haginikitas.

Physical geography

Another clue to the dingo's time of arrival involves looking at the geography of Australia and the land bridges between Australia and surrounding islands at times of low sea level. The dingo is found in all parts of Australia with the exception of Tasmania, where there are no traces of dingoes, either contemporary or archaeological.[38] The island of Tasmania was separated from Australia when the sea levels rose, flooding the land bridge ~12 000 years ago and forming Bass Strait[39] (Fig. 3.2). The presumption is that Tasmania separated from the mainland either before the dingo arrived in Australia or before the dingo could reach the far south-east of Australia, and is used by many as an absolute age for the earliest time point of arrival of the dingo to Australia.

As the largest eutherian mammal present before modern times in Australia (other pre-modern eutherian mammals in Australia include bats and rodents),[40] the dingo seems to have been introduced to this continent by humans. It is unlikely that the dingo found its own way to Australia given there has been no land bridge connecting Australasia to Asia for the last 5 million years, even with successive ice ages. Furthermore, a natural barrier exists, dubbed the Wallace Line, which would have prevented a dingo migration unaided by people. The Wallace Line is a hypothetical boundary line that twists through the middle of the Indonesian archipelago and separates the zoogeographical regions of Asia and Australasia. During his travels through South-east Asia in the 1840s, British naturalist Alfred Wallace (after whom the line

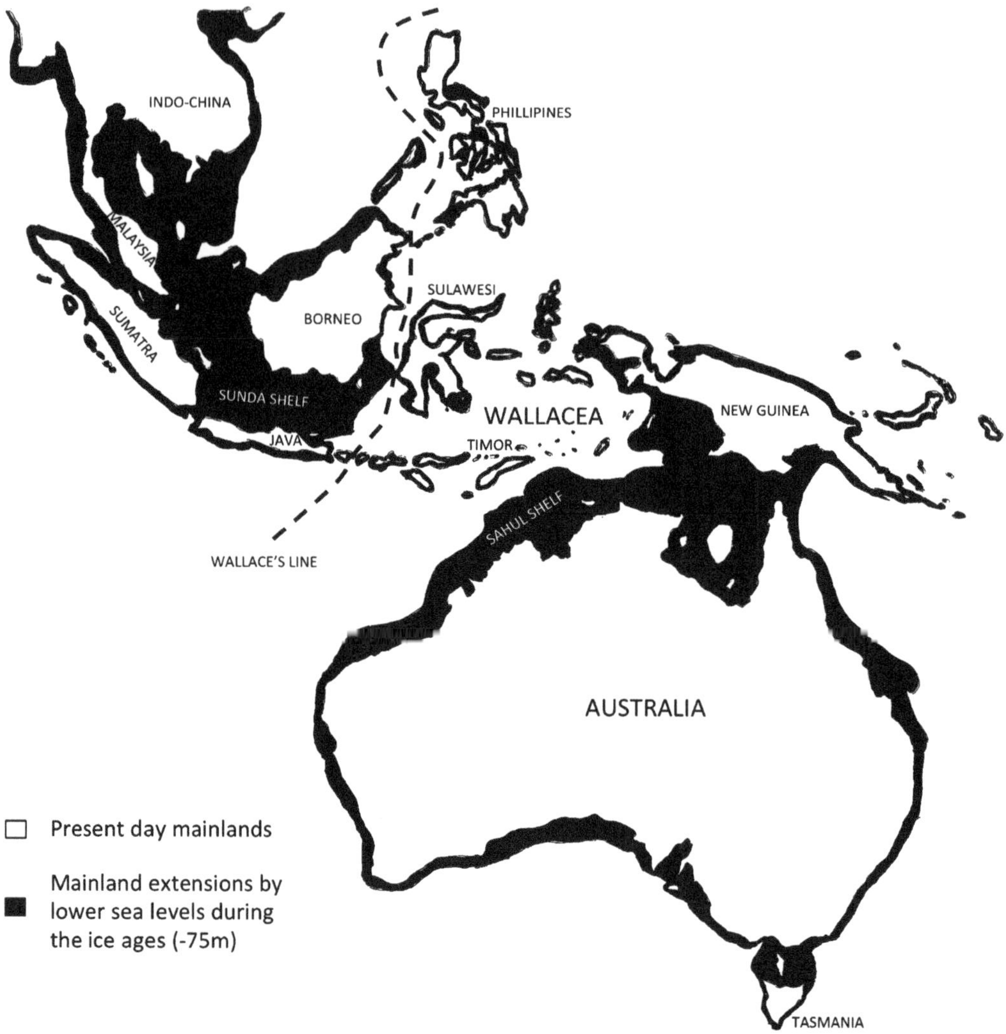

Figure 3.2: Wallace's line separates the fauna of Australia and South-east Asia. The land at the end of the last ice age when the sea level was around 75 m lower than today, is shaded. Given the lack of land bridges the whole way between Australia and Asia, it is likely that humans introduced the dingo into Australia.

is named) discovered a sharp distinction between the species in the north-western and south-eastern part of the archipelago, despite the areas having similar climate and terrain (Fig. 3.2). The reason for the division is a deep oceanic trench that has preserved the gulf between these two islands for millions of years, and prevented the movement of animals via the use of a land bridge connecting Asia to Australia. Wallace's line represents an abrupt difference of distribution for many major animal groups including fish, birds and mammals, many of which are abundantly represented on one side of the line but poorly or not at all on the other side.

This phenomenon helps explain why in ancient times there were no placental mammals in Australia and surrounding islands such as Papua New Guinea (except for rodents and bats).

Instead of the tigers, orang-utans and monkeys that reside in Indonesia and Sumatra, in Australasia you find a wide variety of marsupial mammals such as kangaroos, possums and koalas. Based on these observations, Wallace[41] referred to the dingo 'as probably not truly indigenous… it is in fact difficult to understand how such an animal could, without assistance have arrived in the country, except by means which would have equally admitted the entrance of many other animals. It differs little from the semi-wild dogs of India and other countries and this is an indication that it is geologically speaking a recent immigrant.' Jones[42] agreed, arguing that 'if the dingo had walked here so also could apes and carnivores with the resulting extinction of our marsupial population'.

Basically, any species that could not fly or float (on items such as debris, which would be quite a lot easier for small rodents) stayed in their respective regions. Most probably, the only way that dingoes could have travelled through this barrier would have been with the aid of people. Australia was connected to New Guinea until around 6500–8000 years ago, so it is possible that the dingo could have travelled freely between New Guinea and Australia until this time (Fig. 3.2). Dingoes are not found in New Guinea, but they seem to be closely related to the New Guinea singing dog (*Canis hallstromi*), as explained later in this chapter.

Dispersal time

The fossilised dingo remains are distributed in the far southern and eastern parts of the country, and dated around a similar timeframe. On face value this cluster of dates appears to represent the opening of the dingo record in Australia. Yet this is not considered the first entry date, as it would have taken some time for the dingo to have reached that location if introduction occurred in the north of the continent. Thus, dispersal time (time it took dingoes to spread across the continent) offers some indication of the timing of their arrival, and arguably allows us to calculate an age range from the available archaeological finds. Unfortunately, dispersal time across the Australian continent is hard (in fact almost impossible) to determine, and depends greatly on the initial population and the behavioural and ecological constraints on dingo dispersal.

Since the dingo founder (ancestral) individuals probably had an association with humans, some (but not all) of the introduced population of dingoes may have forged a relationship with local Indigenous Australians (see Chapter 4). This relationship may have hastened the dispersal of dingoes over the mainland as they accompanied or followed Indigenous Australians during trades with neighbouring communities.[43] The many different Indigenous names for the dingo can be traced along clear routes of distribution that may have been influenced by regional differences in dialect where cultural and language barriers prevented the sharing of knowledge. The number of names for the dingo may also be a reflection of the dingo's rapid dispersal across Australia,[44] whereby individual communities encountered wild dingoes before their human neighbours could inform them of its given name. The Indigenous peoples of Tasmania provide another indicator of the speed with which humans could disperse dingoes. Dogs were introduced to Tasmania in 1803, and by 1830 Indigenous people across the island (who had never met a white person) had dogs in their possession.[38]

Tindale[44] estimated the dispersal of the dingo from the north to the south of the country to have taken only 100 years based on the introduced red fox's (*Vulpes vulpes*) rapid spread after it was released in Victoria (~60–80 years to reach the Kimberley coast on the other side of the continent). Others, such as Gollan,[4] have hypothesised dispersal to take no longer than 500 years, but such estimates are highly speculative. It is extremely difficult to devise any real timeline, as there are many factors that influence and limit dingo population growth. These include, for example, social factors (a complex pack structure with breeding limited to the α-pair), environmental factors (minimal access to water sources, periods of drought) and high rates of natural mortality.[8]

Estimated time of arrival

To propose a time of arrival, any calculation must take into consideration all of the available archaeological, geographic, anthropological and genetic evidence and, importantly, adjust for the dispersal time. Given this, it is likely that the dingo reached Australia in the mid-Holocene period around 4000 years ago (based on the cluster of archaeological remains dated around 2800–3500 years old, and allowing for a dispersal time of around 500 years). An introduction is unlikely to have occurred earlier than 12 000 years ago when the land bridge connecting the Australian mainland and Tasmania flooded. Incidentally, 4000–5000 years ago marked a considerable change in human society in Australia, which will be explained in the following section.

A dingo-driven revolution? A time of significant change in Australia

Approximately 5000 years ago there occurred many changes in human lifestyle, migration and trade. This period has been referred to as 'intensification', based on a more intensive utilisation of natural resources by humans, along with innovations in hunting technology, increased population density, expansion of habitat range, more sedentary behaviour[45] and possibly the arrival of the dingo. Such changes were not experienced by Tasmanian Indigenous people who kept a comparatively simpler 'tool kit' and a less dense population.

Flannery[46] described this period as a dingo-driven revolution, suggesting that the establishment of 'man's best friend' was the key factor in triggering such a change. Although this sounds quite a romantic notion, it is not likely to be an accurate representation. Johnson disagreed with this statement in his book *Australia's Mammal Extinctions*,[47] and presented three reasons why it is unlikely to be the case. (1) It is just as plausible that the transportation of dingoes was a consequence of the intensification period, rather than a cause of it. (2) Dingoes are believed to have arrived in the middle of the Holocene population increase, not at the beginning of it. (3) The beginning of the Australian small tool tradition may have preceded the arrival of the dingo (see following section), although both dingoes and tools rapidly spread, suggesting that there may have been formal trade networks. In all, despite any debate over the driving force of this change or what we label this period, there is broad agreement among scientists that several significant changes occurred in Indigenous Australian culture during the late Holocene, and that the changes were intricately tied to a changing relationship between people and their environment.

Origin of new elements

There was a known association between Indigenous Australians and other more seafaring nations such as those from Torres Strait, Papua New Guinea, Indonesia[48–50] and India[19, 51]. Evidence for a network of trade between Asia and Indigenous Australians includes the incorporation of new tools (e.g. backed blades, points, tulas, adzes) into Indigenous Australian society at about the same time as the arrival of the dingo. This is often referred to as 'the small tool tradition'.[52] It is believed that Indigenous Australians acquired them from Asian travellers, [3, 39] but there is much debate surrounding the exact chronology of these tools, and their distribution and origin.[53] In general, though, it is acknowledged that new tools appeared and others that were rare suddenly became common. Some tools, such as the ground axe, were also found all over mainland Australia but not in Tasmania.[3] Dingo bones mentioned above, including those found at Wombah midden, Madura Cave and Fromm's Landing, were associated with backed blades and unifacial pebble tools (flake tools). Although the specific timing of the point technology is debated, most researchers have concluded that small tool technology across Australia dates around 3500–4500 years before present, and that the technology may have arrived together with the dingo.[3, 54]

Extinction of marsupial predators

The timing of the dingo's arrival correlates with the extinction of two major marsupial predators that were present throughout Australia at the time. It is considered more than coincidence that the Tasmanian tiger (*Thylacinus cynocephalus*) and the Tasmanian devil (*Sarcophilus harrisii*) (and incidentally a flightless bird, the Tasmanian native hen, *Gallinula mortierii*)[55] become extinct on the mainland around 3500–5000 years ago, yet survived in Tasmania where the dingo never reached.[8, 56] The demise of these species is often blamed on the dingo because it competed with them for resources and/or preyed on them directly. It is argued by Corbett[57] that the dingoes' 'superior social organisation enabled them to better exploit scarce resources during droughts or after extensive wildfire' (p. 228). However, recent evidence and debate suggests the dingo might not have caused the extinction of these species (discussed in more detail in Chapter 8).

Although the dingo probably didn't help the situation (e.g., dingoes could have killed thylacines, given they were of similar size to male thylacines and considerably larger than females),[58, 59] the mainland extinction of the thylacine is thought by some to be due to human impact and climate change.[45, 60] Scientist Prowse and colleagues[61] built a complex series of models to assess what factors were most influential in the extinction of devils and thylacines. Among other things, they included dingoes, climate and humans, competition between predators, and the influence of climate on vegetation and prey populations. The simulations showed that while dingoes probably had some impact, growth and development in human populations, possibly intensified by climate change, was the most likely extinction driver. In fact, Johnson[47] argued that there was scope for dingoes and thylacines to have co-existed by using different food types. Even though they were both hunters of large mammals, tooth and skull morphology[62] suggest that thylacines were more suited towards larger prey and mainly ate fresh meat, whereas the dingo had a broader diet and was better suited to smaller prey. The diet of the thylacine, however, remains hotly debated.[63]

Seafaring people and their dogs

Domesticated or partly domesticated (tame) dogs could well have been taken by traders in their canoes since 8000 years ago when a trading network in the South-east Asian region is thought to have begun. The method of transport to Australia, and why the dingo was brought into Australia, is not entirely clear. It is most likely that human migrations to Australia were one way. Flood[3] argued that Indigenous Australians could not have exported anything because at that time they lacked watercraft that could travel long distances. Pacific voyagers, in contrast, had large outrigger sailing canoes and reached Tonga and Samoa 5000 years ago. Migrations from Australia, although highly unlikely, are not considered impossible. For example, biting lice (*Heterodoxus spiniger*; Enderlein 1909) found on canid species around the world may have been picked up from Australian marsupials during multiple visitations by non-Indigenous people to Australia.[8, 64] There is still speculation about the origination of this lice and its transference, however, so further investigation is needed to confirm this idea.

So why did these people travel with their dogs? Early seafaring people may have travelled with dogs for many reasons – for protection (guard dogs), companionship, hunting aids or potential sources of food.[8, 65] They could have been taken for spiritual reasons (protector from danger or evil spirits) or as an object for trade. The likeness of the dingo to the dogs of neighbouring islands, such as the New Guinea singing dog,[66] and to dogs across South-east Asia[49] implies that transportation of these domesticated Asian canids extended throughout the Pacific. Margaret Titcomb[67] extensively studied the role of dogs in the islands of the Pacific, which may give further insight. She suggested that the dogs provided much-needed meat, were of great religious significance, and that their teeth, bones and hair were used in rituals of many different kinds as well as in clothing. She wrote:

> It is likely that all Pacific islanders loaded their sea-going craft with food for the voyage, whether a deep-sea fishing voyage, or a planned trip to another island. Ideally for a voyage of any length, the dog, the pig, and the chicken were carried along. On most islands they survived … Most dogs did not live very long. Puppies were more delicious than older dogs. Only those used for breeding could have survived puppyhood … The pacific dog never developed a life of its own, never became feral, for food in the forest was lacking.

Early origin and ancestry theories according to morphological comparisons

Before the advent of molecular biology, the only method available to biologists, archaeologists and zoologists for determining the dingo's origin and relationship with other wild or primitive breeds of dog was by studying their morphology (measurements of the physical structure of the animals, primarily of skulls; Fig. 3.3). Corbett attempted to determine the dingo's ancestry based on almost a decade of research on more than 100 canid skulls from Australia and Asia.[49] He concluded that the Australian dingo was in fact not unique compared to village dogs found across south Asia, and was particularly similar to village dogs of Thailand.[8, 49] Corbett

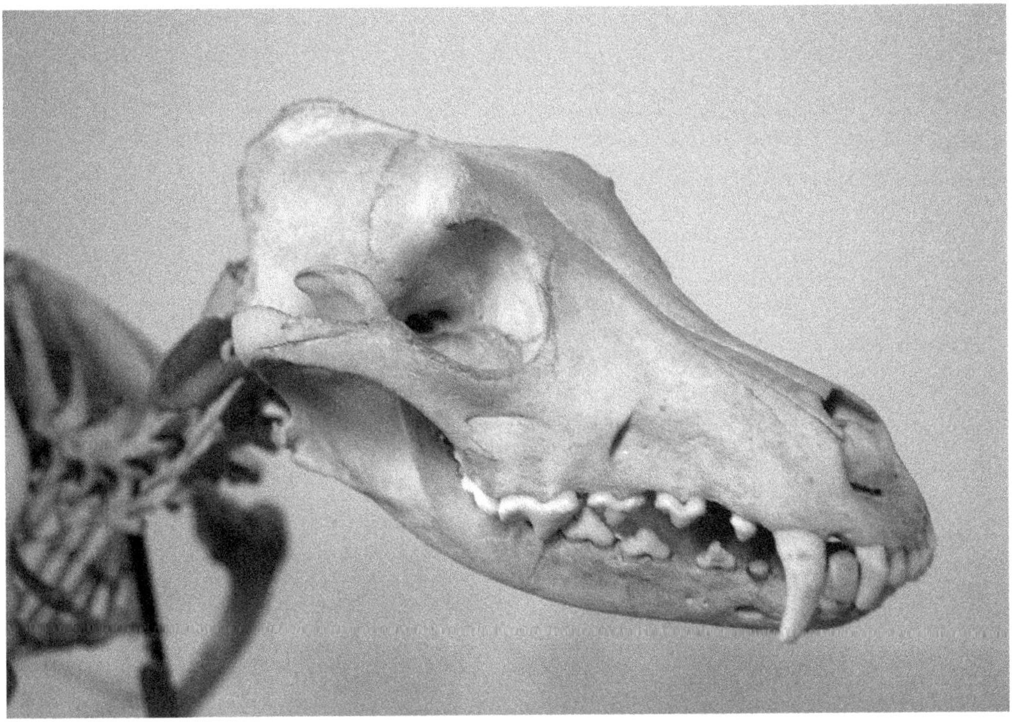

Figure 3.3: A dingo skull housed at the South Australian Museum. Until genetic testing, the skull morphology was used to determine purity and ancestry.
Source: Bradley Smith.

identified potential ancestral species as including the Indian wolf (*Canis lupus pallipes*) and the Arabian wolf (*Canis lupus arabs*).

The possibility of an origin from India has been suggested because the dingo shares similarities in skeletal anatomy with Indian pariah dogs and Indian wolves. Gollan in the early 1980s conducted a study of the Australian dingo and modern and fossilised canids in Asian countries such as Sumatra, Borneo, Java, Papua New Guinea, Thailand and India.[4, 19, 51] He suggested that the dingo resembles the Indian pariah dog much more closely than any from South-east Asia (Fig. 3.4). The closest affinities to the dingo Gollan found were the skeletons of prehistoric dogs from India, such as from the Burzahom in Kashmir, and in particular the domesticated dogs from the Indus civilisation city of Harappa, dating from 3500–4000 years ago. Gollan argued that when the dingo was brought to Australia, it was domesticated or semi-domesticated. Thus, he believed that about 4000 years ago a founder population of domestic dogs was introduced to Australia, and that those dogs were the genetic base for the subsequent feral population of dingoes. An Indian origin of the dingo is not so difficult to imagine, given that there is a strong overlap with important changes in the archaeological and linguistic records in Australia originating from the Indian subcontinent that are likely to have occurred during the Holocene. A genetic link between Aboriginal Australians and Indian people has also been discovered, in studies led by Redd[68] and more recently by Pugach.[69]

Figure 3.4: A modern stray dog from Rajasthan, India. This exemplifies the typical village dog (free-ranging domestic dogs living in urban areas) in parts of Asia that shows some phenotypic resemblance to the dingo.
Source: Judy Dunlap.

There are, however, multiple problems with the theory that the dingo originated in India and was brought to Australia by Indian seafarers.[70] Not all researchers are as confident looking at morphology as Corbett and Gollan. Macintosh, for example, could not come up with an origin for the dingo ancestor after conducting similar comparisons. He did, however, discount the possibility of the Indian wolf ancestry. In a 1974 paper,[71] he wrote: 'I am unable to see the dingo as parent, offspring, or cousin of the pariah dog, and in my opinion there are too many anatomical and behavioural differences, basic and plastic for the dingo to be a derivative of *C. lupus pallipes*' (p. 92). Perhaps most importantly, genetic studies show little relatedness between dingoes and Indian dogs or wolves, as will be discussed in following sections. Lastly, the theories constructed by Corbett and Gollan are limited by the samples that were tested. For example, the sample size and quality of the comparison samples used are not comprehensive. Particularly in the case of Gollan's work, it is difficult to judge his conclusions because the details of the analysis and samples used are not clear.

Morphological comparison can only inform us whether or not the dingo is morphologically similar to other canids to which they are compared. Although inferences can be made, they do not provide definitive proof of ancestry. All morphology can really tell us is that Australian dingoes and South-east Asian dogs are phenotypically similar, and therefore related in some way. It seems that the archaeological records and morphological comparisons have helped little in our understanding of dingo origins and ancestry. Fortunately, molecular biology (genetics) can now provide some of the answers, particularly in terms of relatedness between dingoes and the Asiatic canids, both wild and domesticated.

Modern dingo origin and ancestry theories according to molecular biology

Molecular biology provides a great deal of insight into the history of the dingo. Most importantly, it can reveal the genetic relationships of these canids, and thereby their most likely origin. It can also tell details about the time of arrival and number of founder individuals and, by inference, the route by which they may have been introduced on their way from South-east Asia to Australia. There are several methods that can be utilised in genetic studies, each of which will be discussed.

Origin and ancestry

A seminal study of mitochondrial DNA (mtDNA; maternal genetic material) led by Peter Savolainen, the first comprehensive genetic study of the origins of dingoes, gave strong indications that the dingo originates from an early form of the domestic dog from East Asia.[72,73] The results have been supported by analysis of nuclear genome-wide single nucleotide polymorphisms (SNPs)[74] and of Y-chromosomal DNA (paternal genetic material).[75] Thus, the genetic evidence that dingoes originate from dogs (as opposed to directly from wolves), most probably with an origin in East Asia, is very convincing. The findings from these genetic studies are not entirely unexpected, supporting the assumptions from studies of morphological measurements of dingo, wolf and dog skulls mentioned above.

Savolainen analysed 211 Australian dingoes as well as a large sample of dogs and wolves from worldwide sources. He and his colleagues found that the sole mtDNA founder haplotype detected for the Australian dingo population was a dog haplotype which is commonly found throughout East Asia, but is absent west of the Himalayas.[73] More specifically, a majority of the dingoes sampled had mtDNA type A29, which has been found only in dogs from East Asia, Siberia and Arctic America, whereas 18 of the 19 other types of mtDNA – all very similar to A29 and likely mutated forms of A29 – were unique to dingoes. These dingo sequences were clustered together with domestic dogs in dog clade A (phylogenetic analysis of dog, wolves and other canids has revealed six distinct clades of dog sequences, indicated by the letters A–F. Clade A incorporated more than 71% of the dogs worldwide, and nearly 96% belonged to either clade A, B or C; see Fig. 3.5). This strongly indicates that dingoes originate from these early domestic dogs, carrying haplotype A29. The fact that A29 has been found in southern East Asia, for example in southern China and Thailand, also gives an indication about the geographical ancestry. Importantly, since A29 has not been found in India or anywhere else west of the Himalayas, it contradicts the theories that the dingo has an Indian origin.

A study conducted by Bridgett vonHoldt and colleagues,[74] looking at nuclear genome-wide single nucleotide polymorphisms (SNPs) among dogs, dingoes and wolves, reported similar findings. Most importantly, the dingo and New Guinea singing dog (NGSD) are grouped with domestic dogs, clearly indicating their origin from domestic dogs. Furthermore, dingoes and NGSDs are genetically closest to East Asian dogs, forming a group with breeds such as the chow chow, akita, and Chinese shar pei, distinct from other breeds of dogs from

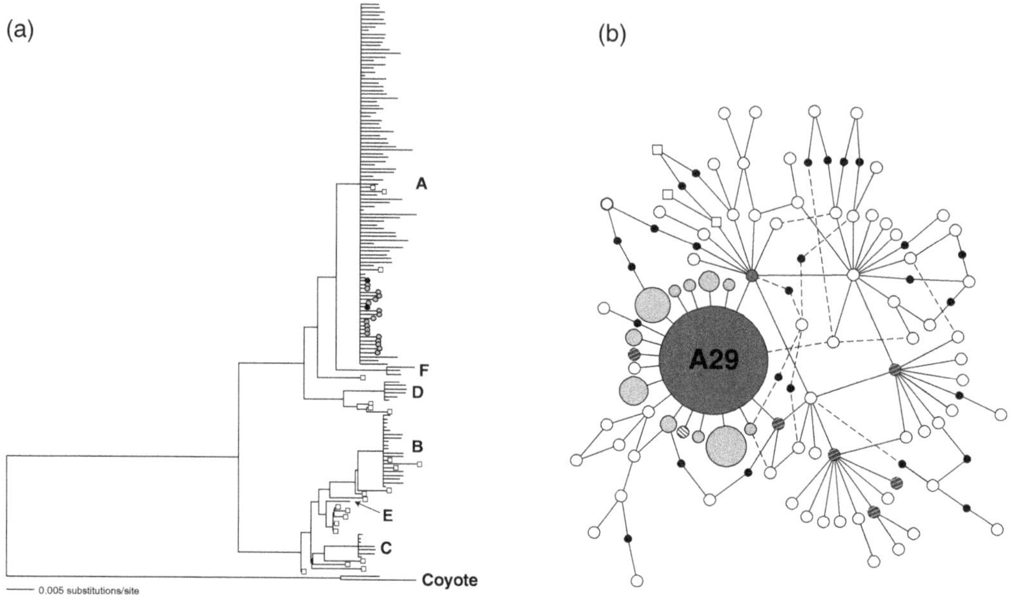

Figure 3.5: Genetic relationships between mtDNA haplotypes (control region sequences 582 bp) from dingoes, domestic dogs and wolves. (A) Phylogenetic tree showing all dog, wolf and dingo mtDNA types. Unlabelled leaves denote mtDNA types from domestic dogs, dog clades A–F are indicated with letters. Small circles (in lower part of clade A) denote types found in dingo, light grey denotes types unique to dingoes and dark grey types found in both dingo and dog. White squares denote wolf mtDNA types. The tree is rooted to coyote sequences. (B) Minimum-spanning network of the main dog clade (clade A). Sequence types (circles) and empty nodes (solid dots) are separated by one mutational step. Grey denotes mtDNA types found among dingoes; dark grey = the two types (including the most frequent A29) found in both dingo and dog; light grey = mtDNA type unique for dingo. Striped circles = type found in dogs in Indonesia, the Philippines, Malaysia or New Guinea. Squares denote wolves. Areas of the grey circles are proportional to mtDNA-type frequencies among dingoes, but the area of A29 is reduced by 50%.
Source: Savolainen *et al.* (2004).[73]

regions such as Europe. Adding further support, a study of Y-chromosomal DNA by Ardalan *et al.*[75] showed that dingoes group with domestic dogs, and that dingoes carry only two haplotypes – one which is common in East Asia, and the other which is unique to dingoes but most closely related to a haplotype unique to East Asian dogs.

To summarise, all genetic markers investigated agree that dingoes are closer to domestic dogs than to any wild canid, and that the dingo originated from a population of early domestic dogs. The opposite – that the dingo would be the ancestor of domestic dogs – is not likely because dingoes originated when domestic dogs had already existed (living with humans) for at least 6000 years.[76] Also, dingoes have a very small portion of dog genetic types in terms of mtDNA, Y-chromosome and major histocompatibility complex. Put simply, the genetic diversity of dingoes can have evolved from domestic dogs, but it could not have been the other way round. That is, the broad diversity of DNA types among domestic dogs could not have originated from the very few dingo DNA types. The genetic studies mentioned in this chapter also show that dingoes are closely related to dogs in East Asia. Interestingly, studies of mtDNA and Y-chromosomal DNA have given strong indications that the domestic dog itself originated in southern East Asia.[11, 75, 77] The question about the geographical origin is still controversial, but it is clear that domestic dogs are related to East Asian wolves, through direct ancestry, admixture or both.

Founder population of dingoes

Studies of nuclear autosomal DNA, mtDNA and Y-chromosomal DNA all indicate low genetic diversity for the dingo population. A study of microsatellite allelic variability revealed much lower genetic diversity among Australian dingoes than among domestic dogs.[78] Similarly, as related above, the mtDNA variation among dingoes is considerably lower than among domestic dogs with a single haplotype, A29, being found in 83% (192/232) of the samples, and all other haplotypes separated from A29 by a single mutation.[72, 73] Diversity of the Y-chromosome DNA sequence was also very low, with only two haplotypes compared to 28 among domestic dogs.[75] Thus, overall genetic evidence suggests a very restricted introduction of the first dingoes into Australia. This indicates that the dingo population was founded from a very small population of animals, or from a population that had undergone numerous bottlenecks, possibly by island-hopping through the South-east Asian archipelago, dramatically reducing genetic diversity.[72, 73, 78] While it is clear that the dingo ancestors experienced a substantial population bottleneck, it is not yet clear if this was caused by several less severe bottlenecks, or if there actually were only a handful of individuals introduced to Australia. The mtDNA and Y-chromosomal data indicates that the founders to the dingo population carried a single mtDNA haplotype and two Y-chromosome haplotypes. It is a fascinating thought that, in theory, dingoes may have originated from only three individuals, one female and two males, possibly introduced at a single occasion. Importantly, the mtDNA and Y-chromosome data give evidence for the long isolation of the dingo since its introduction to Australia. All investigated dingoes carried A29 or very similar mtDNA haplotypes, and there were only two Y-chromosome haplotypes in the dingo population. Introduction of dogs after the first introduction of dingoes should have shown in the form of many other haplotypes among dingoes, since domestic dogs, including those in the Indonesian islands and New Guinea, carry a large number of different haplotypes.

Estimated time of arrival

Calculations based on the genetic divergence among dingoes suggest they were introduced to Australia 4640–18 100 years ago.[72] This estimate is slightly earlier than the time suggested by the archaeological data (3500, or 4000 given time for dispersal). However, this kind of calculation is less reliable than the analyses of origin and ancestry related above, since they are sensitive to a few assumptions, which need to be taken into consideration. For example, it is based on the supposition that A29 was the only haplotype brought to Australia, and that all others (since they are unique to Australia) have later evolved through mutation in Australia. However, there remains a possibility that some of these haplotypes are present in, for instance, neighbouring Indonesian islands, but have not yet been found there. If this were the case, the calculated date would overestimate the time since arrival in Australia. So far, the archaeological data offer the most reliable time estimates, at least of the minimum time for the introduction of dingoes.

Migration route to Australia

In line with the theories described earlier, Savolainen[73] discussed the possibility that dingoes arrived as part of the Austronesian expansion, which spread the Neolithic culture (which included farming, and domestic plants and animals) into island South-east Asia that was previously inhabited by hunter-gatherers. The Austronesian expansion started from Taiwan and spread via the Philippines into Indonesia. However, a later study showed that dogs in mainland South-east Asia, Indonesia and New Guinea (including the NGSD) carry the dingo founder haplotype A29, but that dogs in Taiwan or the Philippines do not. This indicates that dingoes were introduced into Australia via mainland South-east Asia before the spread of the Neolithic into island South-east Asia, possibly through contacts with New Guinea, but not with the expansion of the Neolithic Austronesian culture through Taiwan.[72] This is also supported by microsatellite data from one study that revealed haplotype sharing between Bali street dogs (Fig. 3.6) and Australian dingoes.[79] In a recent article, Sacks and colleagues[80] suggested that the dingoes might actually have been introduced via Taiwan, based on time estimates using the genetic divergence in Y-chromosome microsatellites indicating that the Taiwan dog population and the dingo population have similar age. However, since they did not find any of the dingo-related mtDNA or Y-chromosome haplotypes in Taiwan, and because of the unreliability of genetic time calculations, the basis for this theory seems weak. Figure 3.8 shows the route that, according to mtDNA studies, the dingo may have taken to get to Australia.

Figure 3.6: A stray village dog from Ubud, Indonesia. Stray dogs are a typical sight in many Asian countries. Many of these dogs are looked after by charitable organisations such as the Bali Dog Adoption Rehabilitation Centre.

Source: Shayna Pitch Photography/www.BARC4BaliDogs.org.

There are clear behavioural, morphological and anatomic similarities between the NGSD (see Fig. 3.7) and the dingo.[81] Further, genetic studies have shown extensive haplotype sharing between these two populations, for mtDNA (the founding dingo haplotype A29 was found in two of the three known maternal lineages of NGSD),[72, 73] Y-chromosome DNA (dingo haplotype H-60 is unique to dingoes and NGSDs[75]) and the major histocompatibility complex (all dog leucocyte antigen alleles found in NGSDs were shared with the Australian dingo).[82] A high degree of sharing of the gene pools of Australian dingoes and NGSDs strongly indicates that they originated from the same ancestral canids. There seem to be a few possibilities in terms of their relationship. For example, that they separately originated from the same Indonesian population, or that NGSDs originated from dingoes or vice versa. Because of this strong relationship, Ardalan *et al.*[75] suggested that dingoes were introduced exclusively via New Guinea. They argued that, because all dingoes share a single mtDNA founder haplotype but there are several mtDNA haplotypes present in modern Indonesia and New Guinea, this was the only introduction route of canids to Australia, since introductions of canids from multiple regions into Australia should have brought several different mtDNA founder haplotypes.

Genome-wide sequence studies: a new perspective on dingo origins?

The extraordinary advances of DNA sequencing technology in the last few years now allows the most detailed possible genetic studies in which entire genomes are analysed, though still

Figure 3.7: New Guinea singing dogs, or 'singers' as they are affectionately known, are closely related to the dingo. Unfortunately, because there have been no verified sightings of these dogs in Papua New Guinea since the 1970s, they are considered either rare, or possibly extinct.

Source: Joan Dalgleish.

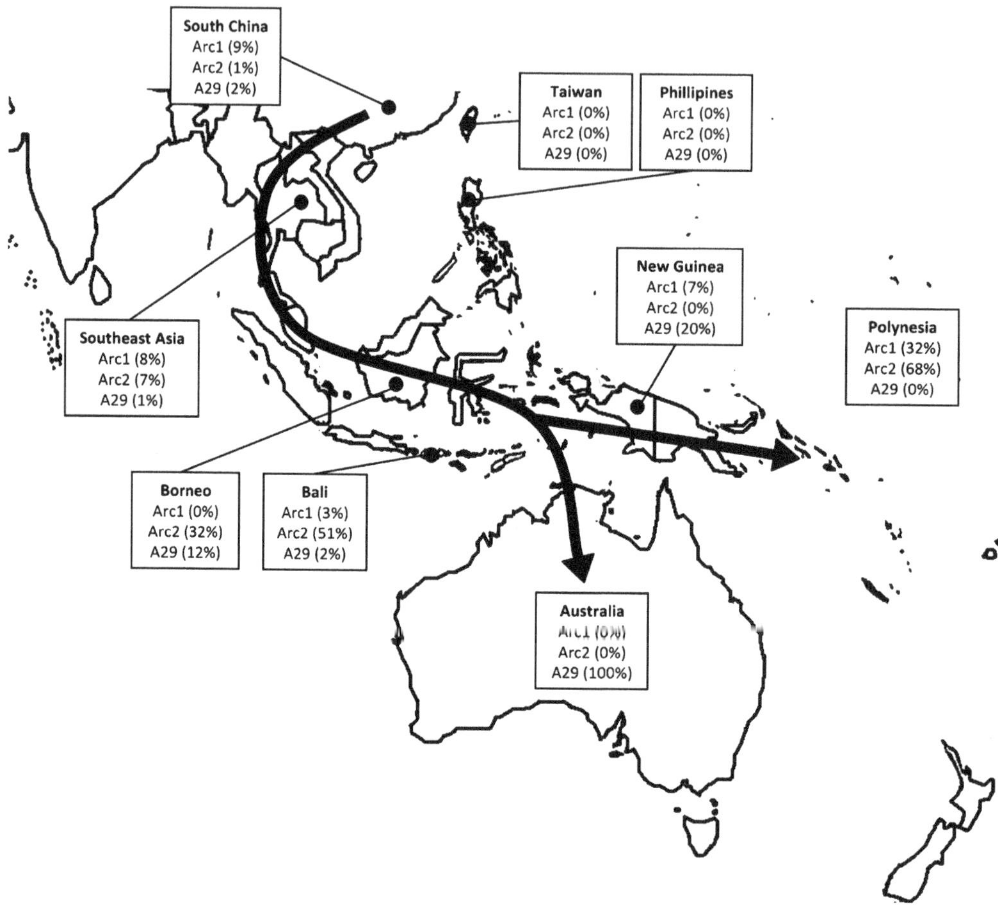

Figure 3.8: Suggested route of the dingo from its suggested origin in South China, through to Indonesia via mainland South-east Asia. According to mtDNA, Australian dingoes and NGSDs carried only haplotype A29 and Polynesian domestic dogs carried two haplotypes, Arc1 and Arc2. Thus, it seems that only a single mtDNA haplotype was introduced to Australia and two others to Polynesia. All three haplotypes are found in mainland and island South-east Asia, among several other haplotypes. Since they are not found in Taiwan or the Philippines it seems unlikely they would have been introduced this way. It is possible that the dingo and Polynesian dog populations were founded from very few individuals from the same Indonesian population, but obtained different haplotypes because of founder bottlenecks. The frequency according to the total number of samples for the region of the Polynesian haplotype (Arc1 and Arc2) and the dingo founder haplotype (A29) are provided according to various geographical regions. Image reproduced from Oskarsson *et al.* (2012).[72]

in only a few individuals, and not yet representative of canids from all areas of the old and new worlds, particularly South-east Asia. A recent study by Freedman and colleagues[9] analysed the genome sequences from seven different canid samples: three wolves representing the broad regions where dog domestication has been hypothesised to have taken place (i.e. the Middle East [Israel], East Asia [North China] and Europe [Croatia]), two ancient dogs (the basenji and the dingo) that represent old populations of domestic dogs, a domestic dog (the boxer) to act as a reference point for modern dogs, and a golden jackal (*Canis aureus*) as a comparison group. Based on their findings, the authors suggested that wolves and all dogs, including dingo and basenji, split from an ancestral population ~15 000 years ago. They proposed that the dingo split from the two other dogs ~12 800 years ago, possibly as dogs dispersed from East Asia to the rest of the world. Of particular importance for the dingo, the authors found that the dingo lineage experienced a dramatic bottleneck ~3000 years ago, which took them from an effective population of 32 500–37 400 to less than 2000 individuals. They also found a significant degree of relationship between the Chinese wolf and the dingo. The research team led by vonHoldt[74] similarly reported evidence to suggest that the East Asian group of dogs, particularly the dingo, were derived from Chinese wolves or admixed with them after domestication (although it is not entirely clear what wolf subspecies Freedman or vonHoldt have sampled). Combined, these findings lend support to the previous genetic evidence that dingoes originated from South-east Asia and that they experienced several bottlenecks as they were transported through the islands of South-east Asia to Australia.

Adaptation to starch-rich diets

There is still some speculation as to whether domestic dogs (including the dingo) originated before or after agriculture. New findings from genome-wide studies led by Erik Axelsson of Sweden's Uppsala University[83] reveal interesting insights about the emergence of domestic dogs; the team studied the genomes of a group of 60 dogs and 12 wolves from around the world. They suggested that the development of agriculture was the catalyst for the domestication of dogs. In other words, the rise of agriculture and the domestication of dogs may have gone hand in hand. They infered this on the basis that domestic dogs possess multiple copies of the AMY2B gene, which is involved in amalyse activity and important for digesting starch (a nutrient found chiefly in the seeds, fruits, tubers, roots and stem pith of plants, notably in corn, potatoes, wheat and rice). In fact, dogs were found to have seven times more copies of the AMY2B locus than wolves. This gene is believed to have rapidly increased in the earliest dog populations and to have been a critical part of the domestication process, facilitating their ability to feed on agricultural garbage around human agrarian settlements. However, Freedman and colleagues[9] found that the AMY2B gene had not spread evenly throughout the early dog population. For example, all the wolves in the study, the dingo and the Siberian husky (historically associated with nomadic hunter-gatherers of the Arctic) had low number of copies of the AMY2B gene, while the basenji and other breeds with histories of long association with humans in agrarian societies had higher numbers. Based on this additional evidence, the authors concluded that domestication of dogs was likely to have occurred before the adoption of agriculture by humans. However, it is equally plausible that all dogs had many copies, but some populations lost them when reverting to more free-ranging meat-based diets. For example, the dingo may have had many copies when it first arrived in Australia, but lost copies when it reverted to a completely wild existence. It is still too early to gauge the significance of such findings from the available data.

Conclusion

We now have a much clearer picture of the origin and ancestry of the dingo than Macintosh was able to draw just 40 years ago. This is largely thanks to the advances in genetic analysis over the last decade. The genetic data provide convincing evidence that the dingo originates from an early form of domesticated dog that existed around 5000 years ago. This is in agreement with the assumptions from previous morphological analyses, based on measurements taken of dingo, wolf and dog skulls. An origin from a population of South-east Asian canines seems most probable. These animals were transported across the Indonesian archipelago, causing several severe bottlenecks (a sharp reduction in population size and subsequent genetic diversity) as they travelled from island to island on their eventual route to Australia. The originating Australian dingo population was probably founded from a small number of animals, possibly even at a single chance event.

The Australian dingo remained effectively isolated from other dog or wolf populations, and from the influence of artificial selection for at least 3500 years[84] (see Chapter 4). Present-day dingoes are homogenous across the continent and appear, at least morphologically, unchanged since this time.[7, 8] Thus, the dingo is both a living fossil of the early forms of the dog that existed 4000–5000 years ago and an animal that, besides having a long history with humans, has lived as a wild animal for an equally long time. In these respects, the dingo is a truly distinctive canid that, according to Savolainen and colleagues,[73] represents a 'unique isolate of early-undifferentiated dogs'. According to Juliet Clutton-Brock,[76, 85] a leading authority on domesticated animals, dingoes are a 'valuable living relic of the early domestic dogs to inhabit southern and eastern Asia in the early prehistoric period'. Consequently, the dingo offers valuable insight into traditional human–canine relationships and is a unique study object for scientific analyses of the domestication, genetics and behaviour of the canids. This is important because researchers are still attempting to understand the process of domestication of wolves, and the complex relationship between canids and humans.

It is believed that dingoes originated when dogs had lived with humans (most of the time, with farmers) for at least 6000 years.[76] These dogs must have been domesticated or tame enough to travel with people in small watercraft, yet retained enough physical, cognitive and behavioural traits to be able to thrive on their own in Australia. It is difficult to speculate what these initial founding dingoes would have been like. That is, the level of domestication they had undergone, and the impact domestication events might have had on their morphology and behaviour. Unfortunately, little is understood about the relationship between peoples and their dogs during this time, and we do not know whether they were actively selecting for (or against) particular traits.

The Australian dingo may have been descended from South-east Asian dogs and bear some physical resemblance to the present-day dogs in South-east Asia, but the dingo is no longer the same animal that arrived in Australia. It is now considered a wild canid that in many respects is more characteristic of a wolf than a domestic dog (as will be covered throughout the book). For example, it can be distinguished genetically,[74, 86] and is phenotypically (Chapter 1), behaviourally (e.g. parental behaviour;[87] Chapter 2) and cognitively[88–92] (Chapter 9) more wolf-like than dog-like. As soon as the dingo arrived, any influence of artificial selection ceased; it has undergone more than 4000 years of natural selection. If modern populations of dingoes were essentially the same as modern domestic dogs, that would suggest a rather remarkable lack of effect for natural selection shaping the dingo since its arrival in Australia. Thus, in the modern context, the dingo should not be classed as a domestic or feral dog (*Canis familaris*), but as a unique species as reflected by the nomenclature *Canis dingo*.

Remarkably, even though the dingo lives as a wild animal, it appears to have maintained an inherent ability to live alongside and communicate with humans – something that is difficult for wolves. For example, dingoes are known to have been in a commensal or at times semi-domestic relationship with Indigenous Australians in several cases,[84] and although they do not make good pets wen compared to dogs they are able to adapt as household companion animals far easier than wolves in a suitable environment[93] (Chapter 10). Compared to wolves, dingoes can more effectively connect to human social cues (Chapter 9). For example, their performance in comprehending human gestures lies between those of wolves and modern breeds of dogs, likely a reflection of their early domestic history and origin from domestic dog populations.[92] The wildness of the dingo, then, is likely a result of more than 3500 years of free-ranging life in Australia but it could also indicate that the founder dingoes were not in a fully domesticated state (in the modern sense). It is likely a combination of both.

Finally, it is worth pointing out that our understanding of the history and ancestry of the dingo matters, particularly in the Australian context where the dingo is managed in most regions as a pest species, and its conservation and heritage status is hotly debated. Macintosh,[5] in an article concerning the origin of the dingo written for an audience of Australian farmers, made the valuable point that 'such a diversity of investigations on an animal may seem very far removed from what the pastoralist particularly requires – some method of effective control, or preferably extermination; but it must be remembered that to control or exterminate any parasitic or other plague it has always first been necessary to find out its identity, life history, and biological pattern'. In sum, the Australian dingo is a unique canid and one of great historical, cultural and genetic importance to both Australia and the world. On these grounds alone, it is well worth conserving.

References

1. Barker BCW, Macintosh A (1979) The dingo: a review. *Archaeology and Physical Anthropology in Oceania* **14**, 27–53.
2. Macintosh NWG (1975) The origin of the dingo: an enigma. In *The Wild Canids: Their Systematics, Behavioural Ecology and Evolution*. (Ed. MW Fox) pp. 87–106. Van Nostrand Reinhold, New York.
3. Flood J (2004) *Archaeology of the Dreamtime: The Story of Prehistoric Australia and its People*. JB Publishing, Adelaide.
4. Gollan K (1984) The Australian dingo: in the shadow of man. In *Vertebrate Zoogeography and Evolution in Australasia: Animals in Space and Time*. (Eds M Archer and G Clayton) pp. 921–927. Hesperian Press, Perth.
5. Macintosh NWG (1956) Trail of the dingo. *The Etruscan* **5**, 8–12.
6. Mulvaney DJ, Lawton G, Twidale C (1964) Archaeological excavation of rock shelter no. 6, Fromm's Landing, South Australia. *Proceedings of the Royal Society of Victoria* **77**, 479–516.
7. Macintosh NWG (1964) A 3,000 years old dingo from shelter 6 (Fromm's Landing, South Australia). *Proceedings of the Royal Society of Victoria* **77**, 498–507.

8. Corbett LK (2001) *The Dingo in Australia and Asia.* JB Books, Adelaide.
9. Freedman AH, Gronau I, Schweizer RM, Ortega-Del Vecchyo D, Han E, Silva PM, Galaverni M, Fan Z, Marx P, Lorente-Galdos B, Beale H, Ramirez O, Hormozdiari F, Alkan C, Vila C, Squire K, Geffen E, Kusak J, Boyko AR, Parker HG, Lee C, Tadigotla V, Siepel A, Bustamante CD, Harkins TT, Nelson SF, Ostrander EA, Marques-Bonet T, Wayne RK, Novembre J (2014) Genome sequencing highlights the dynamic early history of dogs. *PLOS Genetics* **10**, e1004016.
10. Klütsch CF, Seppala EH, Fall T, Uhlen M, Hedhammar A, Lohi H, Savolainen P (2011) Regional occurrence, high frequency but low diversity of mitochondrial DNA haplogroup d1 suggests a recent dog-wolf hybridization in Scandinavia. *Animal Genetics* **42**, 100–103.
11. Pang JF, Kluetsch C, Zou XJ, Zhang AB, Luo LY, Angleby H, Ardalan A, Ekstrom C, Skollermo A, Lundeberg J, Matsumura S, Leitner T, Zhang YP, Savolainen P (2009) MtDNA data indicate a single origin for dogs south of Yangtze River, less than 16,300 years ago, from numerous wolves. *Molecular Biology and Evolution* **26**, 2849–2864.
12. Veth P (1995) Aridity and settlement in north-west Australia. *Antiquity* **69**, 733–746.
13. Veth P (1996) Current archaeological evidence from the Little and Great Sandy Deserts. In *Archaeology of Northern Australia.* (Eds P Veth and P Hiscock) pp. 50–65. Anthropology Museum, University of Queensland, Brisbane.
14. Martin HA (1973) Palynology and historical ecology of some cave excavations in the Australian Nullarbor. *Australian Journal of Botany* **21**, 283–305.
15. Milham P, Thompson P (1976) Relative antiquity of human occupation and extinct fauna at Madura Cave, south-eastern Western Australia. *Mankind* **10**, 175–180.
16. Lowry JWJ, Merrilees D (1969) Age of the desiccated carcase of a thylacine (*Marsupialia, Dasyuroidea*) from Thylacine Hole, Nullarbor region, Western Australia. *Helictite* 7, 15–16.
17. Merrilees D (1970) A check on the radiocarbon dating of desiccated thylacine (marsupial 'wolf') and dog tissue from Thylacine Hole, Nullarbor region, Western Australia. *Helictite* **8**, 39–42.
18. Smith MA (1982) Devon Downs reconsidered: changes in site use at Lower Murray Valley rockshelter. *Archaeology in Oceania* **17**, 109–116.
19. Gollan K (1982) Prehistoric dingo. PhD thesis. Australian National University, Canberra.
20. Kigoshi K, Kobayashi H (1965) Gakushun natural radiocarbon measurements. IV. *Radiocarbon* 7, 10–23.
21. Bermingham A (1966) Victoria natural radiocarbon measurements. I. *Radiocarbon* **8**, 507–521.
22. Lampert RJ (1971) *Burrill Lake and Currarong Terra Australis 1.* Australian National University, Canberra.
23. Attenbrow V (1993) Port Jackson archaeological project: Stage II. In Unpublished Report for Australian Institute of Aboriginal and Torres Strait Islanders (retained in the NSW Department of Environment, Climate Change and Water's Aboriginal Heritage Information Management System: ref. 2864).
24. Attenbrow V (2002) *Sydney's Aboriginal Past: Investigating the Archaeological and Historical Records.* University of New South Wales Press, Sydney.
25. Kigoshi K, Kobayashi H (1966) Gakushun natural radiocarbon measurements. V. *Radiocarbon* **8**, 54–73.
26. McBryde I (1966) Radiocarbon dates for northern New South Wales. *Antiquity* **40**, 285–292.
27. McBryde I (1974) Report on investigation of shell-bed deposits on the property of Mr R. Lancaster, Clybucca, Kempsey, NSW in December 1973.Unpublished report. Retained in NSW Department of Environment, Climate Change and Water, Aboriginal Heritage Information Management System, ref. 252.
28. McBryde I (1982) *Coast and Estuary: Archaeological Investigations on the North Coast of New South Wales at Wombah and Schnapper Point.* Australian Institute of Aboriginal Studies, Canberra.
29. Mulvaney DJ (1975) *The Prehistory of Australia.* Pelican Books, Melbourne.
30. Harney WE (1951) The dingo. Walkabout 1 July, 36–37.
31. McIntosh IS (1999) Why the dingo ate its master. *Australian Folklore* **14**, 183–189.
32. Trezise P (1978) Shadow of the dingo. *Habitat* **5**, 25–26.

33. Horton D, McKenzie K, Bancroft R, De Brabander D (1994) *The Encyclopaedia of Aboriginal Australia: Aboriginal and Torres Strait Islander History, Society and Culture.* Aboriginal Studies Press for Australian Institute of Aboriginal and Torres Strait Islander Studies, Canberra.
34. Pardoe C (1996) Dogs changed the world. *National Dingo News* **Spring**, 19–29.
35. Gunn RG, Whear RL, Douglas LC (2010) A dingo burial from the Arnhem Land Plateau *Australian Archaeology* **71**, 11–16.
36. Macintosh NWG (1965) Dingo and horned anthropomorph in an Aboriginal rock shelter. *Oceania* **36**, 85–100.
37. Bednarik R (2002) The dating of rock art: a critique. *Journal of Archaeological Science* **29**, 1213–1233.
38. Jones R (1970) Tasmanian Aborigines and dogs. *Mankind* **7**, 256–271.
39. Mulvaney J, Kamminga J (1999) *Prehistory of Australia.* Allen and Unwin, Sydney.
40. Strahan R, Van Dyck S (2006) *Mammals of Australia.* New Holland Publishing, Sydney.
41. Wallace A (1893) The Australian Aborigines. In *Stanford's Compendium of Geography and Travel.* pp. 111–157. London.
42. Jones FD (1921) The status of the dingo. *Transactions and Proceedings of the Royal Society of South Australia* **45**, 254–263.
43. Kohen JL (1995) *Aboriginal Environmental Impacts.* University of New South Wales Press, Sydney.
44. Tindale NB (1974) *Aboriginal Tribes of Australia: Their Terrain, Environmental Controls, Distribution, Limits, and Proper Names.* Australian National University Press, Canberra.
45. Johnson CN, Wroe S (2003) Causes of extinction of vertebrates during the Holocene of mainland Australia: arrival of the dingo, or human impact? *The Holocene* **13**, 941–948.
46. Flannery TF (2004) *Country.* Text Publishing, Melbourne.
47. Johnson CN (2006) *Australia's Mammal Extinctions: A 50,000 Year History.* Cambridge University Press, New York.
48. Thomson DF (1949) Arnhem Land: explorations among an unknown people. III. On foot across Arnhem Land. *Geographical Journal* **114**, 53–67.
49. Corbett LK (1985) Morphological comparisons of Australian and Thai dingoes: a reappraisal of dingo status, distribution and ancestry. *Proceedings of the Ecological Society of Australia* **13**, 277–291.
50. Berndt RM, Berndt CH (1981) *The World of the First Australians.* Lansdowne Press, Sydney.
51. Gollan K (1985) Prehistoric dogs in Australia: an Indian origin? In *Recent Advances in Indo-Pacific Prehistory.* (Eds P Bellwood and VN Misra) pp. 439–443. Oxford & IBH Publishing, New Delhi.
52. Gould RA (1969) Puntutjarpa rockshelter: a reply to Messrs Glover and Lampert. *Archaeology and Physical Anthropology in Oceania* **4**, 229–237.
53. Hiscock P (2008) *Archaeology of Ancient Australia.* Routledge, London.
54. Bowdler S, O'Connor S (1991) The dating of the Australian small tool tradition, with new evidence from the Kimberley, WA. *Australian Aboriginal Studies* **1**, 53–62.
55. Baird R (1991) The dingo as a possible factor in the disappearance of *Gallinula morlierii* from the Australian mainland. *Emu* **91**, 121–122.
56. Archer M (1974) New information about the Quaternary distribution of the thylacine (Marsupialia, Thylacinidae) in Australia. *Proceedings of the Royal Society of Western Australia* **57**, 43–50.
57. Corbett LK (2004) Australia and Oceania (Australasian): dingo (*Canis lupus dingo*). In *Canids: Foxes, Wolves, Jackals and Dogs: 2004 Status Survey and Conservation Action Plan.* (Eds C Sillero-Zubiri, M Hoffmann and DW Macdonald) pp. 223–230. IUCN/SSC Canid Specialist Group, Gland, Switzerland, and Cambridge, UK.
58. Letnic M, Fillios M, Crowther MS (2012) Could direct killing by larger dingoes have caused the extinction of the thylacine from mainland Australia? *PLoS ONE* **7**, e34877.
59. Fillios M, Crowther M, Letnic M (2012) The impact of the dingo on the thylacine in Holocene Australia. *World Archaeology* **44**, 118–134.

60. Paddle R (2002) *The Last Tasmanian Tiger: The History and Extinction of the Thylacine.* Cambridge University Press, Cambridge.
61. Prowse T, Johnson C, Bradshaw C, Brook B (2013) An ecological regime shift resulting from disrupted predator–prey interactions in Holocene Australia. *Ecology* **95**, 693–702.
62. Wroe S, McHenry C, Thomason J (2005) Bite club: comparative bite force in big biting mammals and the prediction of predatory behaviour in fossil taxa. *Proceedings. Biological Sciences* **272**, 619–625.
63. Attard M, Chamoli U, Ferrara T, Rogers T, Wroe S (2011) Skull mechanics and implications for feeding behaviour in a large marsupial carnivore guild: the thylacine, Tasmanian devil and spotted-tailed quoll. *Journal of Zoology* **285**, 292–300.
64. Murray MD, Calaby JH (1971) Appendix II: the host relations of the Boopiidae. *Australian Journal of Zoology: Supplementary Series* Supplement 6, 81.
65. Breckwoldt R (1988) *A Very Elegant Animal: The Dingo.* Angus and Robertson, Sydney.
66. Koler-Matznick J, Yates BC, Bulmer S, Brisbin IL (2007) The New Guinea singing dog: its status and scientific importance. *Australian Mammalogy* **29**, 47–56.
67. Titcomb M (1969) Dog and man in the ancient Pacific. *Bernice Pauahi Bishop Museum Special Publication 59.*
68. Redd AJ, Roberts-Thomson J, Karafet T, Bamshad M, Jorde LB, Naidu JM, Walsh B, Hammer MF (2002) Gene flow from the Indian subcontinent to Australia: evidence from the y chromosome. *Current Biology* **12**, 673–677.
69. Pugach I, Delfin F, Gunnarsdottir E, Kayser M, Stoneking M (2013) Genome-wide data substantiate Holocene gene flow from India to Australia. *Proceedings of the National Academy of Sciences of the United States of America* **110**, 1803–1808.
70. Brown P (2013) Of humans, dogs and tiny tools. *Nature* **494**, 316–317.
71. Macintosh NWG (1974) Early man and the dog in Australia. In *Grafton Elliot Smith: The Man and his Work.* (Eds AP Elkin and NWG Macintosh) pp. 83–94. University of Sydney Press, Sydney.
72. Oskarsson MC, Klutsch CF, Boonyaprakob U, Wilton A, Tanabe Y, Savolainen P (2012) Mitochondrial DNA data indicate an introduction through mainland South-east Asia for Australian dingoes and Polynesian domestic dogs. *Proceedings. Biological Sciences* **279**, 967–974.
73. Savolainen P, Leitner T, Wilton A, Matisoo-Smith E, Lundeberg J (2004) A detailed picture of the origin of the Australian dingo, obtained from the study of mitochondrial DNA. *Proceedings of the National Academy of Sciences of the United States of America* **101**, 12387–12390.
74. VonHoldt BM, Pollinger JP, Lohmueller KE, Han E, Parker HG, Quignon P, Degenhardt JD, Boyko AR, Earl DA, Auton A, Reynolds A, Bryc K, Brisbin A, Knowles JC, Mosher DS, Spady TC, Elkahloun A, Geffen E, Pilot M, Jedrzejewski W, Greco C, Randi E, Bannasch D, Wilton A, Shearman J, Musiani M, Cargill M, Jones PG, Qian Z, Huang W, Ding ZL, Zhang YP, Bustamante CD, Ostrander EA, Novembre J, Wayne RK (2010) Genome-wide SNP and haplotype analyses reveal a rich history underlying dog domestication. *Nature* **464**, 898–902.
75. Ardalan A, Oskarsson M, Natanaelsson C, Wilton AN, Ahmadian A, Savolainen P (2012) Narrow genetic basis for the Australian dingo confirmed through analysis of paternal ancestry. *Genetica* **140**, 65–73.
76. Clutton-Brock J (1995) Origins of the early dog: domestication and early history. In *The Domestic Dog: Its Evolution, Behaviour and Interactions with People.* (Ed. J Serpell) pp. 7–20. Cambridge University Press, Cambridge.
77. Ding ZL, Oskarsson M, Ardalan A, Angleby H, Dahlgren LG, Tepeli C, Kirkness E, Savolainen P, Zhang YP (2012) Origins of domestic dog in southern east Asia is supported by analysis of Y-chromosome DNA. *Heredity* **108**, 507–514.
78. Wilton AN, Steward DJ, Zafiris K (1999) Microsatellite variation in the Australian dingo. *Journal of Heredity* **90**, 108–111.
79. Irion DN, Schaffer AL, Grant S, Wilton AN, Pedersen NC (2005) Genetic variation analysis of the Bali street dog using microsatellites. *BMC Genetics* **6**, 610.1186/1471-2156-6-6

80. Sacks BN, Brown SK, Stephens D, Pedersen NC, Wu JT, Berry O (2013) Y chromosome analysis of dingoes and South-east Asian village dogs suggests a Neolithic continental expansion from South-east Asia followed by multiple Austronesian dispersals. *Molecular Biology and Evolution* **30**, 1103–1118.
81. Koler-Matznick J, Brisbin IL, Feinstein M, Bulmer S (2003) An updated description of the New Guinea singing dog (*Canis hallstromi*, Troughton 1957). *Journal of Zoology* **261**, 109–118.
82. Runstadler JA, Angles JM, Pedersen NC (2006) Dog leucocyte antigen class II diversity and relationships among indigenous dogs of the island nations of Indonesia (Bali), Australia and New Guinea. *Tissue Antigens* **68**, 418–426.
83. Axelsson E, Ratnakumar A, Arendt ML, Maqbool K, Webster MT, Perloski M, Liberg O, Arnemo JM, Hedhammar A, Lindblad-Toh K (2013) The genomic signature of dog domestication reveals adaptation to a starch-rich diet. *Nature* **495**, 360–364.
84. Smith BP, Litchfield CA (2009) A review of the relationship between indigenous Australians, dingoes (*Canis dingo*) and domestic dogs (*Canis familiaris*). *Anthrozoos* **22**, 111–128.
85. Clutton-Brock J (2012) *Animals as Domesticates: A World View through History.* Michigan State University Press, Michigan.
86. Wilton AN (2001) DNA methods of assessing dingo purity. In *A Symposium on the Dingo.* (Eds CR Dickman and D Lunney) pp. 49–56. Royal Zoological Society of New South Wales, Sydney.
87. Lord K, Feinstein M, Smith B, Coppinger R (2013) Variation in reproductive traits of members of the genus Canis with special attention to the domestic dog (*Canis familiaris*). *Behavioural Processes* **92**, 131–142.
88. Appleby R, Smith B, Jones D (2013) Observations of a free-ranging adult female dingo (*Canis dingo*) and littermates' responses to the death of a pup. *Behavioural Processes* **96**, 42–46.
89. Smith BP, Appleby RG, Litchfield CA (2012) Spontaneous tool-use: an observation of a dingo (*Canis dingo*) using a table to access an out-of-reach food reward. *Behavioural Processes* **89**, 219–224.
90. Smith BP, Litchfield CA (2013) Looking back at 'looking back': operationalising referential gaze for dingoes in an unsolvable task. *Animal Cognition* **16**, 961–971.
91. Smith BP, Litchfield CA (2010) How well do dingoes, *Canis dingo*, perform on the detour task? *Animal Behaviour* **80**, 155–162.
92. Smith BP, Litchfield CA (2010) Dingoes (*Canis dingo*) can use human social cues to locate hidden food. *Animal Cognition* **13**, 367–376.
93. Smith B (2014) Living with wild dogs: personality dimensions in captive dingoes (*Canis dingo*) and implications for ownership. *Anthrozoos* **27**, 423–433.

4

The role of dingoes in Indigenous Australian lifestyle, culture and spirituality*

Bradley Smith

Australia has been home to Indigenous Australians for at least 50 000 years.[1–3] The majority of this time they were without canine companions. After the dingo arrived, fossil and anthropological evidence suggest it was quickly adopted into almost every facet of Indigenous Australian lifestyle, culture and spirituality. This chapter describes the relationship between dingoes and Indigenous Australians, which is not only fascinating but also quite different from most other human–canine relationships throughout history.

Knowing more about this relationship is important for piecing together the history of dingoes, and in determining whether Indigenous Australians may have shaped dingoes or vice versa. More specifically, it will explore the extent to which any domestication or artificial selection events might have occurred since the time of the dingo's arrival in Australia. While it is difficult to see into the murky life history of the dingo before its arrival on Australian shores, a clear picture is emerging about what happened afterwards. The answer has implications for our understanding and interpretation of dingo behaviour and ancestry, and the way society views the dingo. On a more global level, an understanding of the relationship between primitive dogs (such as dingoes) and traditional societies offers an opportunity to learn more about the early human–canine bond and factors that may have influenced the domestication of early canids. Unfortunately, this is becoming increasingly challenging due to the scarcity of traditional Indigenous communities around the world where the main mechanism for retaining knowledge and traditions is by oral means.

In order to gain an understanding of the Indigenous Australian–dingo relationship we are heavily reliant on the written accounts of early explorers and anthropologists. This requires hours of research and sifting through obscure books and scientific papers for any mention of the dingo or dog. Unfortunately there are quite a few problems with this method. The biggest challenge is that the number of these accounts is limited, and to make things worse they are scattered across various literature sources that are not readily located or accessible. Most include instances of what the observer considered quirky and interesting and therefore worthy of reporting, rather than systematic observations. Although in recent times there has been massive growth in the interest in the lives of animals and the human–animal relationship, when these early observations were made they were not a priority. Most mentions of interspecies interactions are fleeting in nature. Another challenge is that the accounts that do exist are highly anecdotal and typically written by Europeans or uneducated early settlers who often

* Parts of this chapter have been published as: Smith B, Litchfield C (2009) A review of the relationship between Indigenous Australians, dingoes (*Canis dingo*) and domestic dogs (*Canis familiaris*). *Anthrozoos* **22**, 111–128. Replicated with permission from Bloomsbury Publishing.

unknowingly carried some heavy biases (they saw the world through the eyes of a European and not of their human 'subjects'). It is difficult to imagine how much of the relationship we don't know, and will never know, as a result of these biases and misrepresentations. A final challenge is the sheer diversity of Indigenous Australian culture. Each Country, of which there are roughly 400 across Australia, has a different perspective on the dingo. For example, many of the available accounts focus on northern Australian communities but these cannot be generalised to Indigenous perspectives in southern or eastern Australia, which are quite different.[4] For the sake of brevity, this chapter will only summarise the relationship and beliefs of Indigenous Australians towards dingoes in the broadest sense, acknowledging that this might do little justice to the many and varied relationships that existed (and still exist in many places) across Australia.

Yet another confounding problem likely to influence our understanding was that by the time most of the anthropological studies were conducted, many of the communities had domestic dogs. The dog arrived on the scene with the early European settlers in the late 1700s, and resulted in a drastic change in the relationship between Indigenous Australians and dingoes. The more subservient and loyal domestic dog quickly replaced the role of the dingo. Many communities at this time chose the dog because they were more readily obtained and were considered to be 'less troublesome and generally fit in better with living in close quarters with man'[5] (p. 95). However, as you might imagine, before European settlement, because dingoes largely foraged for and took care of themselves, they are likely to have been easier to maintain and less of a drain on resources than domestic dogs.[6] Unfortunately, no real effort was made to distinguish between the two species, so it is highly likely that anthropologists or observers at the time made inferences about dingoes based on observations of domestic dogs. Where possible, this chapter clearly differentiates between dogs and dingoes, and uses the term canid where information pertains to both, in an effort not to make the same mistake.

What should we call it?

Once the dingo arrived in Australia it quickly colonised the mainland and nearby islands and established itself as a functional and integral member of the Australian ecosystem[7] (Chapter 8). Since the first (or founding) population of dingoes may have had previous associations with humans (see Chapter 3), they may have easily and quickly forged a relationship with the local Indigenous people. If this were the case, it would have hastened the dispersal of dingoes as they would have readily accompanied or followed the people, and may have even been used as trade items with neighbouring communities.[8]

The name 'dingo' derives from an Indigenous Australian term believed to describe tame camp dingoes, used by people living in the Port Jackson area (New South Wales) at the time of European settlement.[9] However, the term dingo was not used universally, differing within states (e.g. in New South Wales mirri and warrigal are used)[9] and between states (e.g. ngupanu, papa, parrutju, tjantu, wanaparri, yinura and maliki are used in Western Australia, the Northern Territory and South Australia).[10] In south-eastern Australia (Victoria), there are at least 10 Indigenous place-names that contain the local word for dog.[4]

The diversity of names also gives hints about the difference in behaviour and, in turn, the relationships and roles of dingoes within different communities. For example, definitions of dingo or wild dog in the Anbarra community (Northern Territory) include 'fearful one' (gugurkuja) referring to the dingo being afraid and running away from people on contact in the bush, or 'wild animal or beast' ('an-mugat').[11] Camp dingoes, or those associated with people, are also distinguished from wild dingoes.[9, 12] In Victoria, the language group Ganai called the wild or free-living 'dog' (dingo or feral dog) 'mirigang' and the tame dog 'banj'.[4]

Attitudes towards dogs and dingoes

In many human societies, including some Indigenous Australian societies, dogs appear to fall somewhere between human and non-human animal in status. For example, both dogs and dingoes were afforded certain human privileges,[13] which included being given traditional or mythological names,[14, 15] skin identities or personal kinship names, food and shelter.[16] However, in some cases, dogs were restricted from accompanying men to ceremony but expected to follow the women, in the same way that children accompany women.[14]

To some Indigenous Australians, dogs served as a good reference point for human behaviour. They were regarded as individuals and as members of the community, but ones that did not heed the social laws or rules of kinship.[13] Socially disapproved behaviour such as greed, anti-social or unpredictable behaviour,and sexual promiscuity, for example, are considered dog-like traits.[15, 17] Anyone disregarding marriage 'restrictions' was considered to be acting 'like dogs'.[13]

Today, surveys of Australian households reveal that dog owners consider their dogs to be part of the family. It was no different for Indigenous Australians, to whom in many cases dogs were more than just personal property. So much so that any offence against a person's dog, including accidental or deliberate killing, 'might have violent repercussions'[18] (p. 345). Smyth (1878, as cited in [4]) reported that 'Aborigines are very kind to their dogs, and indeed nothing more offends a black man [*sic*] than to speak harshly to his dogs ... if anyone gave a black man's dog a blow, he would incur bitter enmity'. After the death of a dog (often a deliberate act by non-Indigenous people), Indigenous owners commonly buried it 'with great ceremony, wrapping them in blankets and sheets of bark and lighting fires by their graves after which they decamped and moved up the river' (Rawson 1839, as cited in [4]).

Some authors describe the great affection of Indigenous Australians towards their dogs, particularly puppies, going as far as calling it remarkable.[15] A few accounts have suggested that, like dogs, dingoes were treated as lovingly as the people would have treated their own children.[19] In some cases, dingo genitals were seen to be fondled and petted as a way of soothing tired animals.[5, 20] In some Indigenous communities, such as the Jankuntjara (Central Australia), people were known to groom their puppies' fur searching for scales, sores or parasites. Puppies had a better life than adult dogs. As with many young animals, puppies were particularly appealing to their human caregivers but once they grew older they were largely ignored and left to take their chances with the rest of the canid population.[11] Kolig surmised that dingoes 'quickly progressed from babies to beasts which everybody loathes'[5] (p. 93).

Sometimes a dog remained a favourite with a particular human, who continued to lavish exclusive attention and affection upon it, particularly if it had been hand-reared. For example, Hamilton[15] witnessed favourite puppies being rubbed with a mixture of fat and ochre (just like that used on humans) to make their bodies strong and resistant to evil spirits. Another researcher, Richard Gould[21] found both men and women in Western Desert Indigenous communities had particular favourites, and in some cases provided food for only those animals. Children in Indigenous communities were not always as careful with dogs; a major cause of canine mortality was attributed to children's 'constant' play.[5, 15]

There was a clear distinction in Indigenous Australian attitudes towards camp dogs, with hunting dogs highly regarded and non-hunting dogs less respected.[22] For example, White[22] reported that the superior health and well-mannered behaviour of hunting dogs was in stark contrast to the mongrels that scrounged and stole food to survive. In more recent times, domestic dogs in Indigenous communities have been reported to suffer from chronic eye and skin infections and several other diseases.[11] This is not so much a reflection on community attitudes towards the dogs, as the lack of accessible veterinary resources and care, the unnaturally high dog densities, poor diets and inbreeding as a result of a general lack of intraspecific social cohesiveness.

Although there seems to be little doubt that dingoes were admired and even loved in many cases, quite a few anthropologists observed that dingoes living in some communities often appeared to be in poor physical condition. Gould[21] described the dingoes he encountered as the skinniest dogs he had ever seen. The poor physical condition of dingoes living among humans became an easy way of telling them apart from their healthy free-ranging (wild) counterparts.[23] The dingoes' poor condition reflected the scarcity of food (especially around the camps).[15] This resulted in a 'never-ending battle' for food (p. 65).[24] Dogs faced the same problem of limited food. For example, adult dogs living in the Jankuntjara community mostly survived on raw flour left over from damper-making, or the bones of large game. Only when meat was abundant did the dogs receive 'even a token amount of food' (p. 290).[15] Dogs supplemented their meagre scraps by stealing food around the campsite; they rarely attempted to forage away from camp. As puppies matured, they received glancing blows on the nose to keep them away from food. Hands, sticks or metal bars were used to punish those that crowded around humans at mealtimes.[15]

The spiritual dingo

Soon after their arrival, the dingo was quickly embedded into Indigenous Australian life and mythology. The representations of dingoes are as varied as the Indigenous cultures across the continent. Traditional Indigenous Australian spirituality is often described as the Dreaming. It refers to a period when the world was created and when the 'responsible creational beings' and framework of laws governing Indigenous action were formed.[25] Compared to other animals (with the possible exception of the snake) the dingo is overrepresented in Indigenous Australian mythology, and is associated with the supernatural more than any other animal.[26] The dingo appears in numerous Dreaming narratives across the continent, with Dreaming stories popularised in children's literature such as *The Giant Devil Dingo* (by Dick Roughsey)[27] and *Lasca and her Pups* (by Percy Trezise).[28] For a comprehensive summary of the dingo's role in Indigenous Australian Dreaming (including its role in human and environmental creation, hunting and resurrection), see Parker.[29]

Each group has incorporated the dingoes in a unique way. Deborah Bird Rose, a social anthropologist, spent two years living with the Yarralin people, a remote community in the Victoria River, Northern Territory.[16] This community took note of the physical similarities between dingoes and themselves to such an extent that they believe that humans have dingo origins, the dingo quite literally made us human. By this she meant that dingoes gave humans their characteristic shape with respect to head, genitals and upright stance. The dingo is the only animal that walks now as it did in the Dreaming, and there was a time when dingo and human were one and the same. Each person also has an animal or plant relative, and a spiritual bond with it. 'Old Tim Yilngayari' the 'dog man' of the Yarralin community is known for his 'intimate concern for, and understanding of, dingoes and dogs. His life is connected to dingo life and he has assumed a special responsibility for canines'. Dingo and human death are intricately connected, since 'Human death, too, originates with the dingo. The moon has the Law of eternal life, and he offered it to the dingo'. The dingo is also often associated with the rites of fertility, circumcision and subincision (slitting the underside of the penis with permanent opening of the urethra).[5, 30–32]

In Indigenous Australian rock art, the dingo clearly assumes a mythological rather than a practical role.[30] Some depict the dingo as an intermediary between the living and spiritual worlds, potentially warning humans of approaching evil spirits.[26] Some Indigenous Australians attribute to the domestic dog the ability to sense approaching death in humans[14] and indeed dogs are employed elsewhere around the world as sentinels of epileptic and other seizures and death.[33, 34] By imitating dogs in some ceremonies, young Indigenous men may be endowed with 'whatever good or beneficial qualities that animal might possess'[35] (Fig. 4.1).

This section has focused on the dingo in the Dreaming but some Dreaming narratives also refer to the dog, though it is unclear whether these are domestic dogs, dingoes or both. In general, whenever the dingo/dog appears in the Dreaming it is considered to have detrimental (negative) or uncanny attributes, appearing as a 'vicious', 'murderous', 'bloodthirsty', 'deceitful' and 'wicked creature'. These mythical dingo/dogs wreak havoc through 'acts of destruction; they never obey their masters, are vicious and disloyal; they rape and thieve; and they never act as culture heroes as many other mythical beings do' (p. 123).[26]

Although as a 'fully fledged lawman' the dingo institutes Dreaming laws it also breaks them, appearing as a 'trickster' capable of transcending 'humankind's conceptual boundaries, moving freely between the worlds of gods and humans and playing tricks on both' (p. 184).[25] The dingo, as a sacred totem, provides a reference point for Indigenous Australian customs and social structure, serving as 'a powerful symbol for moderation in behaviour at both individual and group levels'. The 'trickster dingo' threatens the status quo, and Indigenous Australians recognise that this potential trait exists within every member of the community.[25]

The importance of dogs in human culture is well illustrated by the existence of dog burials all around the world, dating to at least 12 000–14 000 years ago.[36] Similar to their human companions, dogs were given the opportunity to take part in the afterlife, viewed as beings capable of true unconditional friendship.[36] In Australia, dingo remains have been found near areas of intense habitation by Indigenous communities, such as mounds and open coastal middens (large deposits containing shells, animal bones and other refuse that indicates the site

Figure 4.1: This image represents a ceremony in New South Wales where men run in circles past boys on their hands and feet imitating dingoes. The swords at times appeared to represent a tail. According to David Collins, by performing this ceremony, it was believed that initiates (the boys) were given power over the dog and whatever good or beneficial qualities that animal might possess.
Source: David Collins.[35]

of a human settlement),[37, 38] suggesting that some dingoes were held in high regard. Pardoe[39] located a human and canid (which he believed to be a dingo) burial site at the Coorong, South Australia. At this site, humans were buried at the centre and surrounded by a large number of hearths (fireplaces), and five dogs were buried at the edge of the site. The intentional positioning of these canids outside the living areas suggests that the canids were protecting their dead human companions in the spirit world. Gunn[40] reported another instance where a dingo was given a special burial in Arnhem Land (northern Australia). 'The dingo burial was placed in a small but locally prominent capped-pillar shelter. As with human bundle-burials, the dingo was wrapped in paperbark and placed on a high ledge, and then closed in with several rocks to prevent disturbance' (p. 12).[40] One of the bones was carbon dated, with the likely result that the dingo died (and was buried) between 1680 CE and 1930 CE.

Raising puppies

Puppies were particularly popular in some Indigenous communities and appeared to serve as substitute children, providing an outlet for nurturing in areas that did not support a large human population.[15] In some communities, childless women were observed to carry dogs around their waist with 'all women who were barren or had ceased bearing children … required to nurse a dog in this way' (White 1915, p. 726, as cited in[15]). Early explorers witnessed Indigenous Australian women breast-feeding young dingo pups.[15] Hamilton recorded the number of dogs kept by individual households and found that 'older persons and those with no resident children have more dogs; younger persons with young families have only one or none at all'. In Anbarra society each woman on average owned about three dogs, with elderly women owning up to 12 dogs.[11] Separate instances of an elderly Indigenous Australian woman fussing over her many dogs in order to compensate for unsatisfactory personal relationships have also been reported.[11, 24]

By all accounts, it was quite easy for Indigenous Australians to acquire dingoes. All they had to do was find a nearby dingo den during whelping season, take a few pups and hand-rear them.[15, 19, 23] They preferred the dingoes as young as possible, usually around two to four weeks old. This suggests an understanding of the critical period of socialisation in canids, where there is only a small window of opportunity to socialise a pup to humans in order for it to be a successful and well-adjusted pet later in life.[41]

Acquisition may have taken place opportunistically during hunting activities when they happened upon dingo dens[15, 32] or have been a targeted or deliberate process.[6, 23] For example, in several eastern Australian and Western Desert communities, during May (heralded by celestial changes) the number of ceremonies increased, reflecting a desire to raise dingo numbers.[32, 42, 43] In this region, healthy dingo pups were kept, weak or deformed pups were killed and perhaps eaten.

Later, Indigenous Australians' knowledge of dingo breeding behaviour (e.g. times and den locations) helped hunters when Europeans started offering bounty payments for dingo scalps (Fig. 4.2). These trained Indigenous hunters were able to make quite a bit of money, but were smart enough to ensure that they didn't exhaust the supply of dingoes. For example, they left alone breeding-age female dingoes, which according to them resulted in sustainable dingo population numbers. In southern Australia, many colonial writers mentioned how skilled Indigenous people were at hunting wild dingoes. In fact, many non-Indigenous farmers actively encouraged Indigenous people to kill the native dogs in exchange for money or goods such as tobacco.[4]

Indigenous Australians had no need to control the population of dingoes that lived with them in their communities because their numbers were never excessive. Despite their close association with humans from puppyhood, dingoes almost always returned to their natural habitat once they reached sexual maturity, never to be seen again. It is possible that dingoes were tied up or somehow restrained as pups and released once acclimated to human environments, but this does not appear to be a well-supported claim. I tend to agree with the assertion made by Macintosh[30] who believed 'the dingo would simply take off once released, and return only to steal food'. It is highly likely that dingoes which failed to adjust to life around humans and became dangerous, or refused to leave once sexually mature, were killed.[16] Rose[16] lived within a remote Indigenous community for some years, and draws interesting parallels between the developmental stages of dingoes and some Indigenous Australian males. Both boys and dingoes were nurtured within the community during their youth but had to leave

Figure 4.2: This image, taken in 1963, depicts an overseer on a cattle property in northern Australia issuing a receipt to Indigenous stockmen for dingo pelts. Dingo scalping became a lucrative venture for Indigenous Australians, who had intimate knowledge of dingoes and proved successful hunters.
Source: National Archives of Australia, A1200, L39260.

upon reaching sexual maturity (at least for a short period in the case of the young men, to fulfil initiation practices).

In contrast, domestic dogs appear never to leave or do so rarely, choosing instead to remain within the community. This is why, in many Indigenous communities, one of the most noticeable things is the large number of dogs, which often outnumber the human inhabitants. For example, at Yatala Aboriginal Reserve (South Australia) in 1971, the ratio was slightly above 1:1[22] and in south-east Arnhem Land (in 2006) there were three dogs per household.[14] The high numbers of domestic dogs may be attributed to their desire to remain near humans rather than leave for natural environments, the greater usefulness of the dog over the dingo, the lack of any attempt to control or suppress breeding, and the capacity for domestic dogs to breed more than once per year compared to the seasonal breeding cycle of the dingo.

Hunting assistants: help or hindrance?

The ability of wild canids to assist hunting is one of the most commonly cited reasons that wolves were domesticated, but the usefulness of such animals is not clear; opinions concerning the effectiveness of the dingo in aiding Indigenous hunting are somewhat contradictory.[30] Hayden,[6] an ethno-archaeologist, is convinced that 'the original and continuing primary reason for the regular adoption of dingoes into Indigenous Australian groups was related to hunting subsistence' and that other uses were 'acquired and accrued advantages, which have sometimes become dominant under settlement conditions'. He cited early accounts that suggested dingoes were trained as pups for hunting, that they hunted on behalf of humans throughout the day and for themselves at night, that they were successful hunters of kangaroos and that their ability to follow scents assisted in the capture of kangaroos, snakes, rats, possums and lizards. Hayden stressed that, regardless of the evidence, we should not doubt the assertions made by Indigenous Australians that confirm the dingo's role in hunting.[6] Cahir and Clark[4] believed that dingoes and dogs are more often used as hunting assistants in southern Australia than in the north (Fig. 4.3).

Figure 4.3: Two Pintupi elders nursing dingo pups before taking them out on a hunt. Photograph taken in 1957 at Lake Mackay.

Source: Douglas Lockwood Collection, Northern Territory Library, PH 0501/0040.

It is important to reiterate at this point that historical reporting often does not specify whether a canid referred to is a dingo or domestic dog. Hayden,[6] for example, may have made this error when he provided examples of domestic dogs to support his assertion that dingoes were used for hunting. This presents a problem, since the traditional use of dingoes may have been different or obscured by later reports focusing on use of domestic dogs that may have been trained or selectively bred for hunting skills. For example, some Indigenous Australian communities in Tasmania immediately recognised the useful hunting skills of domestic dogs and quickly adopted the dogs, despite the fact that the communities had never encountered canids before.[44, 45]

With respect to the Indigenous Australian–dingo relationship, it appears dingoes were rarely utilised or necessary for the successful hunting of large game.[15, 21–24] This may reflect the fact that dingoes were ineffective hunting aids in certain open environments[23] or that the benefit from their assistance was only minimal.[15] Even Hayden[6] admitted that 'dogs vary in their success in game capture and very possibly succeeded less often than hunters using stalking and spearing methods'. Indeed, dingoes were considered a liability for hunting parties that employed strategies based on concealment or stalking.[24] Typically, the hunters hid behind blinds of brush or rock or stalked animals if encountered in the open, but the success of these hunts was compromised if dingoes frightened or chased prey away.[20, 24] Indigenous Australian hunters were skilled trackers who could follow fresh tracks closely without alarming the animals. If dingoes were present, they usually disturbed the prey (e.g. by howling or barking), resulting in a long and unsuccessful chase.[5, 15, 20]

On some occasions, if dingoes did not accompany hunters but were nearby, the dingoes were 'called up' to pursue injured prey animals, tiring and harassing them until the hunter caught up.[46] Kolig,[5] however, suggested that this technique was both tiring for the human hunter and largely ineffective since the dingo was 'notorious for devouring most of the prey before the hunter has a chance to claim it for himself'. Anthropologist Meggitt[23] confirmed this, adding that dingoes that were poorly treated by humans were even more likely to consume game before the hunters got to it, and may have been prompted to return to natural environments in search of more reliable sources of food. It is more likely that dingoes played an active albeit opportunistic role in the foraging activities of women and children.[5, 20] Thus, any meat contributed by the dingo occurred on an opportunistic basis, or during hunts for smaller prey. Furthermore, since the use of fire as a tool for hunting (flushing out prey) and land management was established well before the arrival of the dingo,[8] changing hunting strategies by using dingoes may have been detrimental in terms of hunting success.

Similarly, domestic dogs in many cases were considered such a hindrance to successful hunts that before any hunt the men would drive them back to camp,[15, 24] sometimes with a 'hail of stones' (p. 291).[15] In one case, Gould[24] reported that a favourite dog was tolerated as long as it remained a quarter of a mile behind and caused no interference. At no point in the area visited by Hamilton[15] were dogs used by any serious hunting group specifically attempting to kill large game. These cases, however, may be the exception rather than the rule, with many authors noting the positive effect of dogs (if only a selected few) in hunting, especially in capturing smaller prey items, or larger prey in areas with dense coverage of bush or forest.

Hunting strategies incorporating dogs were used if game was chased rather than stalked,[6] and sometimes dogs were used to assist in driving prey out of hiding.[47] When densities of large game were low, dogs helped capture small items of prey and occasionally brought back kills while accompanying women on gathering outings.[6, 11] Dogs that were left behind by male hunting parties tended to accompany women on their tasks,[21] with small domestic dogs helping dig out rabbit and wombat warrens.[15] It is not known whether dingoes were also used for these purposes.

White[22] described the use of domestic dogs by Indigenous Australians in Yalata Aboriginal Reserve (South Australia) as a significant innovation in hunting technique, greatly enhancing success rates compared with 'the old tribal times'. The range of hunting abilities among domestic dogs varied greatly, but hunting dogs at this site made a substantial contribution to the meat of the community, being considered more effective 'weapons' than spears and clubs. Jones,[44] after summarising the use of dogs in hunting in Tasmania, concluded that they were incorporated into the traditional hunting repertoire replacing men in the role of harriers, 'thus allowing the armed hunters to expend less energy for the same results', a skill 'probably facilitated by observations of Europeans hunting, and by the initial use of European-trained hunting dogs'(p. 267). It must be noted that not all Indigenous communities embraced the use of dogs as hunting aids[14] and, in those that did, not all camp dogs were used in this way.

Trying to neatly summarise this relational aspect is a veritable minefield, with contradiction rife among the writings of an individual researcher, let alone across the literature. It appears that while dogs in many areas were used successfully as hunting aids, dingoes probably were not, perhaps reflecting the dingoes' strong prey drive, independence and inability to be trained (see Chapter 10). Clearer still is that the literature may not accurately reflect the relationship between Indigenous Australian hunters and dingoes before European settlement because accounts were written after domestic dogs had become a major part of Indigenous Australian community life.

Dingoes as dinner

Several cultures around the world see dogs as a useful source of food.[48] This practice, however, doesn't appear to have been widespread in Australia. Corbett,[49] an ecologist who has studied dingoes and dingo-like canids in both Australia and Asia (where the practice of eating dogs is widespread), found it interesting that Indigenous Australians did not adopt the custom of eating canids, since he believed dingoes were used as a source of food by Asians who brought them to Australia. However, given the uniqueness and diversity of Indigenous Australian culture, it is safe to assume that the consumption of dogs varied across the country. In fact, the practice did occur but only in some areas and/or only in specific circumstances. For example, in southern Australia (e.g. Victoria) the practice appears to be more common than in northern Australia.[4] Near Melbourne, escaped convict William Buckley took refuge with the Wathawurrung people for upwards of 30 years.[50] Buckley provided perhaps the most detailed observation of Indigenous Australians eating wild dogs, which they considered a delicacy:

> ...we pitched, or rather erected our bark tents, having killed two immense large wild dogs on our way. The limbs of these animals they broke, and flinging them

> on the fire, they kept them there until the hair was singed, they then took out the entrails, and roasted the bodies between heated stones, covering them over with sheets of bark and earth. After this process, which lasted two hours, they were ready for eating and were considered a dish fit for an Exquisite. They handed me a leg of one, as the best part, but I could not fancy it; and on my smelling it, and turning up my nose, they were much amused, laughing away at a great rate.

In the northern areas, if dingoes were eaten it was most likely during times of emergency, such as famine.[5, 23, 32] As one anthropologist, Tindale, put it, dingoes were 'a kind of last hope meal for people facing abject starvation'.[32] In some cases where Indigenous Australians were observed to eat dingoes even when other meat was available, the meat was not highly prized.[15] One report described in detail the killing of a dog during a food shortage in Central Australia; the humans were unable to eat the dead dog and spent the whole night mourning its passing.[51] Others considered dog to be a taboo as a source of meat, and if desperate enough to eat it the human would be sick.[11] Last, nearly all societies made the distinction between wild and tamed dogs. This may have led to differences in which types could be consumed. For example, camp dogs that were well-known were more likely to be considered sacred and thus inedible, whereas strange or unknown dingoes or domestic dogs might have been fairer game.

A general nuisance

Why keep an animal that is a competitor for the same resources? After all, they both fought for the same prey – particularly those that were young or weak,[8] as well as for food around the camp. Hamilton[15] neatly summarised this competition: 'dogs are significant competitors with man for food, both wild dogs when he goes hunting and camp dogs when he comes home' (p. 290). It is estimated that both wild and camp canids may have been responsible for stealing as much as 10% of the food available, while camp canids contributed only a small portion of the overall edible food taken back to the community.[6, 15, 23, 24] For example, Hamilton[15] calculated that over two six-month periods, dogs living in the Jankunjara Indigenous community contributed 64 kg of meat, largely wallaroo and kangaroo, but consumed 148 kg.

Some reporters described the dingo or dog as a 'camp hanger-on' or scavenger that was not only costly in economic terms but a disruptive nuisance to the community in general.[5, 15] Dingoes wreaked havoc by 'brawling, thieving, barking all night, interfering in ceremonies and tripping up dancers' (p. 293).[15] Communities found it problematic to cache and store food with dingoes or dogs around, sometimes having to construct a table or shelf across branches 2 m off the ground. Children were often forced to eat standing upright to avoid hungry dogs.[15, 52] Similarly, to protect the dead from the 'unwelcome attention of dingoes' rocks or cairns were used on terrestrial graves and 'aerial tombs' were located in trees.[53, 54]

So if the dingo or dog didn't contribute to the food supply (either by assisting with hunting or being the food source themselves), were a significant cause of disruption to everyday life and required some effort to acquire as puppies (in the case of dingoes), why were they kept around? There are many reasons. Canids were able to provide several benefits that cannot be readily measured in economic terms. It seems that pet-keeping was just as popular then as it is today.

Security guards

For communities with no fencing or other type of preventative defences, dogs and dingoes were the perfect watchdogs, a role which they apparently performed with 'great enthusiasm'.[14, 15, 22] Dogs in the Jankuntjara community, for instance, individually protected a territory of ~4.5 m around each human shelter but collectively defended the entire perimeter of the Indigenous community.[15] These dogs greeted returning hunting parties, outsiders or unwelcome strangers and members of 'secret revenge expeditions' with raised tails, snapping jaws and violent barking. Much later, domestic dogs played an important role in protecting Indigenous Australians in Tasmania, serving as essential warning devices: 'countering whites surprising them *en masse* at night and firing musket volleys around their fires' (p. 210).[45] In the early 1800s, sometimes as many as 50 dogs were killed attempting to defend their Indigenous Australian owners from 'roving parties' of European settlers.[45]

Dogs served another important protective role, since it was believed that they could see all 'mamu' or evil spirits that are invisible to humans.[5, 11, 15, 26] The presence of evil forces, capable of being detected only by dogs, explains why women rarely went hunting alone and why they always had dogs with them.[11] So important was this role that even individual dogs considered largely useless served as protectors from human and spirit intruders since they added to the combined power of all the dogs living in the community.[14] The larger the number of dogs, the safer the community felt, with visitors forced to remain at a distance and call out loudly to make their presence known. The dogs' propensity for barking may have been a specifically selected trait in the early domestication process.[48]

While the role of the dog as protector is a focus in the literature, it is less clear whether dingoes served the same role although, like domestic dogs, dingoes are very territorial and thus are likely to have protected the Indigenous community they lived in. It has long been claimed that dingoes do not bark[23, 55] but this is not the case. They emit several vocalisations including snuff-barks and bark-howls (see Chapter 1), which would have alerted their owners to the approach of strangers just as it alerts other dingoes in their natural environments (I have observed this behaviour when interacting with wild dingoes on Fraser Island).

'Three-dog night'

Mutual warming by endothermic (warm-blooded) animals for the maintenance of body temperature is essential for survival in some environments.[48] Given the cold temperatures in desert areas at night, Indigenous Australians probably considered an animal capable of providing warmth as a valuable asset. Early explorers often mentioned the use of canids for this purpose.[23, 32, 56] This is the origin of the term 'three-dog night', which has even found its way into the Macquarie Dictionary of Australian slang.[12, 57] The dictionary states that a 'three-dog night' is a noun for a very cold night, i.e. the number of dogs required to keep warm is a direct reflection of the temperature. On a cold night you will need to sleep alongside one dog, but on a really cold night you will need three. Interspecific huddling can only occur in the absence of fear and aggression, with dingoes capable of serving this role only if raised appropriately from puppyhood, and before sexual maturity when they typically left camp. Dingoes not suited to the role are likely to have been killed or discarded.[48]

Gould[24] witnessed that 'on chilly nights the desert Indigenous people always sleep next to a fire with their dogs huddled around them to keep warm', and this was particularly so for the eldest women. But the use of dogs to keep warm was not always necessary. Hamilton[15] suggested that in most environmentally stable situations sleeping around small fires, with the addition of wind blocks, was sufficient for warmth.[58] Dogs were reported to lie between the fire and their human companions and frequently had to be driven away, thus potentially competing for heat from fires. However, he reported that dogs were clearly useful for warmth during rainy periods or damp heavy morning dews.[15]

Dingo clothing

Dingoes and dogs were also utilised as clothing. Tindale[32] noted that 'where skin clothing was not worn, the dog was of value enough to be kept, a walking blanket for a nomad [sic] unable to carry his wanderings more than a limited number of possessions' (p. 109). Several authors cited by Meggitt[23] observed the use of dingoes as 'living shawls' or lap-dogs, particularly by older members of the community, suggesting that Indigenous Australians depended greatly on their dogs for warmth. Hamilton[15] observed women carrying dogs on their backs for warmth. In colder parts of South Australia it appears that dogs were used as blankets. Indigenous Australians in this area were unable to use any form of animal skin as clothing, since scavenging dingoes tended to eat the garments.

Robert Smyth, honorary secretary to the Board for the Protection of Aborigines in 1860, compiled a series of writings titled *The Aborigines of Victoria*[59] during an attempt to gather information and artefacts of Victorian Indigenous cultures that he saw were fast disappearing. He referred to two interesting uses of native dogs for personal adornment (decoration, clothing etc.). The first was the skin, which was used to make a hunger belt. The hunger belt (Ber-buk) was worn around the waist with the fur positioned outward, and could be tightened when required. It was primarily used when travelling rapidly, or when the wearer had trouble obtaining food or water. When oppressed by hunger, the belt was tightened (Smyth 1878, as cited by [4]). The second was the use of the dog's tail (Wreka baanda, in the Gippsland area), which was suspended from the back of a headband and worn by men (and occasionally women).

Campsite cleaners

Never giving up the possibility of a free meal, both dingoes and dogs constantly scavenged around camp for food scraps and human faeces (that contain pathogens potentially harmful to humans). This role is considered by some authors to be essential in helping to maintain a sanitary and clean site[15, 28, 32, 48] and thereby extending the period the owners could stay in one location. Other authors disagree, suggesting that campsites could be moved if necessary[56] and that dogs and dingoes potentially added to the sanitation and potential health problems as they attracted flies, urinated and defecated around living areas, and were often infested with sores and parasites.[11, 15, 48] The potential spread of pathogens and disease from dogs to humans or wildlife has been widely documented,[60–62] although the threat posed by dingoes as a vector (or transferrer) before European settlement is not well understood.

Was the dingo domesticated in Australia?

Prior to European settlement, less-nomadic communities of Indigenous Australians were known to tame young native animals such as kangaroos, wallaroos, emu and cassowary chicks, and lizards,[5, 32] but the highest numbers of animals within a community were dingoes. In fact, early Europeans most commonly observed dingoes living with Indigenous communities and not in their natural environment. This led them to erroneously believe that the dingo was a domestic animal that turned feral only after European settlement. That is, they believed that dingoes fled to the bush after Indigenous communities were disrupted.[28, 63] The lack of free-ranging dingoes, in conjunction with the intricate relationship observed between Indigenous Australians and dingoes, often led to the latter being labelled as domesticated.[15, 28, 32, 63] However, we now understand the dingo was neither initially nor further domesticated in Australia.[23, 30, 48] After having described the relationship between dingoes, dogs and Indigenous Australians, we are now better placed to discuss the possible reasons why this was not the case.

The apparent lack of attempt to domesticate the dingo puzzled Charles Darwin during his visit to Australia. Darwin had studied dogs extensively and was convinced that even in ancient times attention would have 'been paid to breeding, for the inheritance of good and bad qualities is so obvious'.[64] Other researchers, such as Meggitt,[23] found it remarkable that 'people could associate for several millennia with an economically and ritually significant species without domesticating it completely'. Why did they miss this opportunity, that according to some was so obvious?

To begin to understand, we must first establish what domestication actually is. Domestication implies that the animals have been placed in a captive environment and exposed to certain management practices,[65] eventually losing all contact with their wild ancestral species.[52] Domestication generally fulfils a distinct purpose, need or desire in humans,[64, 66] offers an advantage of some description to the human domesticator and requires strict control or selective breeding of particular individuals within a population of animals. Taming of animals is sometimes incorrectly considered a form of domestication, since by definition tameness is a lack of avoidance responses if approached by a person.[67] Tameness takes place within an individual animal as a result of the learning processes of habituation (where fear of humans or other stimulus is reduced by repeated exposure to them), associative learning (where humans become associated with food, water, shelter and grooming) and/or observational learning (where already-tamed individuals serve as models).[65] Tameness does not take as much effort, but is a vital first step and precursor to the longer and sometimes slower process of domestication.[65, 67, 68] With intense selection, however, domestication can occur rapidly, as was most famously demonstrated in the case of the red fox (*Vulpes vulpes*). Russian geneticist Dmitri Belyaev found that a wild red fox could be kept as a docile pet within 15 generations simply by selecting only for tameness.[69]

In the case of the dingo, some control was evident in terms of puppy acquisition or removal from dens, and the killing of weak or deformed puppies or older dingoes that became dangerous or refused to leave once sexually mature. Thus some dingo traits were clearly preferred by Indigenous Australians but in the context of artificial selection it is impossible to know whether the killing of some dingoes and not others (selecting for preferred traits) made any difference to traits passed on to future generations. Female dingoes that denned and whelped around campsites, potentially increasing their access to resources, may have increased their likelihood of reproductive success.[70] The fact that dingoes were encouraged to leave upon reaching sexual maturity goes somewhat against the principle of domestication, which encourages animals to remain closely bonded with humans.

The later interactions between Indigenous Australians and the domestic dog may provide valuable insight into the lack of intervention or domestication practices imposed on dingoes. Indigenous Australians did not appear to selectively breed domestic dogs, apart from some isolated cases in Tasmania,[44] with most accounts suggesting that there was 'absolutely no evidence of any attempt to restrain free breeding of either category [hunting or pet] of the camp dogs' (p. 202).[22] Any male could breed with any bitch and any mongrel with any hunting dog.[11] Enrich Kolig[5] agreed, adding that the process of artificial selection or domestication was undoubtedly not employed given the low ratio of effective hunting dogs to useless companion dogs. Although puppies showing potential hunting ability were treated more favourably, they were not selected as preferred breeding animals.[22] However, despite the evidence that selective breeding was not deliberately taking place, the fact that dogs with preferred traits were treated more favourably and thereby were more likely to survive and pass on genes, suggests the possibility that there may have been unconscious selection.[64]

Successful domestication is marked by both a cultural change in humans and a biological change in the species being domesticated.[52, 71] Culturally, the animals must become 'incorporated into the social structure of a human community and become objects of ownership, inheritance, purchase, and exchange' (p. 31).[52] Such integration into human society results in the domesticated animals becoming reproductively, genetically and behaviourally isolated from their wild counterparts.

The dingo certainly meets the cultural change criterion of domestication, permeating every aspect of Indigenous Australian society. Dingoes became objects of ownership and items for barter with other Indigenous communities and European settlers, and held deep cultural and spiritual significance for Indigenous Australians. In practical terms, the dingo served as protector, living blanket, food item and companion, with a questionable role as a hunting assistant. Nevertheless, the dingo fails to meet the biological change criterion of domestication, which is caused by selection in response to the human community and environment, and artificial selection for economic, cultural and aesthetic reasons.[52, 71] Domestication of a species inevitably leads to changes in morphology (the structure of an organism) and reproductive traits over time,[69] as well as potential physiological changes to the endocrine system[68] and the hypothalamus.[67] Other changes expected as part of the domestication process include changes in cranial capacity (changes in neocortex), neoteny (the retention of juvenile characteristics such as barking and floppy ears in adulthood) and coat colour (aberrant colours or markings).[71] None of these traits that accompany domestication are apparent in the dingo (aside from the more recent isolated cases below); the modern dingo is essentially unchanged from its prehistoric ancestor of over 3000 years ago.[30, 38]

Most dingoes that had an association with an Indigenous community bred in natural environments, since they typically left when they reached sexual maturity. However, there are two reports of Indigenous communities that maintained a breeding colony of dingoes within the last 1000 years, most likely because their lifestyle could support it.[37] At Kiola and Murramarang Point (New South Wales), dingo remains at some Indigenous burial sites exhibit skeletal markers (particularly cranial) suggestive of selective breeding of specific populations (however, these could result from other factors such as isolation from wild counterparts). These apparently isolated instances are unlikely to have had a significant effect on the entire species. Even if dingo breeding did take place within the community, it would have been difficult to counterbalance the constant loss of tame dingoes through death or running away (p. 23).[23]

The Indigenous Australian–dingo relationship as a whole met only the initial stages of domestication, where the animal maintains a loose contact with humans but breeds freely without human intervention.[72] Such an association describes a relationship between humans and animals that are neither wild nor domesticated, and fall outside the more rigid definitions of domestication that have been proposed.[68] What are some possible reasons for this apparent lack of attempt by Indigenous Australians to domesticate the dingo?

One of the often-forgotten factors that make domestication of wild canids challenging is behavioural barriers or constraints. Wolves, for instance, can be tamed but they cannot be trained to follow commands reliably, resisting 'human direction of their activities and inhibition training' (p. 99).[73] For example, wolves raised by humans typically retain their instinctive behaviour and are therefore very difficult to socialise to humans, are easily frightened and difficult to calm, and show avoidance of humans.[74–76] As discussed in other chapters, dingoes are no different, and exhibit similar behaviour when under human control. Even in modern living conditions, those who have reared dingoes often mention the difficulties of keeping dingoes as pets, suggesting that dingoes are escape artists, very independent and with a powerful prey drive.

Behavioural aspects aside, for Indigenous Australians one of the main limiting factors in trying to domesticate dingoes was their lifestyle, which ranged from nomadic to semi-nomadic. They did not engage in farming or domestication of any species because many clans traversed from one location to the next as a result of changes in season and food availability.[77] Such constant moving provided little opportunity for large numbers of personal possessions to accumulate, and the infrastructure required to successfully domesticate animals was not present. It is more than likely that taming alone may have provided the perceived benefits the dingo could offer the Indigenous community, and further investment in domestication was not required.

Importantly, the key role of the dingo in the Dreaming may also explain why this species was not an ideal choice for domestication. Indigenous Australians have a close kinship with the land and respect for nature, with the Dreaming providing codes of conduct for moral relationships between people and animals. It may have been unacceptable or inappropriate to domesticate a species or to permanently remove an animal from its natural habitat. Thus, despite close association with the dingo, respect for the species is shown by a lack of interference. 'The dingo hunts its own food, makes his own camp, finds his own shelter, and follows his own law' (p. 176).[16] With the dingo viewed as an important being in both the natural and spiritual worlds, there was perhaps even more of an incentive not to disrupt this balance. As the dingo engaged in many behaviours considered unfavourable in humans, it is possible that Indigenous Australians wanted to be closely – but not too closely – associated with the animal. There may have been residual fear and mistrust in some societies as a result of the Dreaming image of the dingo. Consequently, we see in this relationship a unique cultural approach concerning the nature of human–animal interactions, one where domesticating any non-human species was neither appropriate nor necessary.

Conclusion

Soon after the arrival of the dingo, and later the dog, into Australia, those animals were adopted by the Indigenous people, infiltrating many facets of their lifestyle, culture and spirituality. The Indigenous Australian–canid relationship provides a unique example of how a traditional society interacted and viewed an early form of domestic dog over a 4000+ year period, facing some of the difficulties or barriers encountered by earlier prehistoric peoples attempting to domesticate and maintain a population of wild canids. Although some benefits of Indigenous Australians having camp dogs appear obvious – guardians or protectors, companions, bed warmers and mythological creatures – the close association with the animals had disadvantages, including interruptions to hunting, camp life and religious ceremonies, as well as the burden on scarce resources. Behavioural barriers in conjunction with the lifestyle and religious beliefs of Indigenous Australians may have restricted opportunities for domestication as a viable option. It is also possible that tame dingoes provided the necessary perceived benefits, and further investment in domestication was not required. Consequently, it is important that we remain open to other cultural approaches or points of view concerning the nature of human–animal interactions, and whether the need or desire to domesticate any non-human species is always appropriate.

References

1. Flood J (1999) *Archaeology of the Dreamtime: The Story of Prehistoric Australia and its People.* Angus and Robertson, Sydney.
2. Mulvaney J, Kamminga J (1999) *Prehistory of Australia.* Allen and Unwin, Sydney.
3. Rasmussen M, Guo X, Wang Y, Lohmueller KE, Rasmussen S, Albrechtsen A, Skotte L, Lindgreen S, Metspalu M, Jombart T (2011) An Aboriginal Australian genome reveals separate human dispersals into Asia. *Science* **334**, 94–98.
4. Cahir FD, Clark I (2013) The historic importance of the dingo in Aboriginal society in Victoria (Australia): a reconsideration of the archival record. *Anthrozoos* **26**, 185–198.
5. Kolig E (1978) Aboriginal dogmatics: canines in theory, myth and dogma. *Bijdragen Tot de Taal-, Land- en Volkenkunde* **134**, 84–115.
6. Hayden B (1975) Dingoes: pets or producers? *Mankind* **10**, 11–15.

7. Johnson CN, Isaac JL, Fisher DO (2007) Rarity of a top predator triggers continent-wide collapse of mammal prey: dingoes and marsupials in Australia. *Proceedings. Biological Sciences* **274**, 341–346.
8. Kohen JL (1995) *Aboriginal Environmental Impacts.* University of New South Wales Press, Sydney.
9. Ryan JS (1964) Plotting an isogloss: the location and types of Aboriginal names for native dog in New South Wales. *Oceania* **35**, 111–123.
10. Burbidge AA, Johnson KA, Fuller PJ, Southgate RR (1988) Aboriginal knowledge of the mammals of the central deserts of Australia. *Australian Wildlife Research* **15**, 9–39.
11. Meehan B, Jones R, Vincent A (1999) Gulu-kula: dogs in Anbarra society, Arnhem Land. *Aboriginal History* **23**, 83–106.
12. Breckwoldt R (1988) *A Very Elegant Animal: The Dingo.* Angus and Robertson, Sydney.
13. Maddock K (1982) *The Australian Aborigines: A Portrait of their Society.* Penguin Books, Melbourne.
14. Senior K, Chenhall R, McRae-Williams E, Daniels D, Rogers K (2006) Dogs and people in Aboriginal communities: exploring the relationship within the context of the social determinants of health. *Environmental Health* **6**, 39–46.
15. Hamilton A (1972) Aboriginal man's best friend? *Mankind* **8**, 287–295.
16. Rose DB (1992) *Dingo Makes Us Human: Life and Land in an Aboriginal Australian Culture.* Cambridge University Press, Cambridge.
17. Berndt RM, Berndt CH (1981) *The World of the First Australians.* 3rd edn. Lansdowne Press, Sydney.
18. Berndt RM, Berndt CH (1999) *The World of the First Australians.* 5th edn. Aboriginal Studies Press, Canberra.
19. Lumholtz CS (1980) *Among Cannibals.* Australian National University Press, Canberra.
20. Gould RA (1969) *Yiwara: Foragers of the Australian Desert.* Collins, London.
21. Gould RA (1969) Subsistence behaviour among the Western Desert Aborigines of Australia. *Oceania* **39**, 254–280.
22. White IM (1972) Hunting dogs at Yalata. *Mankind* **8**, 201–205.
23. Meggitt M (1965) Australian aborigines and dingoes. In *Man, Culture and Animals.* (Eds A Leeds and P Vayda) pp. 7–26. American Association for the Advancement of Science Symposium Publication.
24. Gould RA (1970) Journey to Pulykara. *Natural History* **79**, 56–66.
25. McIntosh IS (1999) Why the dingo ate its master. *Australian Folklore* **14**, 183–189.
26. Kolig E (1973) Aboriginal man's best foe. *Mankind* **9**, 122–124.
27. Roughsey D (1976) *The Giant Devil Dingo.* Collins, Sydney.
28. Trezise P (1978) Shadow of the dingo. *Habitat* **5**, 25–26.
29. Parker M (2006) Bringing the dingo home: discursive representations of the dingo by Aboriginal, colonial and contemporary Australians. PhD thesis. University of Tasmania.
30. Macintosh NWG (1975) The origin of the dingo: an enigma. In *The Wild Canids: Their Systematics, Behavioural Ecology and Evolution.* (Ed. MW Fox) pp. 87–106. Van Nostrand Reinhold, New York.
31. Macintosh NWG (1965) Dingo and horned anthropomorph in an Aboriginal rock shelter. *Oceania* **36**, 85–100.
32. Tindale NB (1974) *Aboriginal Tribes of Australia: Their Terrain, Environmental Controls, Distribution, Limits, and Proper Names.* Australian National University Press, Canberra.
33. Kirton A, Wirrell E, Zhang J, Hamiwka L (2004) Seizure-alerting and -response behaviors in dogs living with epileptic children. *Neurology* **62**, 2303–2305.
34. Ortiz R, Liporace J (2005) 'Seizure-alert dogs': observations from an inpatient video/EEG unit. *Epilepsy and Behavior* **6**, 620–622.

35. Collins D (1798) *An Account of the English Colony in New South Wales*. Libraries Board of South Australia, Adelaide.
36. Morey DF (2006) Burying key evidence: the social bond between dogs and people. *Journal of Archaeological Science* **33**, 158–175.
37. Gollan K (1984) The Australian dingo: in the shadow of man. In *Vertebrate Zoogeography and Evolution in Australasia: Animals in Space and Time*. (Eds M Archer and G Clayton) pp. 921–927. Hesperian Press, Perth.
38. Gollan K (1982) Prehistoric dingo. PhD thesis. Australian National University, Canberra.
39. Pardoe C (1996) Dogs changed the world. *National Dingo News* **Spring**, 19–29.
40. Gunn RG, Whear RL, Douglas LC (2010) A dingo burial from the Arnhem Land Plateau *Australian Archaeology* **71**, 11–16.
41. Scott J, Fuller J (1965) *Genetics and the Social Behavior of the Dog*. University of Chicago Press, Chicago.
42. Tindale NB (1959) Totemic beliefs in the Western Desert of Australia. I. Women who became the Pleiades. *Records of the South Australian Museum* **13**, 305–332.
43. Radcliffe-Brown AR (1929) Notes on totemism in eastern Australia. *Journal of the Royal Anthropological Institute of Great Britain and Ireland* **59**, 399–415.
44. Jones R (1970) Tasmanian Aborigines and dogs. *Mankind* **7**, 256–271.
45. Boyce J (2004) A dog's breakfast ... lunch and dinner: canine dependency in early Van Diemen's Land. *Tasmanian Historical Research Association* **51**, 194–213.
46. Pickering M (1992) Garawa methods of game hunting, preparation and cooking. *Records of the South Australian Museum* **26**, 9–23.
47. Thomson DF (1949) Arnhem Land: explorations among an unknown people. III. On foot across Arnhem Land. *Geographical Journal* **114**, 53–67.
48. Manwell C, Baker CMA (1984) Domestication of the dog: hunter, food, bed warmer, or emotional object? *Journal of Animal Breeding and Genetics* **101**, 241–256.
49. Corbett LK (2001) *The Dingo in Australia and Asia*. JB Books, Adelaide.
50. Morgan J (1980) *The Life and Adverntures of William Buckley. Thirty-two Years a Wanderer amongst the Aborigines of the then Unexplored Country around Port Phillip, now the Province of Victoria*. Australian National University Press, Canberra.
51. Mountford CP (1948) *Brown Men and Red Sand*. Angus and Robertson, Sydney.
52. Clutton-Brock J (1999) *A Natural History of Domesticated Animals*. Natural History Museum, London.
53. Yarrow HC (2004) *A Further Contribution to the Study of the Mortuary Customs of the North American Indians*. Kessinger Publishing, Montana.
54. Claoué-Long A (2007) *Rural Graves in the ACT: A Historical Context and Interpretation*. http://www.act.nationaltrust.org.au/documents/RuralGravesreport.pdf.
55. Barker BCW, Macintosh A (1979) The dingo: a review. *Archaeology and Physical Anthropology in Oceania* **14**, 27–53.
56. Abbie AA (1969) *The Original Australians*. Frederick Muller, London.
57. Gould RA (1980) *Living Archaeology*. Cambridge University Press, New York.
58. Flood J (2006) *The Original Australians: Story of the Aboriginal People*. Allen and Unwin, Sydney.
59. Smyth R (1878) *The Aborigines of Victoria: With Notes Relating to the Habits of the Natives of Other Parts of Australia*. Victorian Government Printer, Melbourne.
60. Meek PD (1999) The movement, roaming behaviour and home range of free-roaming domestic dogs, *Canis lupus familiaris*, in coastal New South Wales. *Wildlife Research* **26**, 847–856.
61. Jenkins DJ, Andrew PL (1993) Intestinal parasites in dogs from an Aboriginal community in New South Wales. *Australian Veterinary Journal* **70**, 115–116.
62. Jenkins D, Meek P, Ardler A, Hawksby R (1996) Aboriginal community dogs: worms, control and human health. *Dog Health in Indigenous Communities*, 103–125.

63. Walters BJ (1995) *The Company of Dingoes: Two Decades with our Native Dog.* Australian Native Dog Conservation Society, Bargo.
64. Darwin C (1936) *The Origin of Species by Means of Natural Selection and the Descent of Man and Selection in Relation to Sex.* Modern Library, New York.
65. Price EO (1984) Behavioral aspects of animal domestication. *Quarterly Review of Biology* **59**, 1–32.
66. Downs JF (1960) Domestication: an examination of the changing social relationships between man and animals. *Kroeber Anthropological Society Papers* **22**, 18–67.
68. Fox MW (1978) *The Dog: Its Domestication and Behaviour.* Garland STPM Press, New York.
69. Dobney K, Larson G (2006) Genetics and animal domestication: a new window on an elusive process. *Journal of Zoology* **269**, 261–271.
70. Trut LN (1999) Early canid domestication: the farm-fox experiment. *American Scientist* **87**, 160–169.
71. Coppinger R, Schneider R (1995) Evolution of working dogs. In *The Domestic Dog: Its Evolution, Behavior and Interactions with People.* (Ed. J Serpell) pp. 21–47. Cambridge University Press, Cambridge.
72. Clutton-Brock J (1992) The process of domestication. *Mammal Review* **22**, 79–85.
73. Zeuner FE (1963) *A History of Domesticated Animals.* Harper and Row, New York.
74. Koler-Matznick J (2002) The origin of the dog revisited. *Anthrozoos* **15**, 98–118.
75. Fentress JC (1967) Observations on the behavioral development of a hand-reared male timber wolf. *American Zoologist* 7, 339–351.
76. Zimen E (1987) Ontogeny of approach and flight behavior towards humans in wolves, poodles and wolf–poodle hybrids. In *Man and Wolf.* (Ed. H Frank) pp. 275–292. Dr W. Junk Publishers, Dordrecht.
77. Frank H, Frank MG (1982) On the effects of domestication on canine social development and behavior. *Applied Animal Ethology* **8**, 507–525.
78. Kimber RG (1976) Beginnings of farming? Some man–plant–animal relationships in Central Australia. *Mankind* **10**, 142–150.

5
Dingo–human conflict: attacks on livestock

Rob Appleby

'Probably more has been written about the dingo wars than any other aspect of "dingology", yet only two facts stand out: all the effort has been extremely expensive and, by and large, it has not worked' (p. 152).[1]

In the above statement, Corbett, perhaps Australia's best-known 'dingologist', was referring to the wars between livestock producers and dingoes that have raged since the two first came into contact. This chapter is devoted to exploring the most salient aspects of these wars and, with reference to the second part of Corbett's statement, objectively evaluating both the impacts of wild dogs on livestock production and the approaches that have most commonly been employed in an attempt to stop wild dog depredation. The latter part of the chapter then examines how some of these management approaches, particularly in light of some emerging ones, might offer new opportunities to end the wars and move instead towards coexistence.

A history of conflict

Following European settlement, the farming of livestock rapidly expanded in the early and mid 19th century into many reaches of the Australian continent; almost immediately, conflict between livestock producers and dingoes emerged. Dingo numbers are suspected, perhaps ironically, to have grown alongside the expansion of the agricultural industry, particularly in areas where artificial water sources drawn from artesian wells were created.[1] Producers quickly found that sheep and, to a lesser extent, cattle were relatively easy pickings for dingoes (see Fig. 5.1). Breckwoldt[2] recounted several stories from colonists who were among the earliest of Australia's sheep farmers coming into conflict with dingoes. Here is one example, from Michael Hindmarsh (1824, cited in 2):

> The only enemy that attends it is the Native Dog, which commits great havoc. Betimes I have known 100 sheep to have been killed or bitten so as to occasion their death in one night. The bites of these animals are almost sure death to sheep.

Since then a ceaseless battle has been waged against dingoes by pastoralists and the various government agencies charged with supporting the agricultural industry.

Dingoes have been killed through just about any means available, from shooting, 'denning' (the practice of finding den sites and killing any pups found inside) trapping and poisoning (usually through strychnine or compound 1080 – sodium fluoroacetate – baits). The latter has probably been responsible for most dingo deaths, but an accurate tally isn't possible because most poisoned dingoes are never recovered. Somewhat in testament to this, Allen and Sparkes,[3] who examined historical data pertaining to bounties paid for dingo scalps

Figure 5.1: A dingo attacking a sheep in a paddock.
Source: Jean Paul Ferrero/ardea.com.

in Queensland, observed a drop in bounty payments and in the number of professional trappers around the time 1080 was introduced (in the late 1960s), although they noted that other factors may have contributed to a reduction in bounty payments (discussed below).

Bearing this in mind, what is known in relation to the numbers of dingoes destroyed is staggering enough. For example, in Queensland alone, a total of 1.5 million dingo scalps (see Fig. 5.2) have been presented for payment since 1885, with marked variation between particular years, such as 50 000 in 1957–1958 down to 2689 in 1975–1976.[3] In New South Wales, between 1883 and 1930 280 000 bounties were paid.[4] Given the lack of accurate numbers killed by poisoning campaigns, and making an allowance for those dingoes for which no bounties were collected as well as states and territories for which little data is available, it is probably fair to assume that many millions of dingoes have been killed in the interests of protecting livestock in Australia.

Despite some two centuries of persecution, the dingo remains, albeit in various forms in most places due to hybridisation with domestic dogs (see Chapter 2). In fact, only a few scattered places now appear to have resident dingo populations closely resembling those that would have been found before European settlement. What has come to replace dingoes in many areas is probably similar, but just how similar isn't yet clear. Thus, because of ambiguity, the term 'wild dog' is most often used as a catch-all for everything that can't be categorically labelled as a dingo. In the context of control, it also tends to include the dingo. Thus, when the term 'wild dog' is used here it follows this convention, unless otherwise stated.

Figure 5.2: Dead wild dogs hanging in a tree in Benambra, in the Victorian High Country. There are many reasons why dingoes are hung up on fences or trees, including to aid the scalping process and to let farmers in the areas know that dogs have been killed.

Source: Greg Muller.

Despite every effort to wipe them out, the wild dog problem (conflict between pastoralists and wild dogs) is only getting worse, at least according to many media accounts. For example, an article by Townsend[5] painted a decidedly dire picture, with statements such as: 'Wild dogs are killing the wool industry in Queensland. The dingo fence is useless, poison baiting isn't working and the law that says land owners must control wild dogs isn't enforced. Now dog numbers have reached epidemic proportions' and 'A plague of dingoes is killing off the remnants of Queensland's wool industry'. Or take, for example, the following Courier-Mail[6] article entitled 'Sheep slaughter by dingoes threatens flocks', quoting then Cloncurry councillor D.M. Lynch as stating that dingoes are 'working day and night to destroy Australia's sheep industry'. Much of the rhetoric could have easily come from newspaper articles half a century or more earlier.

There are many more examples before and since of newspaper reports with similar sentiments. So what are we to make of such accounts? If they are taken at face value, certainly they suggest an enduring problem when it comes to wild dogs and the sheep industry. However, they also hint at the fact that whatever mechanisms have been used in an attempt to mitigate the conflict, haven't really changed the state of play much. As we'll see below, a closer look at more formal, economic appraisals of the situation doesn't appear to resolve this apparent enigma any more satisfactorily.

Economic costs and other impacts of wild dog depredation

While wild dog depredation clearly has been and remains a problem for many livestock producers, and while a survey of reports relating to the economic costs associated with livestock depredation attests to that fact, inconsistencies between reports makes it difficult to ascertain if the situation is getting worse, better or, for all intents and purposes, hasn't changed much. For example, a report by Gong, Sinden and Braysher[7] listed total costs attributable to wild dogs across Australia at $48.5 million. Five years earlier, McLeod[8] reported a total cost of $66.3 million, suggesting that the situation had improved by a sum of nearly $18 million. Yet in the same year as Gong *et al.*'s[7] report, Hewitt [9] reported a total cost of over $67 million just in Queensland, more than $20 million beyond the total national estimate provided by Gong *et al.*[7]

A good portion of the difference observed between the McLeod[8] and Gong *et al.* reports might be due to a reduction in the sheep industry in the intervening years: McLeod[8] listed the national flock at an estimated 71.2 million, while in the 2009–2010 period the Australian Bureau of Statistics listed it at 68 million. In turn, this might correlate with a lower numbers of recorded attacks (and losses), although this too is unclear. There may be other reasons for any disparity, such as the pricing figures chosen or sources used, as there doesn't appear to be any standardised approach to assessing economic losses, and individual reports vary in the methodology used to collate and calculate losses. Whatever the cause, such disparity makes it difficult to objectively assess economic impacts and, equally importantly, discern any progress in management strategies employed to mitigate losses, at least at a national scale.

Using actual numbers of livestock killed may simplify monetary calculations, yet there can be considerable discrepancy. McLeod[8] assumed that around 0.5% of 71.2 million (~355 000) sheep were taken by wild dogs, yet the year before an ABARE (2003, cited in McLeod)[10] report suggested that closer to 16% (~5.4 million) of sheep deaths could be attributable to wild dogs. This is obviously a considerable difference. Note, though, that the difference is unlikely to be due to actual differences in losses between the two years, and more the result of different methodology in collating and calculating losses. Again, such methodological disparity makes comparisons between periods virtually impossible, such that, even if particular mitigation approaches were having a beneficial impact at a national scale (or indeed if they weren't), they might be obfuscated simply because of differences in the math.

It is likely that certain producers are disproportionally impacted by depredation[1, 2, 11] and therefore for them, the situation has worsened. It isn't always clear why or indeed how often this might be the case, but at local or regional scales it could relate to differences in control practices or husbandry, or the intensity and uniformity of control measures, such that certain locations become hotspots of depredation. In this sense, the situation might be getting relatively worse, in so much as rates of stock losses aren't necessarily changing at a national scale but the impact of losses is more notable for certain producers. However, the impression, at least from certain newspaper reports, is that the situation is worsening in a more ubiquitous sense. If indeed this is true, it makes sense to ask why, especially given the sheer amount of effort put into wiping out wild dogs in many parts of Australia.

Recently, an article by Allen and West[12] posited that dingo depredation has been and continues to be a major cause for a contraction in the sheep industry in Australia over the last 100 years. In exploring the effects of the dingo on sheep production in Australia, Allen and West[12] made some allowances for other influential factors, such as drought, but stated that dingo depredation remains a primary cause of a decline in production. However, as with the overall rates of losses due to depredation discussed above, it is far from clear whether this is the case. For example, in a relatively recent Australian Bureau of Statistics report, sheep and lamb numbers were observed to increase overall by 7% in 2010–2011 (compared to the year before). The report further stated that the sharp reduction observed in sheep numbers in the last 20 years was primarily due to poor market opportunities for Australian wool.[13] An article by East and Foreman[14] assessed the movements of stock around Australia from the point of view of disease spread, and provided some reasoning for the reduction in the sheep industry. Wild dogs weren't mentioned (which is what prompted the Allen and West article),[12] rather, East and Foreman[14] suggested that the reduction was primarily due to a declining demand for wool due to competition from alternative fibres, declining prices, a shift towards the production of lambs and prolonged periods of drought.

Allen and Sparkes[3] noted that in the 1960s and 1970s in Queensland, around the same period which saw a national peak in sheep numbers, there was a simultaneous reduction in sheep numbers and a corresponding increase in cattle production. One figure appeared to demonstrate a correlation between a steady increase in cattle numbers and a decrease in dingo bounties, and a corresponding sharp drop in sheep production approaching the 1970s. The question here is, was the shift to cattle production a direct result of dingo depredation on sheep and therefore a conscious choice on the part of producers to avoid, or attempt to avoid, depredation, or were other factors involved? While this was a possible contributing factor, there are hints that other reasons were involved. For example, Allen and Sparkes[3] described a reduction of 33.8% in the sheep industry in Queensland over a five-year period beginning in 1990, with a higher proportion of the reduction on the southern side of the dingo barrier fence, where protection against wild dog depredation would be expected to be better. North of the dingo fence, the reduction over the same period was actually lower, at 26%. Allen and Sparkes[3] also noted that a major drought occurred between 1991–1996 and there was a large sale of the wool stockpile in the same period, each potentially contributing to decisions to reduce sheep flocks or switch to cattle. Allen and Sparkes[3] concluded that the impact of dingoes on sheep numbers was largely indirect, with producers' perceptions of depredation along the sheep/cattle interface being of major importance, leading producers away from sheep production into cattle production.

McLeod,[8] who specifically included the costs associated with wild dogs to agricultural industries, stated that a major reason for the decline in the national sheep flock in the last 35 years was lower prices for greasy wool. McLeod's[8] report also showed that rabbits were a more costly pest to the sheep industry than wild dogs, with the former estimated to cost $35.5 million and the latter $15.9 million. In fact, foxes could be considered a more significant pest to sheep farmers in relation to costs according to McLeod's[8] estimates, costing $17.5 million, mostly relating to lamb depredation. Rabbits and foxes are both potentially impacted by wild dogs and in the words of Alan Newsome:[15] 'The dingo does a lot of pest control work for us – while we sleep – doing the work we want done' (p. 29). This is another reason to objectively evaluate impacts and mitigation measures and to assess associated costs and benefits as comprehensively as possible.

When the incentives are high enough, sheep flocks rise and so do control efforts, suggesting a correlation between the success of sheep farming and higher control efforts. Glen and Short[4] examined the relationship between bounties paid for dingo scalps and the numbers of sheep in various regions of New South Wales from 1883 to 1930. The relationship between bounties paid and dingo abundance is not necessarily a straightforward one, a caveat noted by the authors and others,[3] but at face value their results suggest a sharp contraction in dingo numbers during the period; where such contractions were highest, sheep numbers were as well. They also observed a decline in the number of bounties paid over the period in most regions, suggesting control measures employed in New South Wales either reduced dingo abundance or otherwise resolved depredation in a practical enough sense to allow wide-scale sheep farming. This is in stark contrast to other states such as Queensland where bounty payments remained relatively stable over long periods. A confounding issue is the introduction and use of large-scale poisoning campaigns (at that time using strychnine) during the period, as this was also likely to reduce bounty payments. There is little information on the degree to which eradication measures worked to reduce attacks or losses *per se*, although it would have been the primary impetus for control effort.[4]

Yet, while arguments in the vein of Allen and West[12] that support widespread control are partly supported by research such as that of Glen and Short,[4] such findings appear to be situation-specific. Many signs point to other causes for a decline in the sheep industry rather than just wild dog depredation. This is what Berger[16] found when undertaking a comprehensive analysis of the relationship between predator control efforts (with a focus on coyotes, *Canis latrans*) and the decline of the sheep industry in the US. There, production costs coupled with market pricing explained most of the variation in models. Berger[16] suggested that predator control had failed to halt decline and that alternative approaches were needed to bolster the industry, as opposed to just killing carnivores. A similar set of complex mechanisms are probably at play in Australia.

However, the impacts of wild dog depredation should not be underestimated. Despite the incongruity in the reports about wild dog damage, this does not mean that wild dogs aren't causing problems in certain areas and sheep producers aren't doing it tough. Dingo depredation may well be, as Allen and West[12] (p. 263) put it: 'the straw that broke the camel's back' for some producers. High rates of depredation on individual properties would obviously take an enormous emotional and economic toll. As Allen and Sparkes[3] noted, for pastoralists living in areas occupied by wild dogs, their mere presence might be disincentive enough when making decisions about whether to raise sheep, particularly when margins are slim. However, as outlined above, depredation has been an ongoing feature of sheep production in Australia since early settlement, including throughout the 1970s when sheep population reached a peak of 180 million,[13] suggesting that the risk or reality of depredation (and other disincentives) is moderated and outweighed at times by higher economic incentives.

Type of livestock

The type of livestock run on a property may play a significant role in both the reality and perception of wild dog depredation. In contrast to sheep farmers, cattle farmers appear to be more tolerant of the threat of depredation events, and even of the events themselves.[2, 17] This is despite an evaluation by McLeod,[8] for example, that the economic losses of cattle farmers are larger than those of sheep farmers when it comes to wild dog depredation. The differences in perceptions and resulting dingo control efforts of cattle and sheep producers in relation to wild dog depredation has been recognised for more than a century, with many cattle producers extolling the virtues of dingoes in reducing the competition from macropods for cattle feed, as opposed to sheep producers, who considers dingoes pose only a threat and no benefits.[3, 17] Losses on sheep farms may be relatively more visible and acute than on many cattle farms. Thus, for various reasons, there appears to be greater tolerance for dingo presence on or near cattle stations (see Fig. 5.3) than on sheep stations, where control objectives tend towards eradication of dingoes.[3]

Why might there be differences between the attack or death rates of sheep and cattle? Allen and Fleming[18] outlined some of the behavioural differences of each type of livestock when confronted by dingoes and during attacks, which demonstrated several vulnerabilities of

Figure 5.3: Cattle and dingoes at Evelyn Downs, a remote pastoral station in South Australia managed by Adam O'Neil and Arian Wallach. The station is a predator-friendly property where no dingo control measures are in place.
Source: Gerrit Schuirmann.

sheep and goats in particular. First, both sheep and goats tend to flee when confronted by a predator, sparking predatory chase behaviour. Sheep are easily outrun by wild dogs, with one study observing high capture success rates (66%) of all chases initiated.[19] Many observed chases ended without capture as wild dogs broke off to chase other sheep.[19] Breed of sheep may play a minor role in relation to losses, as certain breeds may exhibit more effective anti-predator behaviour than others. For example, the mobbing behaviour exhibited by certain breeds of sheep, such as Merinos (the most common breed in Australia), tends to leave the sheep on the outer circle extremely susceptible to attack.[18] Such vulnerabilities are thought to be a primary reason for observations of surplus killing (killing in excess of food requirements) of sheep by wild dogs.[2, 18, 20] In comparison, adult cattle are generally too large for wild dogs to attack with great success, and often defend themselves and their calves when confronted by wild dogs.[18] Thus, direct attacks are the primary reason for sheep losses associated with wild dogs. Diseases spread by wild dogs, such as hydatidosis, are likely to be the primary form of wild dog-associated cattle losses, although direct attacks still account for a considerable proportion.[18]

The role of lethal control in preventing losses

Such differences in losses (or the perception of losses) between cattle and sheep have led some to conclude that sheep and dingoes don't mix,[12, 13] such that where sheep are grazed dingoes will almost certainly need to be lethally controlled (if not eradicated) and/or excluded by fencing. Allen and West[12] contested that in sheep and goat production areas, unless virtually all wild dogs are killed and immigration is largely prevented, dingo numbers will increase to the point where losses to the sheep and goat industries are so high that they are likely to disappear in Australia within several decades. Intuitively, lethal control should act to reduce wild dog depredation events, but the density or abundance of dingoes does not necessarily bear a strong relationship with the amount of loss suffered, as individual wild dogs that manage to avoid destruction appear capable of inflicting considerable damage in isolation.[11, 18] With cattle, a similar problem exists, such that indiscriminate removal of wild dogs fractures packs or opens up vacant territories to dispersing individuals, both of which can actually exacerbate losses.[21, 22] Allen and West[12] concluded that only through broad-scale baiting in rangelands can dingoes be effectively managed, but this seems likely to lead to fracturing and, given the indiscriminate nature of baiting, does not guarantee removal of the most damaging individuals. Impacts on other wildlife are also a concern.

Research involving other canids has clearly shown that particular individuals can be responsible for depredating livestock at high rates. For example, Sacks *et al.*[23] used telemetry collars to study the behaviour of individual coyotes in relation to sheep depredation in the US. They found that breeding coyotes whose territories contained sheep farms were the main predators of sheep. In fact, one breeding male was responsible for 71% of 65 total recorded kills in a single year, and four breeding pairs were implicated in most kills overall. The authors recommended selective removal of individuals in an effort to reduce depredation events, and advised against control beyond a single territory surrounding farms, otherwise vacancies might be created then rapidly refilled by dispersing coyotes. Such a selective process is far from straightforward. Breeding coyotes are often very difficult to capture or target with specialised killing apparatus such as M-44 ejectors, particularly in areas with a long history of control.[24] One hypothesis for this difficulty of capture relates to the familiarity of territorial animals with their territory, and consequently their aversion to or suspicion of novel objects (e.g. traps) or scents.[24–26] Livestock protection collars appear to be one of the few lethal methods capable of targeting individual coyotes that attack sheep.[24, 27] Such collars contain a poison (usually 1080) that kills through ingestion when coyotes seize sheep around the throat, a usual means of dispatching prey. However, the potential need to collar large numbers of sheep makes their use prohibitive[24] and collars can sometimes be punctured by sources other than predators, such as vegetation, rendering them ineffective.

Large-scale and aerial baiting

In one of the first and most comprehensive studies undertaken, Thomson[28] investigated the effects of large-scale aerial baiting campaigns in the Fortescue River region, Western Australia. Unlike many studies of dingoes in Australia, which rely on crude indices of dingo numbers such as counts of dingo tracks, Thomson[28] used radio-collars to directly track and observe a high proportion of resident dingoes within a baited and a control (non-baited) area. Although there was some variation in effectiveness relating to the age and social status of dingoes (younger solitary animals were often completely removed) and the density and distribution of baits, aerial baiting was a cost-effective means for removing high proportions (100% in one trial and 63% and 62% in subsequent trials) of dingoes in such habitat types. Note that 'effectiveness' refers to the numbers of dingoes killed. Thomson[28] said that for total effectiveness to be transferred in terms of a reduction in losses, 100% of dingoes occupying areas running sheep would need to be removed and that this was unlikely to happen very often. He also noted that dingoes in such areas fed almost exclusively on freshly killed prey, making the uptake of meat baits by such individuals less likely, and highlighting the need for other methods.

Coordinated lethal control

Livestock producers sometimes claim that government-controlled lands (such as national parks) which do not necessarily have wild dog control programs are a source of replenishment of dingo numbers in areas that do (i.e. farms) through emigration and dispersal. This difference in tenure is obviously problematic because the respective goals of land managers are in stark opposition in some cases, with farmers seeking to control or eradicate wild dogs and national park managers often seeking to conserve native wildlife such as dingoes. An experiment of sorts on what has been referred to as a 'nil tenure approach' has been underway for some time in the Brindabella Wee Jasper region of New South Wales. In this area, which constitutes ~150 000 ha of mixed tenure land, a cooperative has formed between the various managers, both public and private, to address wild dog and fox issues.[29] The program, which is primarily reliant upon broad-scale baiting and to a lesser extent on trapping, was able to reduce stock losses by an average of 75% across a three-year period. It was awarded an acknowledgement from the Australian Department of Family and Community Services for helping farming families deal with the emotional toll of wild dog attacks on sheep.[29] The approach is now being more widely adopted across the state. This is somewhat reminiscent of the historical approach highlighted by Glen and Short[4] in the cooperative management of wild dogs throughout much of the sheep country in New South Wales. While losses haven't been completely halted, it would seem that the Brindabella Wee Jasper approach has been a success in the eyes of the farmers involved, not least because the emotional as well as the economic costs have been lessened.

This example highlights the potential success of coordinated lethal predator control across large areas in reducing sheep losses. This is consistent with the recommendations of McIlroy, Gifford and Cooper[30] and Allen and Fleming.[18] It is also in keeping with the idea of buffer zones espoused by Thomson.[28] However, its success is largely reliant upon a compromise whereby managers of national parks abutting sheep properties actively destroy resident dingoes that would otherwise be protected. Another success of the Brindabella Wee Jasper approach is the recognition of the wild dog problem (including the emotional costs) to sheep farmers by managers across a variety of tenures, who now share the burden of reducing losses. Perhaps the success of a cooperative approach can be achieved without a heavy reliance upon lethal approaches; this will be discussed later.

Even in the Brindabella Wee Jasper example, where control efforts are more coordinated and cooperative than might commonly occur, it is apparent that total eradication of wild dogs is almost impossible under most circumstances. Although something close to local or even regional eradication has been achieved at times in the history of wild dog management in Australia,[4] the more usual scenario is some level of persistence in populations of wild dogs. This is especially the case where control is haphazard and where dog-proof fencing is largely absent or has fallen into disrepair.[12] Calls for baiting, the reintroduction of bounties or the adoption of other methods focused on killing wild dogs might seem justifiable to those whose livelihoods are deeply affected by wild dog depredation but it appears that the best this might achieve is often only a limited and/or temporary reprieve and that, at worst, it may even exacerbate the situation.

Alternatives to lethal control and eradication

Along with concerns over efficacy, lethal methods are more likely to be regarded as publicly unacceptable[31, 32] and concurrently there may be an increasing regard for non-lethal approaches.[33] The potential impacts of baiting[34, 35] and trapping[36] on non-target native animals has also been highlighted. Put another way, while lethal control in some contexts is likely to continue, there appears to be a growing impetus to find and apply alternative non-lethal solutions where possible.[37]

Much of this book fosters an accord with a relatively new paradigm in the history of dingo management in Australia: that of conservation. For dingo conservation to become a reality, a shift is required in certain key places, away from lethal control as the dominant approach and towards non-lethal techniques to managing conflict with livestock producers. Thankfully, a growing suite of approaches are either being put into practice or may be employed if their efficacy is experimentally determined. The identification of key places where dingoes can or should be conserved (or indeed reintroduced) remains a hot topic of discussion,[38, 39] further complicated by issues such as the ecological role of dingoes and hybridisation – particularly how hybridisation can be detected and managed and whether the ecological roles of hybrids are different from those of dingoes.[1, 40–42] Prime candidate sites will almost certainly be conservation reserves and national parks, though there may be calls for consideration of other places as well. A good example of one such reserve is Sturt National Park.[15] Interfaces between grazing properties and areas potentially important to preservation of dingoes present acute challenges that require considerations beyond conventional lethal-centric practices for managing wild dogs. Beyond reserves, alternative approaches to lethal control, or a mixture of non-lethal and lethal strategies where warranted, might offer some respite to producers suffering depredation.

The dingo fence and fencing

Fencing has been and continues to be a major enterprise in Australia in relation to protecting livestock from wild dogs. Without fencing, lethal control practices may be relatively futile, particularly when such control effort is largely uncoordinated or piecemeal, as areas where control measures are not enacted become potential sources for repopulating the areas where control has occurred.[1, 18] Where fencing is a sufficient barrier to wild dogs and is maintained and properly managed, lethal control practices may be unnecessary or the necessity may be reduced. In accordance with this, some producers appear more confident with fencing as opposed to baiting campaigns, as Allen and Sparkes[3] found with certain Queensland graziers. Allen and Sparkes[3] determined that in large open areas such as semi-arid pasture lands and where control effort was wide-scale and cooperative, baiting and fencing were equally effective. Fencing was more effective in smaller forested areas, where control efforts were less coordinated or supported. In either case, fencing appeared to be a useful approach. A potentially major advantage of effective fencing is that it is a barrier to wild dogs regardless of fluctuations in their density, dispersal or movement patterns, foraging activities or behaviour generally, all of which are likely to affect the success of conventional lethal approaches.

Glen and Short[4] highlighted a historical relationship between bounty payments for dingo scalps and sheep numbers in New South Wales, suggesting that if bounty payments reflected dingo abundance in any reasonable way, then many areas in the state had come close to achieving eradication by the 1930s. While much of this was due to comprehensive baiting, shooting and trapping, another major influence noted by the authors was the establishment of the iconic dingo barrier fence (Fig. 5.4). In its current state, it stretches 5614 km across three Australian states, from the Darling Downs in Queensland to the Great Australian Bight in South Australia. The utility and effectiveness of the dingo barrier fence is controversial and remains the subject of much debate, although it is likely to continue in some capacity well into the future. Breckwoldt[2] discussed in detail the use of barrier fences in Australia, including the history of the dingo fence, as did authors Holden[43] and Woodford.[44] More recently, Bradshaw and Ritchie[45] reignited discussion on the dingo barrier fence and dingo management in a more general sense.

Fencing has its drawbacks: it is not a barrier only to dingoes and wild dogs, but also to virtually any other medium-sized mammals. Thus, differences in the densities and

Figure 5.4: A section of the dingo fence near Coober Pedy in South Australia. Its length, its remoteness and the terrain make maintenance of the fence difficult and expensive.
Source: Bradley Smith.

assemblages of animals (and plants) on different sides of fences can occur. Fencing is expensive to erect and maintain, but it is likely to remain a pillar of wild dog management in or around livestock production areas in the future.

Similar to fencing but more temporary and mobile, fladry, which entails the use of rope or cord with pennant-like flags at regular intervals, can be erected around livestock pastures. Perhaps surprisingly, this appeared to be considerably worrying to wolves during experimental trials, enough that it effectively formed a barrier and prevented their ingress into pastures, at least for relatively short periods of time.[46] Where ordinary fladry begins to fail, an approach nicknamed 'turbo fladry'[47] offers another level of aversion in the form of electric shock, providing a powerful reinforcing disincentive to wolves bold enough to attempt to cross the barriers.[48] The advantage of fladry (electrified or otherwise) is that is relatively mobile, meaning that night pens can be established around flocks, wherever they graze across pastures. Usually such night pens also have people stationed to shepherd the sheep (discussed below). Emerging evidence from experimental trials suggested this combination can largely halt depredation events.

Given the high costs of fencing and maintenance and the disparity of cost-sharing often borne by the relatively unlucky producers closest to large collective barrier fences such as the dingo barrier fence,[3] it would potentially be of great value to producers for dingo conservation groups to contribute to these costs where possible. This could be particularly beneficial at the interfaces between sheep or cattle properties and national parks and reserves where dingo protection may be paramount. Local environmental, bush care and conservation groups could provide in-kind assistance with fence construction and maintenance in their areas, with larger-scale involvement by state or national groups that can better coordinate such efforts. Such a practical (and perhaps metaphorical) enterprise to mend fences could foster an increased awareness and acceptance of the positions of stakeholder groups that might otherwise see themselves at odds, and encourage the development of other avenues in the pursuit of common ground.

Shepherding

Allen and West[12] suggested that in the past it was the coordinated effort to eradicate dingoes that made sheep farming sustainable in Australia, but there are other differences in management that might have been equally if not more important. For example, human shepherds actively patrolled and protected flocks from predators, especially at night. It was only after the introduction of strychnine poisoning, coupled with improved and more consistent fencing, that the practice of shepherding began to wane.[2, 11] In modern farming practices, shepherding might seem antiquated, but a return to physical shepherding, coupled with other non-lethal approaches discussed below, might have merit in some circumstances. For stations where depredation is high, shepherding, particularly at critical times of year that match seasonal peaks in depredation, could help curtail such events.[42] As with fencing programs, there could also be room for conservation groups to become involved in shepherding and gain valuable experience of the challenges involved in farming where wild dogs are present. The cost-effectiveness of such approaches obviously needs to be taken into account and many farms may not be well-suited to the practice, particularly large holdings with flocks that roam over hundreds of kilometres or more. In some cases, though, shepherding might represent an additional tool worth exploration (or re-exploration).

Modifying dingo behaviour

Much of what is argued below aims to develop an idea that the intelligence of dingoes (see Chapter 9) could actually be advantageous when it comes to their management, rather than burdensome. Animals that are capable of learning, remembering and passing on lessons to young are well-suited to certain non-lethal management approaches, and indeed success may be reliant on such traits. For example, learning that certain areas are very difficult to breach or that certain animals are difficult to capture or kill or are otherwise well-protected might serve to redirect predatory effort elsewhere and away from stock. Dead dingoes learn nothing and pass along no lessons. Therefore, unless eradication is achieved, a reliance on lethal control inevitably becomes cyclical. Non-lethal methodology that targets and subsequently modifies dingo behaviour potentially has an advantage in that it works in unison with evolved cognitive mechanisms designed to steer animals away from potentially dangerous or harmful events, encouraging them to look for an easier meal elsewhere.

Livestock-guarding animals

The saying 'fight fire with fire' has some relevance to managing wild dog depredation, albeit with a twist. One alternative means of defending flocks from wild dog depredation is through the use of livestock guarding animals. Such animals naturally exhibit aggressive or defensive behaviours when confronted by predators, and may actively deter predators from approaching or pursuing protected livestock. Jenkins[49] reviewed and summarised many of the challenges and opportunities associated with the use of livestock-guarding animals in Australia, concluding that while their use was limited at the time there was nevertheless considerable potential. Jenkins[49] provided briefs on many of the most relevant options including alpacas, llamas, donkeys and a range of livestock-guarding dog breeds such as maremmas and great pyrenees. Several case studies were presented that showed compelling, albeit preliminary, evidence of a reduction in losses and conflict from the use of livestock-guarding animals.

More recently, van Bommel and Johnson[50] conducted a survey with 150 participants, all of whom had used or were using livestock-guarding dogs (most often maremmas) in Australia. Results were encouraging, with 65.7% of those surveyed stating that depredation events had ceased after obtaining and training the dogs and a further 30.2% stating that depredation events had been reduced. A potential limitation related to the number of dogs required to protect stock. For example, flocks greater than 100 individuals to one livestock-guarding dog are unlikely to be spared depredation altogether (although some exceptions were noted). In a compelling case study, a sheep station suffering around 15% losses before the introduction of maremmas reduced this to 4% three years after the dogs were purchased. Most of the remaining 4% loss was not considered to be due to depredation events. Costs associated with the dogs were recouped within the first 12 months.

The benefits of using guard animals such as maremmas appear to be stark in some circumstances. An example, not specifically related to wild dogs but that highlights the potential value of guard animals, comes from Middle Island in south-western Victoria. A population of little penguins had been decimated by foxes, from around 600 penguins in 1999 to just 10 in 2005. In 2006 two maremmas were introduced on to the island, and there has since been little to no recorded predation by foxes. The penguin population has increased to over 100 individuals as a result. Prior to the introduction of the dogs, all manner of lethal control approaches had been attempted in an effort to eradicate foxes, including baiting and shooting, yet predation continued unabated.

Aversive conditioning

In a similar vein to livestock-guarding animals, there are emerging approaches to conditioning escape and avoidance responses in predators attempting depredation, collectively labelled here as aversive conditioning. In wildlife management situations, aversive conditioning usually refers to punishing specific undesirable behaviours with stimuli that induce fear, are uncomfortable, painful or otherwise noxious, in an effort to establish a relationship between such behaviour and a negative consequence.[11, 51–55] This process is essentially an attempt to apply aspects of learning theory[56, 57] and in particular, operant (or instrumental) conditioning.[58] In practice, this can prove challenging.[54] However, as shown below, several important studies have begun to demonstrate the promise of such an approach.

One promising arm of aversive conditioning in the context of depredation management involving canids is the use of aversive (shock) collars. Linhart *et al.*[59] and Andelt *et al.*[60]

demonstrated dramatic avoidance learning in coyotes subjected to a small number of shock events when confronted with prey species that they had habitually killed (respectively, lambs and rabbits). In fact, many authors (Schultz *et al.*,[61] Hawley *et al.*[62] and Rossler *et al.*)[63] all reported evidence that collared wolves (and, in the latter case, their pack mates) avoided certain areas after receiving an electric shock on approach. Aversive conditioning might therefore have a role in mitigating livestock-related conflict (and other conflict types, see Chapter 6), particularly in circumstances where conflict involves high conservation priority populations of dingoes. It may also be worthy of experimentation in cases where other management approaches are failing.

Along with intelligence, other behavioural traits exhibited by canids such as dingoes might help with non-lethal management. For example, a potential advantage of the territoriality exhibited by wolves and other related canids is that they tend to fend off would-be rivals. This borrows from ideas first considered in lethal control programs, where buffer zones[28] are established around sites, such that bordering territories with intact social groups might limit the ability of transient or dispersing social canids from further afield to enter the vacant buffer zones and create new territories. In theory, such a concept transfers well to a non-lethal paradigm, especially in cases where territorial residents learn not to depredate livestock (e.g. through conditioning) and in turn prevent ingress from naïve rivals that might be more inclined to attack stock.[64] These territorial effects can potentially be retained over multiple years, increasing the cost-effectiveness of non-lethal approaches.[64] However, as Rossler and Gehring[63] and Schultz and Jonas[61] noted, conditioned wolves may sometimes shift away from livestock areas altogether, leaving a vacant territory near pastoral sites that may become occupied by naïve wolves, potentially perpetuating the depredation cycle. The effects of aversive conditioning on territorial behaviour need to be more rigorously investigated.[63] Different experimental conditions also need to be explored, for example, protecting an area such as a paddock, as opposed to protecting the livestock animals themselves.

As previously discussed, lethal livestock protection collars were designed to target individuals that are actually attacking livestock. The principle of targeting the individuals directly involved in depredation attempts is equally sound when it comes to non-lethal approaches. In such cases, this could be more akin to collared animals learning that certain former prey now exhibit biologically relevant (albeit artificial) anti-predator strategies. Well-known natural examples are skunks and echidnas, armed with noxious and potentially painful stimuli (respectively, scent and quills) that predators learn to avoid after a small number of negative interactions. Retrospectively arming animals with an anti-predator strategy such as a shock collar on a dingo that is activated by close proximity to a livestock animal (i.e. during an attack) simply borrows from more naturalistic 'arms races' exhibited the world over by predators and prey. This mirrors what Linhart *et al.*[59] and Andelt *et al.*[60] found in their experiments, where the major discriminatory stimuli were the potential prey animals themselves. Could similar learned responses be expected in dingoes?

There are some hints that they could. For example, on Fraser Island, despite high numbers of cane toads, very few dingoes have been reported as exhibiting the signs of poisoning usually associated with cane toads (QPWS, unpublished data). In fact, dingoes have regularly been observed to ignore cane toads (Appleby, unpublished data). There might be several reasons for this. One is similar to emerging research that indicates young quolls[65] and planigales[66] can learn to avoid cane toads, through non-lethal exposure to an analogue to the toxin produced by toads. This is known as conditioned taste aversion. Thus, it is possible that young dingoes ingest a small amount of poison after interacting with toads, experience some of the negative consequences of the toxin and learn to avoid toads. Another element may be due to some form of cultural transmission, similar to collared and non-collared wolves alike exhibiting avoidance behaviour. Whatever the reason, although cane toads are having a devastating impact on other species, dingoes appear to be largely unaffected, at least on Fraser Island.

Land use

Newsome[15] maintained that part of the 'kill or conserve' dilemma when it comes to dingoes stems from historical decisions about land tenure. He suggested that if cattle graziers rather than sheep graziers had been the first to discover western New South Wales, dingoes would not have been as much of a problem in the state. Breckwoldt[2] contended that because agricultural enterprise was so important during European settlement of Australia, it was always the dingo that was blamed for any conflict that arose and not the settlers. More than 200 years later, decisions about land are almost impossible to change. He went further to say that, wherever land use planning could serve to keep dingoes and sheep separate, it should be undertaken. As was discussed above, conservation of dingoes is more likely to be accepted within reserves and national parks bordered by cattle or other agricultural industries than in those bordered by sheep farming enterprises. It would seem very challenging at the moment to attempt to conserve dingoes in areas where land use is dominated by sheep grazing; any exceptions would need to be compelling and undoubtedly subject to considerable management endeavours. Neither would it seem prudent to continue sheep farming close to key dingo conservation areas, unless agreement can be reached in relation to managing or preventing dingo depredation primarily through non-lethal means. Perhaps such areas could pursue the Brindabella Wee Jasper model discussed above, but shift from lethal to non-lethal management. Perhaps this could even be pursued in the Brindabella Wee Jasper area itself. As a final point, Corbett[1] suggested the areas that ultimately offered the greatest hope of dingo conservation are island refugia. These include famous islands such as Fraser Island, but also mainland areas that can be sufficiently protected. Here, 'protected' means not just an absence of human persecution, but the prevention of hybridisation with domestic dogs (see Chapter 2).

Conclusion

Conservationists must recognise that a major impediment to dingo conservation is providing convincing evidence for the benefits of dingoes to counterbalance their very real negative impacts on livestock producers. Divides between these and other stakeholder groups all too often appear to be unbridgeable. Yet the rise in environmental awareness in Australia in the last few decades, including within the farming industry, brings a new set of opportunities to allocate money and effort to resolving conflict in ways that are demonstrably beneficial. Ultimately, innovative approaches underpinned by scientific research are likely to lead to mutually beneficial outcomes for stakeholders on both sides of the fence.

Lethal control must be evaluated on a cost–benefit basis, as would any approach that is proposed. Here, benefits are not a reduction in wild dog numbers, but a reduction in depredation events along with other economic and social benefits.

In some cases, such an evaluation would likely support lethal control and in such cases the pragmatic view would be to accept that dingo conservation efforts are better directed elsewhere. At the same time, where populations of largely pure dingoes are found, especially where there are reasonable prospects for their preservation (e.g. hybridisation can be minimised), a pragmatic view would be that farming sheep adjacent to such areas is likely to lead to significant conflict unless non-lethal management approaches can be employed effectively. If pastoralists continue to graze sheep close to such areas then some tolerance of losses might be required, perhaps mediated to some degree through compensation schemes. Agreements in relation to sharing the burden of erecting or maintaining barrier fencing between producers and conservation authorities, as well as other private and public stakeholder groups, could also be pursued. A good portion of the notable success of schemes such as that employed in Brindabella Wee Jasper region stems from landholders across a variety of tenures coordinating efforts to resolve the problem of wild dog depredation.[29] There is no reason why these principles of strategic and coordinated management can't be applied in other areas where dingo conservation has a high priority. A shift away from lethal control is likely to be the major difference. Perhaps a transitional process can occur whereby non-lethal management options are adopted as and when they are shown to be effective. Likewise, reintroduction of dingoes based upon their ecological roles and broader restoration principles must be weighed against the likely costs to sheep producers. Where restoration is viewed as a high priority, a conflict management plan should be developed before any reintroduction. This should include strategies to prevent and resolve conflict.

No one approach is likely to be a panacea, and even bullets aren't magic. All approaches must be weighed on merit and should be implemented on a cost–benefit basis, or trialled in experimental settings that allow for proper evaluation and comparison. Arming pastoralists with techniques to minimise stock losses and not simply to kill dingoes indiscriminately, and prioritising areas for dingo conservation or livestock production, are likely to progress the situation from one steeped in conflict towards one of coexistence.

References

1. Corbett LK (2001) *The Dingo in Australian and Asia.* JB Books, Adelaide.
2. Breckwoldt R (1988) *A Very Elegant Animal: The Dingo.* Angus and Robertson, Sydney.
3. Allen LR, Sparkes EC (2001) The effect of dingo control on sheep and beef cattle in Queensland. *Journal of Applied Ecology* **38**, 76–87.
4. Glen AS, Short J (2000) The control of dingoes in New South Wales in the period 1883–1930 and its likely impact on their distribution and abundance. *Australian Zoologist* **31**, 432–442.
5. Townsend I (2013) The dogs that ate a sheep industry. *ABC Radio National.*
6. Courier-Mail (1947) Sheep slaughter by dingoes threatens flocks. *Courier-Mail.*
7. Gong W, Sinden J, Braysher M, Jones R (2009) *The Economic Impacts of Vertebrate Pests in Australia.* Invasive Animals Cooperative Research Centre, Canberra.
8. McLeod R (2004) *Counting the Cost: Impact of Invasive Animals in Australia, 2004.* Cooperative Research Centre for Pest Animal Control, Canberra.
9. Hewitt L (2009) Major economic costs associated with wild dogs in the Queensland grazing industry. Agforce, Brisbane.

10. McLeod PJ (1996) Developmental changes in associations among timber wolf (*Canis lupus*) postures. *Behavioural Processes* **38**, 105–118.
11. Fleming PJS, Corbett LK, Harden R, Thomson PC (2001) *Managing the Impacts of Dingoes and Other Wild Dogs*. Bureau of Rural Sciences, Canberra.
12. Allen BL, West P (2013) Influence of dingoes on sheep distribution in Australia. *Australian Veterinary Journal* **91**, 261–267.
13. ABS (2013) *Historical Selected Agriculture Commodities, by State (1861 to Present), 2010–11*. Australian Bureau of Statistics, Canberra.
14. East IJ, Foreman I (2011) The structure, dynamics and movement patterns of the Australian sheep industry. *Australian Veterinary Journal* **89**, 477–489.
15. Newsome AE (2001) The biology and ecology of the dingo. In *A Symposium on the Dingo*. Royal Zoological Society of New South Wales, Sydney.
16. Berger KM (2006) Carnivore–livestock conflicts: effects of subsidized predator control and economic correlates on the sheep industry. *Conservation Biology* **20**, 751–761.
17. Hrdina FC (1997) Marsupial destruction in Queensland 1877–1930. *Australian Zoologist* **30**, 272–286.
18. Allen LR, Fleming PJS (2004) Review of canid management in Australia for the protection of livestock and wildlife: potential application to coyote management. *Sheep and Goat Research Journal* **19**, 97–104.
19. Thomson PC (1992) The behavioural ecology of dingoes in north-western Australia. II. Activity patterns, breeding season and pup rearing. *Wildlife Research* **19**, 519–530.
20. Short J, Kinnear JE, Robley A (2002) Surplus killing by introduced predators in Australia: evidence for ineffective anti-predator adaptations in native prey species? *Biological Conservation* **103**, 283–301.
21. Allen LR (2013) Wild dog control impacts on calf wastage in extensive beef cattle enterprises. *Animal Production Science* **54**, 214–220.
22. Allen LR, Gonzalez T (1998) Baiting reduces dingo numbers, changes age structures yet often increases calf losses. In *Australian Vertebrate Pest Control Conference*.
23. Sacks BN, Jaeger MM, Neale JC, McCullough DR (1999) Territoriality and breeding status of coyotes relative to sheep predation. *Journal of Wildlife Management* **63**, 593–605.
24. Jaeger M (2004) Selective targeting of alpha coyotes to stop sheep depredation. *Sheep and Goat Research Journal* **19**, 80–84.
25. Harris CE, Knowlton FF (2001) Differential responses of coyotes to novel stimuli in familiar and unfamiliar settings. *Canadian Journal of Zoology* **79**, 2005–2013.
26. Sequin ES, Jaeger MM, Brussard PF, Barrett RH (2003) Wariness of coyotes to camera traps relative to social status and territory boundaries. *Canadian Journal of Zoology* **81**, 2015–2025.
27. Burns RJ, Zemlicka DE, Savarie PJ (1996) Effectiveness of large livestock protection collars against depredating coyotes. *Wildlife Society Bulletin* **24**, 123–127.
28. Thomson PC (1986) The effectiveness of aerial baiting for the control of dingoes in north-western Australia. *Wildlife Research* **13**, 165–176.
29. Hunt R (2005) Brindabella Wee Jasper wild dog/fox working group: the nil tenure approach to a landscape issue (wild dogs). In *3rd NSW Pest Animal Control Conference*. NSW National Parks and Wildlife Service, p. 16.
30. McIlroy J, Gifford E, Cooper R (1986) Effects on nontarget animal populations of wild dog trail-baiting campaigns with 1080 poison. *Wildlife Research* **13**, 447–453.
31. Fox CH, Papouchis CM, Hirsch K, Lamont G (2005) *Coyotes in our Midst: Coexisting with an Adaptable and Resilient Carnivore*. Animal Protection Institute.
32. Reiter DK, Brunson MW, Schmidt RH (1999) Public attitudes toward wildlife damage management and policy. *Wildlife Society Bulletin* **27**, 746–758.
33. Fitzgerald G (2009) *Public Attitudes to Current and Proposed Forms of Pest Animal Control: A Summary and Review of the Australasian and Selected International Research*. Invasive Animals Cooperative Research Centre, Canberra.

34. Glen AS, Dickman CR (2003) Effects of bait-station design on the uptake of baits by non-target animals during control programmes for foxes and wild dogs. *Wildlife Research* **30**, 147–149.
35. Glen AS, Gentle MN, Dickman CR (2007) Non-target impacts of poison baiting for predator control in Australia. *Mammal Review* **37**, 191–205.
36. Newsome A, Corbett L, Catling P, Burt R (1983) The feeding ecology of the dingo. I. Stomach contents from trapping in south-eastern Australia, and the non-target wildlife also caught in dingo traps. *Wildlife Research* **10**, 477–486.
37. Johnson CN, Crowther MS, Dickman CR, Letnic MI, Newsome TM, Nimmo DG, Ritchie EG, Wallach AD (2014) Experiments in no-impact control of dingoes: comment on Allen *et al.* 2013. *Frontiers in Zoology* **11**, 17.
38. Allen BL, Fleming PJS (2012) Reintroducing the dingo: the risk of dingo predation to threatened vertebrates of western New South Wales. *Wildlife Research* **39**, 35–50.
39. Dickman CR, Glen AS, Letnic M (2009) Reintroducing the dingo: can Australia's conservation wastelands be restored? In *Reintroduction of Top-order Predators.* (Eds MW Hayward and MJ Somers) pp. 238–269. Wiley Blackwell, London.
40. Daniels MJ, Corbett LK (2003) Redefining introgressed protected mammals: when is a wildcat a wild cat and a dingo a wild dog? *Wildlife Research* **30**, 213–218.
41. Elledge AE, Allen LR, Carlsson B, Wilton AN, Leung LK (2008) An evaluation of genetic analyses, skull morphology and visual appearance for assessing dingo purity: implications for dingo conservation. *Wildlife Research* **35**, 812–820.
42. Purcell B (2010) *Dingo.* CSIRO Publishing, Melbourne.
43. Holden P (2002) *Along the Dingo Fence.* Hodder and Stoughton, London.
44. Woodford J (2003) *The Dog Fence: A Journey Across the Heart of Australia.* Text Publishing, Melbourne.
45. Bradshaw C, Ritchie E (2012) Can Australia afford the dingo fence? *The Conversation.* http://theconversation.com/can-australia-afford-the-dingo-fence-7101.
46. Musiani M, Mamo C, Boitani L, Callaghan C, Gates CC, Mattei L, Visalberghi E, Breck S, Volpi G (2003) Wolf depredation trends and the use of Fladry barriers to protect livestock in western North America. *Conservation Biology* **17**, 1538–1547.
47. Shivik JA (2006) Tools for the edge: what's new for conserving carnivores. *Bioscience* **56**, 253–259.
48. Lance NJ, Breck SW, Sime C, Callahan P, Shivik JA (2010) Biological, technical, and social aspects of applying electrified Fladry for livestock protection from wolves (*Canis lupus*). *Wildlife Research* **37**, 708–714.
49. Jenkins D (2003) *Guard Animals for Livestock Protection: Existing and Potential Use in Australia.* NSW Agriculture.
50. van Bommel L, Johnson CN (2012) Good dog! Using livestock guardian dogs to protect livestock from predators in Australia's extensive grazing systems. *Wildlife Research* **39**, 220–229.
51. Kloppers EL, St Clair C, Hurd TE (2005) Predator-resembling aversive conditioning for managing habituated wildlife. *Ecology and Society* **10**, 31.
52. Rauer G, Kaczensky P, Knauer F (2003) Experiences with aversive conditioning of habituated brown bears in Austria and other European countries. *Ursus* **14**, 215–224.
53. Shivik JA, Martin DJ (2000) Aversive and disruptive stimulus applications for managing predation. In *Proceedings of the 9th Eastern Wildlife Damage Management Conference.*
54. Shivik JA, Treves A, Callahan P (2003) Nonlethal techniques for managing predation: primary and secondary repellents. *Conservation Biology* **17**, 1531–1537.
55. Smith TS, Herrero S, Debruyn TD, Wilder JM (2008) Efficacy of bear deterrent spray in Alaska. *Journal of Wildlife Management* **72**, 640–645.
56. Mackintosh N (1997) Has the wheel turned full circle? Fifty years of learning theory, 1946–1996. *Quarterly Journal of Experimental Psychology: Section A* **50**, 879–898.
57. Wasserman EA, Miller RR (1997) What's elementary about associative learning? *Annual Review of Psychology* **48**, 573–607.

58. Staddon J, Cerutti D (2003) Operant conditioning. *Annual Review of Psychology* **54**, 115–144.
59. Linhart SB, Roberts JD, Schumake SA, Johnson R (1976) Avoidance of prey by captive coyotes punished with electric shock. In *Proceedings of the 7th Vertebrate Pest Conference.* pp. 302–306. Available at digitalcommons.unl.edu/vpc7/30/.
60. Andelt WF, Phillips RL, Gruver KS, Guthrie JW (1999) Coyote predation on domestic sheep deterred with electronic dog-training collar. *Wildlife Society Bulletin* **27**, 12–18.
61. Schultz RN, Jonas KW, Skuldt LH, Wydeven AP (2005) Experimental use of dog-training shock collars to deter depredation by gray wolves. *Wildlife Society Bulletin* **33**, 142–148.
62. Hawley JE, Gehring TM, Schultz RN, Rossler ST, Wydeven AP (2009) Assessment of shock collars as nonlethal management for wolves in Wisconsin. *Journal of Wildlife Management* **73**, 518–525.
63. Rossler ST, Gehring TM, Schultz RN, Rossler MT, Wydeven AP, Hawley JE (2012) Shock collars as a site-aversive conditioning tool for wolves. *Wildlife Society Bulletin* **36**, 176–184.
64. Shivik JA (2004) Non-lethal alternatives for predation management. *Sheep and Goat Research Journal* 19**,** 64–71.
65. O'Donnell S, Webb JK, Shine R (2010) Conditioned taste aversion enhances the survival of an endangered predator imperilled by a toxic invader. *Journal of Applied Ecology* **47**, 558–565.
66. Webb JK, Brown GP, Child T, Greenlees MJ, Phillips BL, Shine R (2008) A native dasyurid predator (common planigale, *Planigale maculata*) rapidly learns to avoid a toxic invader. *Austral Ecology* **33**, 821–829.

Plate 1: Dingoes are extremely flexible and agile, climbing trees, rocks and sand banks with ease.

Source: Bradley Smith.

Plate 2: Living in alpine regions requires a thick double coat and a bushy tail to keep dingoes warm and dry in the winter months.

Source: Bradley Smith.

Plate 3: The dingo varies little across the mainland of Australia. In fact, it has remained unchanged for the past 3500–5000 years.

Source: Bradley Smith.

Plate 4: An adult ginger dingo in a protected Indigenous area in the Simpson Desert.

Source: Arian Wallach.

Plate 5: A black and tan coloured dingo.

Source: Bradley Smith.

Plate 6: White or cream is the rarest of the three coat colours.

Source: Bradley Smith.

Plate 7: Pure dingoes have been exported to zoos and wildlife parks in the US. Two of these dingoes and a wolf are seen playing together in a private wildlife park there, a rare opportunity for the two species to interact.

Source: John Williams.

Plate 8: The dug-out base of a large tree provides an ideal location for a dingo family to raise a litter of puppies.

Source: Jean Paul Ferrero/ardea.com.

Plate 9: The typical dingo greeting or group ceremony, in which the subordinate juvenile dingoes all press together to lick the muzzle of a dominant adult. This is known as active submission, and resembles a food-begging ritual in pups.

Source: Bradley Smith.

Plate 10: A juvenile dingo passively submitting to an adult. Deference behaviors are offered by the dingo in response to a perceived threat. In this case the dingo has rolled over on its back to expose vulnerable underparts.

Source: Bradley Smith.

Plate 11: At around three to four weeks during weaning, the mother and father regurgitate food for their pups. To solicit regurgitation, puppies lick the muzzle of the adult.

Source: Jean Paul Ferrero/ardea.com.

Plate 12: An iconic Fraser Island dingo walking along the beach at Eurong. Dingoes on the island are given ear-tags to aid in identifying individuals.

Source: Bradley Smith.

Plate 13: Two males play-fighting on a beach in the Nadgee Nature reserve, far south coast New South Wales. This type of play can be useful practice for the real thing, and for establishing dominance hierarchies.

Source: Jean Paul Ferrero/ardea.com.

Plate 14: The dingo is considered by many ecologists to be a keystone predator. That is, they play an integral role in the contemporary ecology of Australia, and are essential to biodiversity and the conservation of other native species.

Source: Jean Paul Ferrero/ardea.com.

Plate 15: It is easy to see why dingoes are attractive as pets when they are young, but they soon outgrow puppyhood.

Source: Bradley Smith.

6
Dingo–human conflict: attacks on humans

Rob Appleby

Chapter 5 reviewed conflict between graziers and dingoes, but in more recent years direct conflict has emerged between humans and dingoes. Or at least, in more recent years stories about such conflict have begun to appear with some regularity in the media, mostly originating from Fraser Island. The management of the dingoes on Fraser Island has been highly controversial and publicised, particularly since the death of a young boy in 2001. Various stakeholders have raised concerns about aspects of the management and research of dingoes on Fraser Island. These include capture and handling procedures, the use of collars for tracking or aversive conditioning, the use of ear-tags for identification, a policy against feeding and interaction, the condition of dingoes, the use of fencing to exclude dingoes from certain human-use areas, and the lethal control of dingoes deemed to be a risk to human safety. Several research projects have been conducted on Fraser Island in relation to dingoes and interactions with people, but most have been brief or piecemeal in nature and highly speculative when it comes to dingo behaviour. Yet it is knowledge of the latter, and perhaps of the behaviour of human participants as well, that is likely to lead to better management decisions. This chapter will review the conflict between tourists and dingoes, with a major focus on the idyllic tourism destination of Fraser Island. It concludes by suggesting ways in which humans can learn to coexist with dingoes in such places.

The death of Azaria Chamberlain

Most interactions between humans and dingoes are either positive or benign, but these are overshadowed by much rarer negative interactions. The most conspicuous place where negative interactions occur in Australia is Fraser Island, although conflict of a similar nature occasionally occurs elsewhere (e.g. Uluru, Kakadu, Myall Lakes). The most famous case of human–dingo conflict occurred thousands of kilometres away from Fraser Island at another iconic Australian national park, Uluru. The story has been told and retold many times, so in the interests of brevity only the most salient aspects will be covered here. Readers are directed to *The Chamberlain Case: Nation, Law, Memory*[1] and *The Chamberlain Case: The Legal Saga that Transfixed the Nation*[2] for more details. However, in reviewing negative human–dingo interaction dynamics, it certainly seems an appropriate enough place to start.

On the evening of 17 August 1980, Azaria Chamberlain, who was born a little more than two months earlier, went missing from a tent in a campground at Uluru (commonly referred to at the time as Ayers Rock) in the Northern Territory. The body was never recovered, however, some of the child's clothing was found nearby, with further evidence to suggest signs of a struggle. After more than three decades of speculation, the exact truth about what happened on that fateful night remains a mystery. What took place afterwards is infamous in Australia's legal history.

The following are extracts of the lengthy and numerous legal proceedings that followed Azaria Chamberlain's disappearance, drawn largely from Biber.[3] The first of four coronial inquests, which began on 16 December 1980, concluded that a dingo had taken Azaria Chamberlain but the Attorney-General of the Northern Territory, Chief Minister Everingham, later moved to quash the findings. A second inquest, which began on 14 December 1981, concluded with the coroner's report stating that 'there is positive evidence linking Mrs Chamberlain to the homicide' (p. 120).[3] This sparked a court case in the Northern Territory Supreme Court, beginning on 13 September 1982. At its conclusion, Lindy Chamberlain (now Lindy Chamberlain-Creighton) was found guilty of murder and her then husband Michael Chamberlain was found guilty of being an accessory after the fact. Lindy Chamberlain was given a mandatory life sentence and Michael Chamberlain was sentenced to 18 months hard labour; he was immediately released on a bond, under conditions of good behaviour for a period of three years.

Lindy Chamberlain commenced her life sentence at Darwin Penitentiary and almost immediately lodged a notice to appeal to the Federal Court. Only 19 days after hearing her verdict, she gave birth to a daughter in Darwin Hospital. She had been told that she would be permitted to keep the baby in prison. On 18 November 1982 she applied to the Federal Court for her release on bail pending the appeal, stating 'exceptional circumstances' due to the psychological trauma of not bonding with her baby. A day later, the Federal Court granted her bail and she was released from prison. Her appeal judgement came on 29 April 1983, with the Federal Court unanimously dismissing the Chamberlains' appeal. Lindy Chamberlain was subsequently ordered to return to prison.

The Chamberlains then appealed to the High Court, with a request for bail for Lindy Chamberlain similar to that of her earlier appeal. On 2 May 1983 the application for bail was refused. Lindy Chamberlain remained in prison and after a reasonably lengthy process the High Court of Australia did not overturn the original convictions.

On 2 February 1986, a matinee jacket believed to belong to Azaria Chamberlain was found at Uluru, after an English tourist fell to his death and was subsequently dismembered, apparently by dingoes. This initiated a search for his missing body parts, during which the jacket was found. The jacket was shown to Lindy Chamberlain and she immediately identified it as the jacket Azaria was wearing when she disappeared. Five days after the discovery of the jacket, Marshall Perron, then Attorney-General of the Northern Territory announced an inquiry into the case. Special legislation was created for the inquiry to occur. Justice Trevor Morling of the Federal Court was appointed as Commissioner. In a very comprehensive review of the evidence, Justice Morling concluded that there were now serious questions as to the validity of the original evidence and in light of new evidence such as the discovery of the jacket, considerable doubt had arisen in relation to the Chamberlains' guilt. On 2 June 1987 Justice Morling's report was tabled and on the same day the government granted a formal pardon to the couple. On 15 September 1988 the Northern Territory Court of Criminal Appeal formally adopted the findings of the Commission, removing all convictions and pronouncing the

Chamberlains innocent. The Chamberlains received compensation for wrongful convictions from the Northern Territory government, together amounting to $1.3 million.

A third coroner's inquest began on 29 November 1995. It returned an open finding in relation to the cause of Azaria Chamberlain's death, officially concluding 'the cause and manner of death unknown'.

The most recent coronial inquiry, which took place in mid 2012, concluded that Azaria had been killed by a dingo and the death certificate was amended to reflect this finding.[4] However, no new evidence directly demonstrating this was presented at the inquiry. Instead, incident reports were drawn from other places, such as Fraser Island, by the coroner in making the decision.[4] It is for this reason that, while it remains a distinct possibility that a dingo was involved in the Chamberlain tragedy, the coronial findings fail to resolve the issue satisfactorily. After all, a presentation of evidence from elsewhere suggesting that a dingo was capable of being involved is not the same as directly demonstrating that in fact one was. This perhaps raises more questions in relation to the precedent set by such a finding, than it does in regard to dingo behaviour and human–dingo conflict. As this chapter is devoted to the latter , further discussion of this point might best be left to those with the legal knowledge to do it justice. Regardless of whether this particular finding is or is not considered the final word on the matter, the Chamberlain case will undoubtedly remain the best-known example of human–dingo conflict and one of the most famous legal cases in Australia's history.

Apportioning blame in attacks: who is responsible?

After attacks from wildlife, questions naturally arise in relation to root causes and prevention. But for some, questions are more about 'who' or 'what' is to blame, dragging the discussion away from sensible reflection and the development of a carefully considered response. Abrupt responses at the behest of a vocal minority or because of a politically motivated agenda usually do not lend themselves to long-term solutions, particularly if rigorous science has fallen by the wayside in the process. The threat of litigation also undoubtedly hangs over management agencies, potentially influencing decision-making processes.

When this threat becomes a reality, the law sometimes seems to offer little help and perhaps just makes the situation more difficult. Wading through cases of legal suits against state or other government wilderness/wildlife management agencies in North America, where wildlife attacks generate a great deal more litigious activity than appears to be the case in Australia, is a lesson in semantics as much as it is a lesson in the law itself. Take the case of a family that successfully sued the state of Utah following a bear attack that killed an 11-year-old boy.[5] The case was successfully appealed in the Utah Supreme Court, after a District Court judge dismissed a wrongful death case and another District judge awarded the plaintiffs $1.95 million. A major point of contention during the Supreme Court trial revolved around whether a bear could be considered a 'natural condition of the land' and therefore, whether the state could or could not be considered culpable for any damages wrought by bears. The Supreme Court reversed a decision by a District Court judge and decided that bears that scavenged for food in human-use areas were not a 'natural condition' on publicly owned or controlled land, making the state at least partially culpable. Again, what kind of precedent this result sets is interesting enough, and it is noteworthy that two of the five Supreme Court judges did agree that 'indigenous wildlife' was a natural condition and therefore the state was immune. However, what is most interesting about this case is that it serves as an example of how the law interprets responsibility, even going so far as to assign percentages. The previous case at the District level had found the Forest Service was responsible for 65% of the tragedy, because it owned the land that the attack occurred on. The Supreme Court case found that the State Division of Wildlife Resources was responsible for a further 25% and the remaining 10% was the responsibility of the family, primarily because a granola bar wrapper and a can of soda (that may have attracted the bear) were found within the boy's tent.

Apportioning blame in such a way is immediately arresting, as it suggests quite a deep understanding of a situation that would otherwise seem difficult to defend, at least going by incident dynamics as they are reported in the scientific literature. In Australia, similar lawsuits are, perhaps thankfully, fewer and farther between, but the threat of litigation may influence some policy and management decisions in relation to wildlife, including dingoes.[6]

Dingoes on Fraser Island

Interactions between humans and dingoes occur at a variety of places around Australia, with incidents reported in national parks including Uluru and the Kimberleys in the Northern Territory, and Mungo Lakes and other New South Wales national parks and reserves, as well as at interfaces between such reserves and human endeavours such as mining sites or residential developments. However, by far the most commonly referred to location for interactions is Fraser Island, where the most visible efforts have been employed to manage such interactions. It is also where our attention now turns, although much of what is discussed is equally applicable to other areas of the country.

Background

Before discussing human–dingo interactions, it is important to provide a little background. Fraser Island, at approximately 166 000 ha, is the world's largest sand island, lying just off the coast of Queensland and attracting more than 300 000 visitors annually[7] (see Fig. 6.1 and Plate 12). It was listed as a World Heritage Area in 1992. The traditional owners of the island are the Butchulla people, who have probably occupied K'gari (Fraser Island) for millennia.[8]

Appleby and Jones[9] analysed data from a preliminary mark–recapture study conducted by QPWS (beginning in late 2009) and subsequently estimated a population range of 104–197 dingoes, with a rather rubbery point estimate of ~130 individuals as of the beginning of the study.

The diet of the island's dingoes is a much-discussed and debated topic, with some members of the public concluding that there is inadequate food available for dingoes and that this may be causing negative interactions with people because dingoes are hungry.[10] The role food might play in interactions will be discussed at length later in this chapter, but diet and food availability are likely to remain hot topics for some time. Unlike many other parts of Australia, large macropods such as kangaroos are largely absent from the dingo's diet on Fraser Island. However, there are numerous potential substitutes and, as a result, considerable

Figure 6.1: The dingo is a major tourist drawcard on Fraser Island. Visitors on a day tour get an opportunity to observe dingoes on the eastern beach.
Source: Bradley Smith.

breadth to the diet nonetheless. Angel[11] found 18 species of mammals represented in 126 dingo droppings (scats) collected from the southern and central parts of the island. She also compared items found in dingo scats with prey abundance estimated by trapping and other detection methods. She found that the long-nosed bandicoot (*Perameles nasuta*) as well as the pale field rat (*Rattus tunneyi*) featured heavily in both capture records and scats. It was concluded that bandicoots, as well as fish, were an important constituent of dingo diet on Fraser Island. A longer-term scat study is underway on the island, which should provide a more up-to-date profile of dingo diet. Follow-up population studies of likely prey items, such as bandicoots, are required.

Baxter and Davies[12] reported on a GPS collar study of 18 dingoes. Home ranges varied from 547 ha to 57 459 ha, with males generally having much larger home ranges than females. However, the data requires a more detailed examination to better ascertain dingo movement patterns, particularly in relation to individuals likely to be living in stable territories in comparison to dispersing or transient individuals. The latter are likely to have had a substantial influence on home-range size estimates.

History of conflict: the death of Clinton Gage

More than 20 years after the disappearance of Azaria Chamberlain dingoes would again hit the headlines, implicated in the death of a person, a nine-year-old boy named Clinton Gage.

This time though, and unlike the Chamberlain case, there could be little doubt of their involvement. The following is a selective summary of the Queensland Police Service report detailing the circumstances surrounding the fatal mauling of Clinton Gage by two suspected dingoes in April 2001.[13] Clinton and a friend of similar age (name withheld) had left their families and been walking along a track towards an area known as the Binngih Sandblow, which can be accessed from the Waddy Point campground where the Gage family were camping, when two dingoes appeared. According to a statement given by Clinton's friend, the dingoes came from behind the children. Clinton began running and went ahead of the other child, who did not run but kept walking. While running, Clinton apparently tripped and fell to the ground; the other child kept walking past him, with the dingoes following. The other child then stopped briefly and the dingoes returned to where Clinton was lying on the ground and looked as if they began to sniff at him. The other child then walked a considerable distance around to the campground where his parents were.

At around the same time, Clinton's father and younger brother had begun looking for Clinton together. After walking for some time up the track towards the sandblow, they came across Clinton's body. No time of death is given in the report but, based on statements of those interviewed, Clinton died at some stage between his friend leaving the scene and his father arriving, a timespan of ~20–30 min. While Clinton's father was retrieving the body, Clinton's younger brother, who was standing a short distance away, was harassed by two dingoes, presumably the same ones that had mauled Clinton. The boys' father 'shooed' the most aggressive dingo away. As they made their way back to the campground the dingoes kept harassing the younger boy, especially if he moved too far in front of his father.

A QPWS ranger was called to the scene and assisted the family. Sometime later, the ranger was informed that two dingoes had arrived at the campground and, suspecting that they were the dingoes (one male and one female, ages not specified) involved, shot them in the presence of other authorities. A post mortem was later carried out on the destroyed dingoes. The male dingo's stomach was found to contain no items of human origin, and although the female dingo's stomach did contain suspicious material no conclusion in relation to origin could be ascertained. Dental impressions of the male dingo's teeth revealed patterns consistent with the injuries sustained by Clinton Gage. The first and arguably most controversial management action taken was the destruction of 31 dingoes in the month that followed the attack.

In reviewing this tragic event, several important questions have arisen that have considerable management implications. For instance, was this an act of predation by the dingoes, or was there some other reason for the attack? This is important, as an answer potentially leads to quite different conclusions about risk and risk management given different situations. For example, in his comprehensive review of bear attacks, Herrero[14] established that in many cases black bears engaged in negative interactions with humans were actively depredating upon them, while grizzly bears were more often acting defensively (e.g. mothers protecting cubs from a perceived threat). This led to very different recommendations or 'rules of thumb' in relation to people's reactions when facing an attack: people should generally react aggressively when attacked by black bears, and generally react passively (e.g. play dead) when attacked by grizzly bears, particularly adult females with cubs.

In the case of the Clinton Gage incident, it is difficult to reach robust conclusions about the nature of the attack. There are, however, some details of the incident that stand out. That the two boys were unsupervised by an adult may have had considerable bearing on the outcome, and supervision now constitutes one of the primary recommendations reinforced by QPWS. Notably, it would be surprising if similar situations of a lack of adult supervision of children encountering dingoes had never occurred before (and indeed several have been directly witnessed since – Appleby, unpublished data), yet this was the first that ended in a fatality.

The difference between the boys' reactions to the dingoes highlights the importance that human behaviour may play during interactions. It is intriguing that although Clinton Gage tripped and fell after running away, the dingoes appeared to initially walk past him, instead following the other boy, who was walking. The act of walking and running may trigger an inquisitive response or pursuit from dingoes,[15] though it is unclear whether such responses are due to predatory or other motivations. Indeed, there is much about this and other serious incidents that are yet to be properly ascertained, an objective that should be considered a priority if such events are to be effectively minimised. This remains an important part of an ongoing study (Appleby, unpublished data).

Incident dynamics

The most up-to-date summary of incidents is provided in the most recent review of dingo management on Fraser Island, conducted by an independent commercial company after a successful tender bid (the Ecosure report).[16] Unfortunately, raw figures were not provided, so the brief rundown given here has been derived from graphical representations in the report and therefore lacks precision. From such graphs[16] (p. 83) it is clear that, overall, incidents have risen in recent years, approaching 300 in 2012. Most incidents are of relatively low management concern (deemed 'category C' by QPWS – a full description of the categories is given in the Ecosure report). Incidents appeared to peak in or around 2003 at more than 500 category C incidents. The more serious incidents (categories D and E) appear to occur at far lower rates, at fewer than five of each category annually between 2005 and 2012 and a peak, again around 2003, in category D incidents of around six incidents.

Appleby *et al.*[15] suggested that incident reports may not accurately reflect incident rates, as reporting rates themselves might vary. This was also recognised in the Ecosure report.[16] A skew in reporting consistency might therefore be expected, such that incidents of less concern are possibly reported less often, while more serious incidents might be reported more often.[15]

The relationship between incident categories was not explored in detail in the Ecosure report, so any possible progression from less serious to more serious categories is unclear. However, such trends are apparent in the main graph provided in the report (p. 83) and this is worthy of further investigation.

There appear to be seasonal elements to incidents. For example, Appleby *et al.*[15] identified some peaks in incident reporting around March/April, which coincides with the Easter holiday period (and consequently a surge in visitor numbers) as well as the mating component of the dingo breeding season. This trend appears to have held over time.[16] Other peaks have

Figure 6.2: Serious incidents on Fraser Island can often involve children. Moments before this picture was taken, these children were playing unsupervised when a dingo approached.
Source: Bradley Smith.

also been observed to coincide with holiday periods.[16] Location too appears to have some relationship with incidents, with the Ecosure report identifying six areas where the majority (90%) of category D and E incidents occurred (although one of the areas encompasses almost a third of the island). See Fig. 6.2 for an example of interaction between children and a dingo.

Types of interactions and potential causes

Most human–dingo interactions that occur on Fraser Island are probably positive or benign with respect to the human participants,[6] although there is no mechanism in place to properly quantify this. Sightings, which could possibly serve as a point of comparison, are currently recorded in a relatively opportunistic manner. If such a comparison were possible, it would serve as a useful contrast to the rarer but more conspicuous negative interactions. The severity of incidents ranges from fairly innocuous activity on the part of dingoes such as loitering in human-use areas, to more serious events in which dingoes threaten or injure people (e.g. biting). The death of Clinton Gage is an unfortunate reminder of the catastrophic conclusions some incidents can have, albeit on the rarest of occasions.

The search for answers behind such attacks remains complex and ongoing. The Fraser Island Dingo Management Strategy[8] refers to some 20 factors that may be directly or indirectly associated with negative dingo–human interactions, many of them behavioural in nature. Corbett[17] summarised three major potential reasons for agonistic behaviour by dingoes directed at people.

1. Regarding humans as competitors (for space or resources).
2. Regarding humans (mostly children) as prey.
3. Young dingoes playing with humans (mostly in the form of practising hunting and aggressive behaviours).

These were, in effect, untested hypotheses in relation to incidents of the greatest management concern that can occur. The first two remain untested because they probably occur quite rarely and are difficult to test. Nonetheless, they remain feasible explanations in some circumstances and shouldn't necessarily be discounted, even if they are very challenging to explore. The 'play' hypothesis might be an exception in terms of testing, at least observationally speaking, and will be discussed a little later.

There are other potential hypotheses that might be worth considering, drawn from, for example, Moyer's[18] seminal discussion of the types of aggression exhibited by animals, which he divided into the following seven categories: predatory, inter-male, fear-induced, irritable, territorial, maternal and instrumental. Moyer's[18] categorisations are incomplete and represent more defensive than aggressive behaviours.[19] Moyer[20] came to see such categorisation as too restrictive and questioned some of the kinds of aggression he had originally delineated, such as territorial aggression. Predation may not fit well as a specific category of aggression, as it is more appropriately thought of as an entirely different behavioural state. Nonetheless such classifications serve as a useful starting point. Some of these categories of aggression are obviously quite context-specific and might not appear at first glance to be of great relevance in relation to human–dingo interactions.

For example, inter-male fighting is dependent on the co-presence of unfamiliar male conspecifics[19] and probably would not serve as a useful descriptor of dingo aggression directed at humans (even that involving adult male dingoes), or at least could prove very difficult to discern. At the same time, given that there are incident spikes around the Easter holiday period, which aligns with the mating component of the dingo breeding season, and that more incidents involve male dingoes,[16] there is some possibility of a redirection of aggression to humans that might stem from inter-male competition. Support for this idea came from Corbett,[17, 21] who described observations of dingo behaviour directed at cattle, particularly calves, in the Barkly Tableland, Northern Territory. During the breeding season, Corbett[21] observed that calves and dingoes were often at water sources together, with calves sometimes becoming 'frisky' – dingoes subsequently pursued and attacked them, perhaps because hormones associated with breeding were high at the time.[21]

Corbett[21] suggested that because calves were unable to appease or subvert such aggressive behaviour, this meant that the behaviour would continue, often in conjunction with other dingoes, with the result being immobilisation or death of the calf. In Corbett's view, the motivation of the dingoes was probably not related to hunger (calves that were killed were often not consumed), but more a displacement activity due to frustration over competition between males for access to breeding females, and resultant inter-male aggression. Corbett[21] did not provide details such as gender (although the fact that they were mostly male can be surmised from statements in the text) or age, but it is an interesting illustration of motivationally ambiguous behaviour. Another, perhaps related example will be presented later.

Certain kinds of aggression do serve as more straightforward anchor points through which human-oriented aggression can be explored and there is clearly overlap with some of the hypotheses Corbett[17] proposed. These expand on the possible trajectories of aggressive behaviour and are worthy of inclusion in discussing or indeed modelling serious incidents, which remains a work in progress. Future research must take a closer look at aggression by dingoes towards conspecifics and during predation events, and how this may or may not relate to aggressive interactions with humans. The purpose here is to simply introduce and expand upon Corbett's[17] original propositions and highlight the complexity of such an exploration.

Another attempt to summarise the sequence of events that might lead to a negative dingo–human interaction was given, in relation to dingo participants, as follows.[22]

Attraction → Habituation → Interaction → Aggression

Although a useful conceptualisation, the model is not altogether robust as it was not accompanied by an adequate explanation of how or why each component might lead to the next. Nor does it consider possible feedback or interchange between components, or how context might influence responses.[16] An example is the unqualified link between an interaction occurring, and the exhibition of dingo aggression, possibly the most important step in quantifying negative interaction dynamics and subsequent risk. The processes leading to aggressive interactions are yet to be properly explored, but what follows is a brief discussion of some pertinent considerations.

Habituation

Habituation is a major element in Masterson's[22] original model, and the idea remains a commonly cited feature of human–wildlife interactions (i.e. not just dingo–human interactions).[24] As we'll see though, the concept and its use in exploring human–wildlife interaction dynamics is not without serious complications.

A relatively precise definition of habituation is the reduction of a response to a repeated stimulus that is neither positively or negatively reinforced.[25] Habituation is a non-associative learning mechanism that allows an animal to passively discriminate between useful and useless biological stimuli. In much of the literature pertaining to negative wildlife–human interaction, this definition is relaxed somewhat to infer that habituated animals have lost their fear of humans.[24] A typical example of the thought process leading to this relaxed view in the context of dingoes on Fraser Island is as follows.[7]

> Although warned by signage and literature not to feed the dingoes, animals have been regularly fed at picnic and camping grounds, barge landings, resorts, and island residences. Inadvertent feeding opportunities were available at garbage dumpsters and bins, from careless campsite hygiene, fish offal, and bait disposed

> of on beaches, as well as food stolen from tents and food containers. As a result, dingoes became habituated to human presence and lost their fear of humans because of regular, ongoing, and rewarding contact (p. 39).

There are considerable discrepancies with, and in some cases misuses of, concepts such as habituation in the wildlife–human interaction literature.[24] This includes the theoretical gap between the precise and relaxed definitions, inappropriate interchange with other learning concepts (e.g. food conditioning – see below), a general lack of empirical support for its almost canonical usage, considerable challenges in preventing and reversing habituation in some contexts, and the nature of its role in aggressive interactions involving humans. It is clear from the example above that confusion can arise in relation to the concept. The notion of 'ongoing, rewarding contact' suggests something more akin to conditioning or associative learning featuring positive reinforcement (where behaviour that is positively rewarded is more likely to be repeated), than it does habituation. Animals that have learned that a stimulus is biologically irrelevant (i.e. habituation) tend to ignore it, not seek it out.

Also, the idea that there has been a 'loss of fear' obviously implies that there was fear to begin with. Animals can be born with what might be regarded as an innate 'fear', or a related response to certain stimuli, primarily in cases where their survival might be limited without it. For dingoes to be inherently fearful of humans, arguably, selection for fearful individuals would need to be continuous and strong. Perhaps this was partly the impetus behind the initial cull of animals deemed to be 'habituated' on Fraser Island immediately following the death of Clinton Gage in 2001. But an innate fear of humans should not therefore be considered a 'natural' state for dingoes.[23] In certain locations on the mainland, almost two centuries of artificial selection pressure from lethal control practices has undoubtedly led to a skew in some resident populations in relation to responses to human stimuli, away from boldness and towards something suggestive of a fear response. There is good reason to consider dingoes undergoing less or no persecution as more often being naturally inquisitive (or bold), as there were early accounts of such behaviour soon after European settlement.[26]

Experiences with very young dingoes on Fraser Island suggests that they often take their lead from older dingoes; if the older dingoes are calm, so are they (Appleby, unpublished data). Provided observers were quiet and relatively still, rarely was any overt fear response observed in pups that were in close proximity, unless older dingoes were agitated for some reason (Appleby, unpublished data). Great care was taken during this study to limit the exposure of young dingoes to humans. Nonetheless, such occasional observations were useful, particularly as behavioural responses of dingoes to humans might be different elsewhere.[27] These could serve as useful points of comparison in relation to the effects of lethal control and any associated skew in behavioural responses.

Boldness–wariness spectrum

Most behaviour occurs on the 'boldness–wariness spectrum' (or shy–bold gradient/continuum, as it has been put by some authors).[28–30] Individual dingoes will lie somewhere on this gradient, and even change position depending on context. Our expectations of how animals should behave may have a lot to do with how we manage them, so if innate 'fearfulness' is desirable or required in Fraser Island dingoes then, inevitably, lethal control practices will need to continue in order to remove any individuals that lie closer to the boldness end of the spectrum. However, this hardly seems in keeping with the values we might expect to be upheld in World Heritage wilderness areas, and in fact the management authorities responsible for the island have stated that they do not support such an approach.[8]

As suggested above, boldness exhibited in the right contexts may well be a trait that ends up making a maturing dingo a great parent or pack leader (see[31] for a general discussion) and therefore retention of such traits within the population is desirable. For example, Reale *et al.*[29] found that bolder North American bighorn ewes (*Ovis canadensis*) reproduced earlier and had greater weaning success than shy ewes, and that variability in temperament was influenced by life-history trade-offs and annual changes in selection pressures. Put another way, a natural population requires variation in temperament across individuals in order to cope with environmental and other stressors that vary from year to year: in some years or contexts, shy animals might survive or reproduce more successfully, while in other years or contexts, bolder animals are more successful.[28] Plasticity in such traits can have similar advantages conveyed in different contexts or at different times. Such plasticity may largely circumvent the selective pressure exerted by humane destructions on the island unless the pressure is overwhelming, which again does not match the principles of conservation management. An alternative to destruction therefore is to seek to modify behaviour exhibited by dingoes which is deemed to represent a risk to human safety, but not at the cost of the dingo's life, through aversive conditioning and other forms of behavioural management.

Feeding: setting the stage for interactions

As was discussed above, habituation is often confused with the concept of conditioning, or more specifically, so-called food conditioning. In their comprehensive review of the nomenclature used in bear management, Hopkins *et al.*[32] had this to say, 'Habituation and conditioning are commonly but erroneously used interchangeably, and may be the two most complex and misapplied terms in human–bear management' (p. 161). Hopkins *et al.*[32] defined a food-conditioned bear as one that has 'learned to associate people (or the smell of people), human activities, human-use areas, or food storage receptacles (e.g. bear-lockers, trashcans, dumpsters, backpacks, vehicles) with anthropogenic food', citing Herrero *et al.*[33] Exchanging bears for dingoes in the context of Fraser Island, this is probably a reasonable definition of a food-conditioned dingo. Note that this is distinct from habituation, as described above. Note also that there is a clear link between food, humans as a stimulus complex and other important elements such as location. In this sense, we can see that anthropogenic food sets the stage for human–dingo interactions to occur. Importantly, food is unlikely to be the only impetus for interactions, particularly when it comes to younger dingoes (discussed below). Nonetheless, food can be regarded as a primary reinforcer for many interactions and, as conditioning theory would suggest, it probably doesn't take many occasions of access to food, particularly in the absence of counter-consequences (e.g. aversive events), for a strong association to be built and maintained.

Hopkins and colleagues[32] suggested that food conditioning occurs along a spectrum such that bears can, based upon behavioural assessment, be categorised as highly food-conditioned when they habitually seek food over relatively long periods of time in human-use areas despite repeated hazing and wide availability of natural foods. This is opposed to a lower level of food conditioning in bears that seek food only when natural food availability is low, in low human-use areas and on rarer occasions. This too probably has cross-over with dingoes on Fraser Island, and it might be useful to borrow such concepts in progressing management on the island.

In some circumstances, bears that display aggression towards humans in an effort to obtain food can learn an additional association: that aggression equals food. This is akin to Moyer's[18] 'instrumental aggression', in which the exhibition of aggressive behaviour is reinforced, causing the frequency of such exhibitions to increase. There have not been many documented cases of similar behaviour exhibited by dingoes, although the suggestion of it is not unheard of. As with habituation, considerably more research is required to better understand the role of food in producing or maintaining aggressive behaviour by dingoes towards humans.

Restricting access to anthropogenic food remains one of the primary management strategies for dingoes on Fraser Island. It's fair to say that, while QPWS employs what they call the three Es (education, engineering and enforcement – see below), much of the management of human–dingo interactions on the island is actually summed up as the three Fs (food, fencing and fines). The latter two are covered below. Not everyone agrees, however, that restricting access to anthropogenic food is a good thing.

Immediately following the death of Clinton Gage, QPWS and various other government departments received an influx of public submissions in relation to dingo management on the island, a common idea being that of so-called 'feeding stations'.[7] The idea seemed to be that aggressive behaviour directed at humans by dingoes was related to hunger and that feeding stations would therefore supposedly satisfy dingoes and possibly keep them away from areas of high human use. Corbett[17] and the Ecosure[16] report rejected such arguments, stating that feeding stations would introduce an unnatural influence in relation to population dynamics and artificially inflate the carrying capacity of the island. It's worth adding that there are a plethora of other challenges with the idea, not least of which are what food would be provided, where stations would be located and how many would be established. Further, how would food be delivered (e.g. would people be visible during supply and, if not, how would they be obscured), how often would food be supplied to achieve the goal of sating dingoes, what would it cost, and who would pay for and manage the stations? Feeding stations would need to be managed in terms of people seeking to view dingoes, which counteracts the premise of keeping dingoes away from human-use areas and limiting opportunities for interactions. It's highly unlikely that feeding stations will ever eventuate on the island, but if the idea were ever taken seriously such questions would need answers.

Obviously the issue of anthropogenic food is complex and worthy of far more discussion than can be entertained here. One final point, though, relates to the motivations of people who deliberately provide dingoes with food. Much of the impetus for this appears to come from observations (and subsequent interpretations) of dingoes that are in poor condition.

Indeed, some dingoes are considerably thinner than others and consequently some people might think that providing them with food is an act of kindness. However, this is a particularly slippery slope. There are many animals that reside in wilderness areas, all of which suffer hardships and endure the trials of life in an effort to survive and ultimately reproduce. Not all will, or can, be successful. If humans were to intervene so that many more survived and reproduced than would naturally be the case, then the situation is no longer natural. As unpalatable to us as starvation and suffering in nature may be, such occurrences are a part of nature and it is not our place to overrule natural selection processes – the very same processes that have given rise to all the diversity in wilderness areas. Surely dingoes deserve more credit as the consummate survivors that they are, and that we can all admire them for being, than those choosing to feed them appear to give?

Play behaviour/dominance displays

Another, rather conspicuous and motivationally ambiguous behaviour that is sometimes directed at humans by dingoes has been loosely considered to be related to dominance behaviour.[8] The term 'dominance' carries a particular connotation of an individual that supplants another, usually when competing for food or breeding opportunities.[34] In the context of interactions between dingoes and people, this notion of dominance is questionable. However, as such ambiguous displays often feature behavioural events that are reminiscent of aggression, such as lunging and snapping, it can be very difficult to ascertain the intent behind these displays.[35, 36] Although the intent may be more playful than aggressive, human participants tend to perceive it as the latter.

There are some clues that playful behaviour is common in nature. For example, most of such behaviours are enacted by young dingoes (Appleby, unpublished data). Play is most common among juveniles, and has often been suggested to provide a mechanism through which they learn and practise the social and foraging skills necessary for survival (see Plate 13).[37–40] If play is a motivating factor behind some interactions with humans it is not clear what exactly is being practised, although similar displays are sometimes directed at other dingoes and objects such as trees and bushes (Appleby, unpublished data). One possibility is that humans represent a surrogate 'play mate' or even just an object against which to practise, akin to training.

If indeed play is occurring, there are some differences between this type of play and the play behaviour dingoes exhibit with each other. The latter often involves much more rough-and-tumble contact, including biting, although this is probably somewhat inhibited. Regardless, it can elicit whines and yelps from the receiver. Like most mammals that exhibit social play, dingoes undoubtedly have mechanisms so that when they play with one another, either they can effectively communicate that what is about to happen is play, or can retrospectively communicate that intent.[36] This might involve subtle or contextual cues[36] that humans can easily miss. Indeed, if a human is bitten by a dingo even in a playful context it will possibly do some damage, consistent with some of the injuries recorded in incidents. Regardless of the possibly playful intent, injury to people is unlikely to be tolerated by management.

Perhaps of greater concern is if such play behaviour were to escalate to something similar to that described by Corbett[21] in relation to dingo interactions with calves. If indeed Corbett[21] was right and an escalation or continuance of aggression is due to a lack of appeasement displayed by calves, then what might start out as playful intent might progress to something more serious because human participants wouldn't be expected to exhibit appeasing behaviours. Presumably this would entail something akin to active or passive submission[21] (see Chapter 2). This is the main concern of managers. Whether the behaviour is playful or not, if an interaction concludes in injury to a person the dingo's intention becomes of marginal importance.

A theme worth reiterating here is that a better understanding of dingoes' motivations might give clues of how humans can respond to prevent escalation. In the course of many purportedly playful interactions with dingoes where the human observers remained passive and reacted as little as possible, the situation never escalated to aggressive behaviour such as biting. This was regardless of time, location or context (e.g. whether the person was sitting, squatting or standing: Appleby, unpublished data). Perhaps this is a useful analogue to appeasement, at least in relation to adult human behaviour when confronted by a dingo exhibiting such a display. However, an alternative explanation might be that a lack of interest shown by a person simply translates to 'I am not interested in playing'. This might mirror the behaviour of a parent or other dingo uninterested in playing. More detailed analysis is required in order to better inform management procedures and any recommendations in relation to human responses to dingoes.

Other causes and sources of correlation

All reported incidents that occur on Fraser Island are kept in a database, compiled and collated by QPWS. Many dingoes are ear-tagged for identification (Fig. 6.3). Appleby *et al.*[15] examined an initial set of incident reports. Although models were often very simplified, several variables stood out in terms of the more serious incidents examined (e.g. where biting occurred). These included the human participants walking or running, or being in water at the time of the incident. In the latter case it was not always clear whether human participants were in the water at the beginning of the interaction or moved into the water after the interaction began, but for now it would seem prudent to recommend visitors avoid running or moving into water when dingoes are in close proximity. This is in keeping with current recommendations from QPWS. Much greater exploration of the causal factors and motivations behind dingo interactions with humans is required.

Managing interactions between dingoes and humans

Past and current management: the QPWS' three Es

Earlier, the three Es of the QPWS management strategy were mentioned – education, engineering and enforcement. These represent the three cornerstones of the overall strategy and are pivotal in the ongoing management of dingoes and people on Fraser Island.

Education

Comprehensive conflict resolution on Fraser Island involves management strategies targeting both dingoes and humans.[10, 41] Education and interpretive programs endeavour to teach

Figure 6.3: Many Fraser Island dingoes are given ear-tags so that they can be more easily identified and tracked over time. An ear-tag in the left ear indicates a male, in the right ear it indicates a female.
Source: Bradley Smith.

visitors, island staff and residents, in an effort to reduce dingo–human conflict. Early programs were favourably reviewed. Beckmann and Savage,[41] for example, singled out the 'Be Dingo Smart!' campaign, stating that it was considered by international experts to be very innovative and best-practice. Education strategies remain a major management priority. A wide variety of interpretive approaches are utilised in the education campaign including widespread signage, personal contact (i.e. rangers and staff presenting information directly), brochures, web pages, an information DVD and press releases.[16] The Ecosure[16] report recommended several additional strategies to maximise reach, including brochure handouts to emphasise that 'some dingoes can be dangerous' and flyers with multiple languages strategically placed at areas of high visitor traffic such as arrival and departure points. Ongoing formal education of tourist operators is also essential.

Elements of some of these education strategies have been criticised. For example, Hytten and Burns[6] examined the representations of Fraser Island dingoes in various governmental discourses. They concluded these were often inconsistent, have unrealistic expectations of acceptable dingo behaviour and failed to state explicitly that dingoes have killed. They added that vigilant supervision of children should be the central message of all educational materials. They questioned the strength of the association of food, habituation and aggression in dingo behaviour, calling for more research to explore this association and indeed many of the other factors that have been identified as potentially contributing to aggressive behaviour. Several revisions have now been made to interpretive materials.

The Ecosure[16] report also flagged some concerns in relation to a lack of credible scientific evidence in support of claims, including the provision of food leading to aggression in dingoes, the applicability of the notion of dominance testing, and the best course of action for people who are approached by dingoes. Bringing the messages in line with the results of up-to-date scientific studies will help to ensure that such concerns are largely abated, as will further engagement with stakeholders.

Engineering

A major factor in preventing interactions is the strategic use of fencing. Of particular importance is separating dingoes and people in areas where relatively large numbers of people congregate and where food is likely to attract dingoes, such as campgrounds. Where habitual feeding has been suspected in or around residential or resort areas these too have been fenced,

Figure 6.4: Many camping areas and townships are now fenced to limit the interaction between dingoes and people, and prevent dingoes' access to human food or waste.

Source: Bradley Smith.

including Kingfisher Bay Resort on the western side of the island and the Eurong resort and Eurong Second Valley townships on the eastern side (Fig. 6.4). There are 12 fenced areas on the island.[16]

The Ecosure[16] report analysed associations between incidents and the absence/presence of fencing, with equivocal results. On the face of it, it appears that fencing may not have had meaningful effect in reducing incidents at five major sites on the island. However, the report did suggest that several factors, including small sample sizes, may have confounded results. Importantly, the report concluded, no serious incidents had occurred within any fenced area.

These results raise questions about the relationship between dingo access to food and incident rates, as almost all areas were primarily fenced in an effort to reduce dingo access to anthropogenic food, along with keeping people and dingoes separated in a more general sense. It is perhaps too early to draw conclusions in this regard and, as the Ecosure[16] report contended, the absence of serious incidents post-fencing is probably justification enough for fencing, although there is some contention otherwise (see below). In principle, keeping people and dingoes separated, particularly families with young children, is an obvious incident prevention strategy and where justifiable is likely to remain a key engineering approach on the island. Nevertheless, there remain unanswered questions relating to the role of anthropogenic food in incident dynamics and, most importantly, incidents of significant risk to human safety. Fences are not much help much if interactions occur outside their boundaries.

Fencing is not without controversy, as some residents and members of the public have questioned its necessity in certain places, and its ecological impacts.[16] Any future fencing projects will need to take such considerations into account, and consultation with affected stakeholders is likely the best way to resolve concerns.

Other engineering projects have potentially aided in resolving aspects of conflict dynamics. Most serious incidents involving biting that were reviewed by Appleby *et al.*[15] occurred at Indian Head, which for some time had a campground without toilet or waste disposal facilities. Large collections of poorly disposed-of human waste may have attracted dingoes and this could have led to higher rates of interactions. Many of the most popular camping areas, walking tracks and day-use areas on the island now have such facilities, in addition to lockers in which to store food or rubbish (Fig. 6.5). Not all do, however, as this is logistically (due to remote locations with difficult access) and economically challenging. Visitors are encouraged to bring their own chemical toilets, but it is unclear how this initiative is faring.

Several large open waste disposal sites formerly existed on the island, and large numbers of dingoes were often seen at them (Fig. 6.6).[42] In an effort to reduce dingo access to food waste and discourage such congregation, such dumps have been phased out in favour of fenced waste disposal sites with industrial bins for visitor, residential and commercial waste.[7]

As might be expected, there are only a limited number of places where waste disposal stations can be constructed, consequently some camping areas are distant from a convenient waste disposal location. Unfortunately, if waste is not disposed of responsibly it attracts dingoes to certain areas. Such behaviour can be reinforced with only a small number of rewarding experiences. Waste disposal is a major challenge on an island, particularly one that is a World Heritage listed site. Visitors must dispose of their own waste responsibly and, as

Figure 6.5: Campers are given areas to store food items safely.
Source: Bradley Smith.

with food, keep it safely out of reach of dingoes. When this doesn't occur, the final E – enforcement – becomes a requirement.

Enforcement

Despite broad education campaigns warning people about the possible risk dingoes may pose, along with advice on keeping food and waste out of reach, not everyone adheres to the legal requirements. An initial step in enforcement is issuing a warning notice (a Notice to Comply) in an effort to get the people involved to correct the situation. Where this fails or where breaches are more problematic, a more formal approach is taken. Most often, penalty infringement notices (PINs) are given to people who failed to secure food or waste properly, but were probably not deliberately attempting to feed dingoes. From 2003 to mid 2012, 453 PINs were served, of which only eight were for unauthorised feeding.[16] A further 10 infringement notices were dealt with by the courts, nine of which concluded with fines of between $500 and $34 000.[16] In an exceptional and highly publicised case, wildlife photographer and dingo campaigner Jennifer Parkhurst was fined $40 000 for multiple offences.

There are signs that PINs are decreasing in frequency, but the reasons for this are not entirely clear.[16] A possible and positive reason is that compliance is improving, but there could be a reduction in monitoring compliance by staff or an increasing hesitation to issue compliance-related notices.[16] Prevention remains the ideal target of management strategies, as

Figure 6.6: Waste disposal areas are a great source of food for dingoes. New fencing infrastructure prevents dingoes from accessing such areas.
Source: Bradley Smith.

enforcement procedures such as PINs obviously indicate that undesirable human behaviour has already occurred and therefore undesirable dingo behaviour may arise. Nevertheless, enforcement will likely need to remain a feature of management, serving as a disincentive to those who might otherwise ignore recommendations or warnings, and directly punishing those who do ignore such directives.

Lethal control

When the management strategies stemming from the three Es described above fail, dingoes that are persistently involved in serious incidents are often destroyed. Ballard and Fleming[43] argued that once a dingo had been consistently fed (or otherwise supplemented with food by humans) and displayed aggression, the only course of action available to management authorities was lethal control. This is an unfortunate reality in many cases, but it doesn't have to be.

The Ecosure[16] report found that at the time of writing and since 1992, 135 dingoes had been humanely destroyed, eight of which were sick or injured, leaving 127 destroyed because of negative human–dingo interactions. Most of those animals were destroyed since 2001, when Clinton Gage was killed, with 32 dingoes killed in that year alone. If 2001 is classed as an aberrant year, then on average approximately nine animals per year were destroyed between 2002 and the time of the Ecosure report. Of the 76 animals destroyed between 2002 and the release of the report in 2012, 24 (31%) involved animals that had some link (e.g. were resident) in the areas between Dilli Village and Eurong and a further 11 (14.5%) at Hook Point. Almost 80% of all dingoes destroyed under these circumstances were less than 18 months of age. Further, 69% were male and 31% were female. This suggests that the majority of individuals likely to be destroyed for incidents with humans are young males residing in (or otherwise connected to) the area from Hook Point to Eurong. This 'model' is rather crude; building useful predictive models that incorporate incident rates, destructions, the impacts of various management approaches and other important factors is challenging, and the subject of an ongoing study (Appleby, unpublished data). However, simple as the model is, it highlights aspects of incident dynamics that might represent an opportunity to trial alternative approaches to lethal control which, at present, is the almost inevitable outcome of dingoes being involved in progressively more serious incidents.

The Ecosure[16] report contended that current and past levels of lethal control were unlikely to have adverse impacts at a population level, although this was qualified with reference to a need to properly determine population metrics and the appropriate methodology with which to do so. However, the removal of 108 individuals (including the 32 culled in 2001) in little more than a decade from a population likely not to comprise a great deal more individuals than this is not trivial. Consequently, the impacts might not be either. The island's dingo population is isolated (geographically closed) and the prospect of immigration from the mainland seems remote. Island populations have been recognised as more susceptible to extinction processes because of such isolation, a major proposed mechanism being inbreeding depression.[44] There is currently no information pertaining to the level of inbreeding in the island's dingo population and this, along with several other genetic metrics, needs to be rectified. However, even in the absence of such information, it would pay to maximise the natural levels of genetic diversity. There are other reasons why genetic diversity is important in the long-term conservation of the population, such as the potential threat posed by disease when relatedness is high. Disease might be a more insidious route through which extinction could occur. In this way, high rates of destruction might be an indirect contributing factor to population decline.

As was the case with livestock conflict discussed in Chapter 5, there are questions about the effectiveness of lethal control in relation to reducing risk. Certainly the removal of an individual that poses a risk might seem to result in an immediate reduction in risk, but because lethal control fails to target the reason for the risk in the first place such a reduction might be relatively short-lived. This means that control inevitably becomes cyclical, and if it is employed as a last resort (as is most often the case, with exceptions such as the initial cull immediately following the death of Clinton Gage) it is a reactionary procedure, meaning other individuals may have developed, or be developing, associated problem behaviour. This is somewhat evident in annual statistics associated with incidents, which exhibit some fluctuation (mostly involving upward spikes) but have tended to remain stable across the record-keeping period (e.g. see Fig. 22 of the Ecosure report).[16] Treating the root causes in a proactive and preventative way, as much of the management strategy is directed at doing, aids in breaking this cycle of incidents leading to destructions.

A commendable course of action in the management plan is minimising reliance upon lethal control. This is evident in the decreasing rates of destruction in recent years (see[16],

p. 83). There are additional measures, such as aversive conditioning and strategic area management (discussed below) that might assist in this regard, as will innovative approaches to non-lethal management and visitor education. Encouragingly, only one individual was destroyed in the last 12 months,[45] which is considerably lower than the established average. This is the best way to ensure that lethal control will have as little impact on the population as possible.

Future management strategies for minimising human–dingo interactions

The problem and possible causes of negative dingo–human interactions have been discussed, as well as the current strategies employed by the managing body. There are, however, many other techniques and methods that could be used to better manage and conserve populations of dingoes in areas of high human traffic.

Aversive conditioning

A good deal of the relevant content in relation to aversive conditioning was covered in Chapter 5 on livestock conflict, but there are additional points more specific to the situation on Fraser Island (and the broader context of human–dingo interaction conflict) worth covering here. The original management strategy for Fraser Island, implemented in 2001 after the Clinton Gage incident,[8] recommended the need to reduce and even reverse dingo habituation, which was and perhaps still is to some degree seen as one of the major constituents of conflict. Previous concerns regarding habituation aside, such rhetoric reflects an interest in modifying dingo behaviour that is of concern to managers. A result was the implementation of hazing, which is a fairly loose behavioural modification concept in which aversive stimuli are directed at dingoes in an effort to discourage them from visiting or loitering within human-use areas. This most often involved rangers using slingshots to shoot clay balls at dingoes. Little in the way of detailed records was kept in relation to hazing practices and outcomes; consequently, evaluating the success or failure of the approach is problematic. Both the original Fraser Island Dingo Management Strategy[8] and the most recent review[16] discuss hazing interchangeably with aversive conditioning. Although there is undoubtedly conceptual cross-over, there are also considerable differences between the terms. Shivik *et al.*[46] stressed that an understanding of the basic tenets of animal behaviour are necessary to avoid the misuse of non-lethal methods (see also[47]) and a focus on animal learning theory can help distinguish flawed from fruitful methodology.

For example, a major limitation suggested by rangers' observations when conducting hazing was that instead of learning a desired association between aversive events and particular locations (place avoidance), dingoes instead learned an association between aversive events and the presence of rangers (e.g. based on vehicle sound or uniform). This is perhaps easy to understand. Place avoidance might be particularly difficult to achieve in such circumstances, where the most salient stimuli during aversive events (the presence of rangers) overshadows the relevance of the location. In addition, such places have most likely become associated with something positive (e.g. the prospect of food) and if aversive events can be avoided by simply avoiding rangers, then this association and its effects will be retained. At any rate, dingoes continued to visit campgrounds but avoided rangers, possibly leading some rangers to think that hazing might be working.

Hazing is no longer practised on the island. As flawed as it might have been, with nothing to replace it, there are no intermediate steps for dealing with problem dingo behaviour. In an effort to explore an alternative approach, much more in line with conventional aversive conditioning theory, Appleby (unpublished data) conducted a series of trials of aversive (shock) collars on dingoes that had been previously identified as exhibiting behaviour of major concern to managers. During four trials, each involving a different individual, some useful preliminary results were garnered. Two dingoes ceased their problem behaviour immediately after receiving an aversive stimulus (shock). After exposure to two shocks on subsequent days, a third dingo became hesitant to approach a backpack or a vehicle that it had previously approached without hesitation. This aversion was apparent even when food was presented as an incentive to approach. The fourth dingo consistently fled after receiving a shock no matter what target behaviour was involved, including, in one case, close approach to an unattended child. No aggression was directed at humans during or immediately following any aversive stimuli, with dingo subjects tending to move rapidly away from human observers.

Aversive conditioning need not be limited to shock collars. Another possibility worth consideration is the use of something akin to inert bear spray (essentially, pressurised cans of water) carried by visitors and deployed upon close approaches by dingoes. The use of conventional bear spray in North America has been instrumental in reducing negative outcomes that might otherwise have been mitigated with lethal force.[48] As is the case in North American studies, the approach has a major advantage in that it puts a behavioural modification tool in the hands of visitors, dramatically widening the prospect of consistent stimulus presentation to individual dingoes displaying target behaviour. Such a spray would be difficult to misuse against dingoes as the active constituent usually associated with bear spray is absent, and the effective range of the spray would be limited. As with shock collars, the goal is to find a stimulus that is aversive enough to halt the unwanted behaviour but not so aversive as to have long-lasting or detrimental effects. Stimuli that are not sufficiently aversive are inherently incapable of modifying behaviour in any beneficial way, as Edgar, Appleby and Jones[49] found. Use of the inert spray cans could be monitored: cans could be distributed by QPWS through the various centres that service the island or through the visitor bottlenecks that occur when travelling to the island via barges and ferries. Returns policies upon completion of visits could be undertaken using similar means, such as drop-off bins along with incident report forms, allowing usage to be estimated. A pilot study could directly monitor usage and explore efficacy.

Where individual dingoes are deemed too great a risk to human safety to remain free during aversive conditioning, an alternative is to carry out the procedure in an enclosure. For example, in trials in the US, nicknamed 'bear school', brown bears were captured, temporarily contained within an enclosure and aversively conditioned for four to six days.[50] Although expensive, such an approach has several significant advantages. It removes possible risk to visitors during the conditioning process, and provides the opportunity to test release protocols or decide what to do if release is not deemed possible.

However, support for aversive conditioning is not unequivocal, with certain detractors labelling it 'ethically questionable'.[51] There is indeed an ethical quandary in relation to the use of aversive conditioning, in that animals have to be subjected to an unpleasant experience. Yet in the absence of such approaches, animals involved in conflict with humans, particularly where significant risk to human safety is a concern, are almost inevitably going to be destroyed. In this sense, aversive conditioning, can be thought of as quite literally life-saving for the animals involved. This perhaps parallels life-saving medical or veterinary procedures that involve unavoidable pain or discomfort.

The real dilemma is not whether aversive conditioning is ethically questionable, as Gunn[51] originally framed it, but whether aversive conditioning is more ethically questionable than lethal control. By no means should aversive conditioning be considered a panacea, and its use might be restricted to very specific situations, but neither should it be disregarded as an alternative to lethal control. Modern tools used in aversive conditioning such as shock collars allow for precise and controllable exposure to the intensity and duration of aversive stimuli. In the course of a conditioning experiment, it is likely that any animals involved would be exposed to such stimulation for only seconds at a time and on a limited number of critical occasions (e.g. during the exhibition of target behaviour). Without aversive conditioning, managers, stakeholders and interest groups must resign themselves to the fact that lethal control will often be the only remaining option. Aversive conditioning represents a potentially viable alternative. If successful, it could allay human safety concerns while allowing dingoes to remain alive and play their part in the long-term survival of this most important population. The ethical quandary posed by aversive conditioning versus lethal control urgently requires resolution.

Strategic area management

Perhaps as another parallel to livestock conflict mitigation, strategically managing the usage of certain areas on Fraser Island might lead to a reduction in incidents. QPWS already closes campgrounds where chronic problem behaviour by dingoes (or presumably humans) is observed. With reference to some of the key areas where incidents occurred,[16] there might be an opportunity to explore other types of site use management. For example, the beach area in front of the Eurong and Eurong Second Valley townships is popular with visitors, but also with fishers. If the prospect of a meal of fish scraps is attracting dingoes to the area, then consideration should be given to closing this and other popular beaches to fishing. In addition, well-signed, vandal-proof surveillance cameras could be trialled at key locations (e.g. Eurong beachfront) in an effort to discourage inappropriate behaviour such as feeding dingoes, and encourage proper behaviour, as well as allow review of any incidents that might occur.

Young dingoes may be particularly susceptible to learning strong associations, both negative and positive, and associations with humans and anthropogenic stimuli are likely to be no different. Limiting exposure to such stimuli may have merit and could be experimentally tested using standard control and treatment comparisons. In management terms this might include closing campgrounds, or restricting human use of campgrounds and indeed some beach areas in locations close to dens or in areas where very young dingoes have been observed. At early stages of rearing, it would be expected that the movement of young animals would be restricted somewhat, meaning that there would be a need for relatively small areas of human exclusion and possibly only for short periods. However, it is noteworthy that dingo families are capable of moving reasonable distances on occasion, possibly in relation to food availability (both natural and anthropogenic) or other factors (Appleby unpublished data).

Public education and participation

Although contentious in some respects, strategies to inform visitors and others about dingoes, their behaviour and the risks they may pose are already comprehensive. An idea worth consideration is establishing what could be called 'Dingo Week', to intensively promote education on human safety and responsible behaviour around dingoes, dingo ecology and behaviour more generally, and the research being done in this regard. To increase impact, such an event would be best timed to coincide with the Easter holiday period when management concerns are usually high. Various activities including public presentations by rangers, researchers and traditional owners, speciality brochures and competitions and giveaways could be developed in an effort to target key demographics, such as families with small children or international visitors. One idea would be a public photographic competition to encourage submissions of photographs of dingoes, following recommended guidelines (e.g. impromptu photographs taken at a safe distance without encouraging dingoes to approach). Submissions could be useful for population and general management but, importantly, could also provide an opportunity to foster and reinforce appropriate behaviour when viewing and photographing dingoes.

Conclusion

Fraser Island represents one of the last and certainly best places to conserve dingoes in Australia, at least in a form similar to what would have existed before European settlement. At the same time, the death of Clinton Gage underlined the risk that dingoes pose to human safety in some circumstances. This poses a conservation conundrum, which Woodroffe *et al.*[52] phrased in the form of a challenging question: what to do when a species vulnerable to extinction causes loss of human life or affects human livelihoods? Given the importance of the Fraser Island dingo population, this is a particularly pressing question. From the information presented in this chapter, it is clear that a great deal of effort and expenditure has gone into attempting to resolve human–dingo conflict on Fraser Island. Despite this, conflict continues and lethal control of dingoes remains a management mainstay. There are, however, promising signs that reliance on lethal control might be lessening. While some have concluded that the rate of lethal control is unlikely to affect the population in any lasting detrimental way, it seems at best premature to argue this given that the population is closed, dingo abundance is relatively limited and there is virtually no information on genetic relatedness and inbreeding. Neither, though, can attacks by dingoes be underplayed, occasional as they might be. Non-lethal approaches such as aversive conditioning, coupled with strategic area management, might aid in striking the balance between human safety and conservation of dingoes. Thompson *et al.*[7] put it well when they said that resolution would come not from removing dingoes but from a partnership with all stakeholders taking collective responsibility for actions that lead to negative interactions. Put another way, the onus, and indeed the responsibility, is and should always be on the human component of human–dingo interactions.

In Chapter 5, we saw that conflict between dingoes and livestock producers is complex, persistent and often enigmatic. Even after more than 200 years, neither producers nor dingoes have yielded. Conflict directly involving people, such as in the form of attacks, is a relatively recent phenomenon yet we've seen that it is no less complex or enigmatic. Both types of conflict pose considerable challenges to the emerging paradigm of dingo conservation, such that any progress in relation to conservation can only occur hand-in-hand with progress towards resolving conflict.

References

1. Staines D, Arrow M, Biber K (Eds) (2009) *The Chamberlain Case: Nation, Law, Memory.* Australian Scholarly Publishing, Melbourne.
2. Crispin K (2012) *The Chamberlain Case: The Legal Saga that Transfixed the Nation.* Scribe Publications.
3. Biber K (2009) Judicial extracts. In *The Chamberlain Case: Nation, Law, Memory.* (Eds D Staines, M Arrow and K Biber) pp. 113–151. *Australian Scholarly Publishing*, Melbourne.
4. Morris ES (2012) *Inquest into the Death of Azaria Chantel Loren Chamberlain.*
5. Morgan E (2013) *Family of Boy Killed by Bear Supported by Utah Supreme Court Ruling.* KSL.com.
6. Hytten KF, Burns GL (2007) Deconstructing dingo management on Fraser Island, Queensland: the significance of social constructionism for effective wildlife management. *Australasian Journal of Environmental Management* **14**, 48–62.
7. Thompson J, Shirreffs L, McPhail I (2003) Dingoes on Fraser Island: tourism dream or management nightmare? *Human Dimensions of Wildlife* **8**, 37–47.
8. EPA/QPWS (2001) *Fraser Island Dingo Management Strategy.* Environmental Protection Agency/Queensland Parks and Wildlife Service.
9. Appleby RG, Jones D (2011) *Analysis of a Preliminary Dingo Capture-mark-recapture Experiment on Fraser Island Conducted by the Queensland Parks and Wildlife Service.* Environmental Futures Centre, Griffith University, Brisbane.
10. Burns GL, Howard P (2003) When wildlife tourism goes wrong: a case study of stakeholder and management issues regarding dingoes on Fraser Island, Australia. *Tourism Management* **24**, 699–712.
11. Angel DC (2006) *Dingo diet and prey availability on Fraser Island.* University of the Sunshine Coast.
12. Baxter G, Davies N (2013) *Tracking Dingoes on Fraser Island.* University of Queensland Press, Brisbane.
13. QPS (2002) *Death – Clinton Troy Gage – Dingo Attack Fraser Island: Police Report.* Queensland Police Service, Maryborough District Criminal Investigations Branch, Hervey Bay Office, Queensland.
14. Herrero S (2002) *Bear Attacks: Their Causes and Avoidance.* Lyons Press, Guildford, Connecticut.
15. Appleby RG, Burns GL, McBroom J, Jones D (2004) *The Fraser Island Dingo Incident Database: A Preliminary Examination.* Griffith University, Brisbane.
16. Ecosure (2012) Fraser Island dingo management strategy review. Report to Department of Environment and Heritage Protection, Brisbane.
17. Corbett LK (1998) *Management of Dingoes on Fraser Island.* ERA Environmental Services.

18. Moyer KE (1968) Kinds of aggression and their physiological basis. *Communications in Behavioral Biology. Part A:* **2**, 65–87.
19. Popova NK, Nikulina EM, Kulikov AV (1993) Genetic analysis of different kinds of aggressive behavior. *Behavior Genetics* **23**, 491–497.
20. Moyer KE (1973) The physiological inhibition of hostile behaviour. In *The Control of Aggression: Implications from Basic Research.* (Ed. JF Knutson) pp. 9–38. Aldine, Chicago.
21. Corbett LK (2001) *The Dingo in Australian and Asia.* JB Books, Adelaide.
22. Masterson S (1994) *Dingo Monitoring Report.* Queensland Parks and Wildlife Service, Central Station, Fraser Island.
23. Dept of Environmental and Heritage Protection (2013) *Fraser Island Dingo Conservation and Risk Management Strategy.* Ecosystem Services, Department of Environmental and Heritage Protection, Brisbane.
24. Whittaker D, Knight RL (1998) Understanding wildlife responses to humans. *Wildlife Society Bulletin* **26**, 312–317.
25. Eibl-Eibesfeldt I (1985) *Ethology: The Biology of Behaviour.* Holt, Rinehart and Winston, New York.
26. Breckwoldt R (1988) *A Very Elegant Animal: The Dingo.* Angus and Robertson, Sydney.
27. Thomson PC (1992) The behavioural ecology of dingoes in north-western Australia. II. Activity patterns, breeding season and pup rearing. *Wildlife Research* **19**, 519–530.
28. Wilson DS, Clark AB, Coleman K, Dearstyne T (1994) Shyness and boldness in humans and other animals. *Trends in Ecology and Evolution* **9**, 442–446.
29. Réale D, Gallant BY, Leblanc M, Festa-Blanchet M (2000) Consistency of temperament in bighorn ewes and correlates with behaviour and life history. *Animal Behaviour* **60**, 589–597.
30. Stuber EF, Araya-Ajoy YG, Mathot KJ, Mutzel A, Nicolaus M, Wijmenga JJ, Mueller JC, Dingemanse NJ (2013) Slow explorers take less risk: a problem of sampling bias in ecological studies. *Behavioral Ecology* **24**, 1092–1098.
31. Sih A, Bell A, Johnson JC (2004) Behavioral syndromes: an ecological and evolutionary overview. *Trends in Ecology and Evolution* **19**, 372–378.
32. Hopkins JB III, Herrero S, Shideler RT, Gunther KA, Schwartz CC, Kalinowski ST (2010) A proposed lexicon of terms and concepts for human–bear management in North America. *Ursus* **21**, 154–168.
33. Herrero S, Smith TS, DeBruyn TD, Gunther K, Matt CA (2005) From the field: brown bear habituation to people – safety, risks, and benefits. *Wildlife Society Bulletin* **33**, 362–373.
34. Martin P, Bateson P (1986) *Measuring Behaviour.* Cambridge University Press, Cambridge.
35. Panksepp J, Siviy S, Normansell L (1984) The psychobiology of play: theoretical and methodological perspectives. *Neuroscience and Biobehavioral Reviews* **8**, 465–492.
36. Pellis SM, Pellis VC (1996) On knowing it's only play: the role of play signals in play fighting. *Aggression and Violent Behavior* **1**, 249–268.
37. Bekoff M (1995) Play signals as punctuation: the structure of social play in canids. *Behaviour* **132**, 419–429.
38. Bekoff M, Allen C (1997) Intentional communication and social play: how and why animals negotiate and agree to play. In *Animal Play: Evolutionary, Comparative and Ecological Perspectives.* (Eds M Bekoff and JA Byers). Cambridge University Press, Cambridge.
39. Dugatkin LA, Bekoff M (2003) Play and the evolution of fairness: a game theory model. *Behavioural Processes* **60**, 209–214.
40. Bekoff M (1974) Social play and play-soliciting by infant canids. *American Zoologist* **14**, 323–340.
41. Beckmann EA, Savage G (2003) *Evaluation of Dingo Education Strategy and Programs for Fraser Island and Literature Review: Communicating to the Public about Potentially Dangerous Wildlife in Natural Settings.* Environmental Protection Agency.

42. Moussalli A (1994) A preliminary field study of dingoes inhabiting the Waddy Point region, north-east Fraser Island: diet, social organisation and behaviour. BEnvSc thesis. Griffith University, Brisbane.
43. Ballard G, Fleming PJS (2012) Non-violence has its place, but let's give the dingoes due credit. *The Conversation*. http://theconversation.com/non-violence-has-its-place-but-lets-give-dingoes-due-credit-8635.
44. Frankham R (1998) Inbreeding and extinction: island populations. *Conservation Biology* **12**, 665–675.
45. Belcher R (2014) Pers. comm.
46. Shivik JA, Treves A, Callahan P (2003) Non-lethal techniques for managing predation: primary and secondary repellents. *Conservation Biology* **17**, 1531–1537.
47. Caro T, Durant S (1995) The importance of behavioral ecology for conservation biology: examples from Serengeti carnivores. In *Serengeti. II. Dynamics, Management, and Conservation of an Ecosystem*. (Eds ARE Sinclair and P Arcese) pp. 451–472. University of Chicago Press, Chicago.
48. Smith TS, Herrero S, Debruyn TD, Wilder JM (2008) Efficacy of bear deterrent spray in Alaska. *Journal of Wildlife Management* **72**, 640–645.
49. Edgar JP, Appleby RG, Jones DN (2007) Efficacy of an ultrasonic device as a deterrent to dingoes (*Canis lupus dingo*): a preliminary investigation. *Journal of Ethology* **25**, 209–213.
50. Jonkel C (1994) Grizzly/brown bears. In *The Handbook: Prevention and Control of Wildlife Damage* (Eds SE Hygnstrom, RM Timm and GE Larson) pp. C17–C23. Cooperative Extension Service, University of Nebraska, Lincoln.
51. Gunn I (2011) Death of the Fraser Island dingo. *The Conversation*. http://theconversation.com/death-of-the-fraser-island-dingo-793.
52. Woodroffe R, Thirgood S, Rabinowitz A (2005) The impact of human–wildlife conflict on natural systems. In *People and Wildlife: Conflict or Co-existence?* (Eds R Woodroffe, S Thirgood and A Rabinowitz). Cambridge University Press, Melbourne.

7

Chasing the yellow dog's tale: the science of studying dingoes

Damian Morrant

Still ran Dingo – Yellow-Dog Dingo – very much bewildered, very much hungry, and wondering what in the world or out of it made Old Man Kangaroo hop.

Much like Yellow-Dog Dingo in Rudyard Kipling's classic children's story,[1] scientists often wonder what makes wild animals do the things they do. Information about dingoes is sought for a broad variety of reasons, ranging from determining effective methods of eradicating them because of their impacts on the livestock industry, through to developing ways in which they can be conserved and protected for their role as a biodiversity regulator. Dingoes have been the subject of considerable scientific interest and research over the past half-century; despite this work we still have a lot to learn about this iconic predator.

Some people believe that dingoes pose a major threat to human safety, livelihoods and biodiversity conservation. Of particular concern is their predation on domestic animals and native fauna, and their potential to transmit disease (e.g. hydatid disease). Dingoes are often considered a native animal since they have existed in Australia for thousands of years, yet others see them as an introduced pest that does not belong in Australian ecosystems. The perception that dingoes do not belong can make their deleterious impacts seem more serious. For these reasons, much of the early scientific research into dingo biology focused on understanding their behaviour so that they could be controlled more effectively using lethal methods.

Dingoes, however, play an important role as the apex predator in many Australian ecosystems, and their interactions with other organisms have far-reaching implications for ecosystem health. For example, dingoes may regulate the distribution and activity of smaller non-native predators such as cats and foxes, play an important role in controlling populations of invasive species, and regulate populations of their prey species. Recent improvements in our understanding of the dingo's role in Australian ecosystems have shifted much of the research focus towards investigating the benefits that may stem from their ecological role.

Conflicting perceptions of dingoes have led to considerable debate among the general public and scientists about the costs and benefits of dingoes in the Australian landscape.[2–6] These divergent views indicate an urgent need for scientific research to improve our understanding of the interactions between dingoes, humans, domestic animals and native fauna. Armed with the results of such research, policy-makers and land managers will be able to make management decisions based on scientific knowledge rather than on anecdote and superstition. Many authors have already studied aspects of dingo biology, and I will focus on some of these studies in this chapter.

Examples of dingo research in action

Alan Newsome, in association with Laurie Corbett and other researchers, conducted much of the early scientific research on dingoes for the CSIRO.[7–27] They investigated many aspects of dingo biology including social behaviour, prey selection and movement patterns. Much of their work is summarised in Corbett's seminal text *The Dingo in Australia and Asia*.[22] The conclusions from such early dingo research provided valuable insights into dingo ecology and behaviour that have informed and guided much of the work that has followed.

Dingo researcher Peter Thomson conducted a nine-year investigation into the ecology, behaviour and social organisation of dingoes on the lower Fortescue River in Western Australia.[28–36] Thomson sought to provide information which would guide dingo management decisions in a region where rugged undeveloped land adjoined sheep-grazing areas. Much of the focus of Thomson's research was therefore on gaining insights that would allow effective dingo control strategies. Data were collected using several methods including analysis of faeces and stomach contents, visual observation, radio-tracking, and measurements taken from trapped animals.

Ecologists Chris Johnson (University of Tasmania) and Euan Ritchie (Deakin University) investigated the effects of dingo control on herbivores and feral cats at two sites within several different Australian environments. Each site included an area where dingoes were controlled and one where they were not. The pioneering work showed that dingoes play an important role in helping to maintain biodiversity in Australian ecosystems. Johnson and Ritchie teamed up with Mike Letnic (University of New South Wales), Arian Wallach (James Cook University) and Adam O'Neill (Evelyn Downs Station), who conducted related dingo research, in a successful bid for the prestigious 2013 Australian Museum NSW Office of Environment and Heritage Eureka Prize for Environmental Research.

My own dingo research, as part of my PhD through James Cook University and in association with CSIRO and the Centre for Tropical Environmental and Sustainability Science, has investigated the potential impacts of dingoes on the fauna of the coastal lowlands of Queensland's Wet Tropics. My work is supported by advice from experts in social science and ecology (Brad Congdon from James Cook University, Chris Johnson, and James Butler from CSIRO). I have used methods such as GPS tracking, analysis of dingo DNA, determination of diet from faeces and stomach contents, stable isotope analysis of hair and vibrissae (whiskers), and surveys of local residents. Throughout my research I have worked closely with another PhD student, Felicity Smout, a veterinarian who is investigating the disease ecology of wild dogs in the Wet Tropics. Her work has revealed much about the diseases and parasites of dingoes and other dogs in the region.

The studies mentioned above are only a few examples of research involving dingoes. This chapter provides an overview of some of the ways in which scientists collect information about wild dingoes. It will not focus on knowledge acquired through captive studies, anecdotal evidence or the traditional knowledge of Australian Aboriginal people, nor will it detail work which has involved reviews of dingo-related literature, critiques of research methods or modelling using pre-existing data. While such resources have provided valuable insights into many facets of the lives of dingoes and have informed and inspired field researchers, they are beyond the scope of this chapter.

This is by no means an exhaustive review or a how-to manual. The chapter deliberately avoids detailed descriptions of some methods because, if poorly applied by untrained users, they have great potential to cause unnecessary (and unacceptable) suffering in animals. Readers who wish to learn more about the methods discussed in this chapter should consult the scientific literature, and/or speak with experts who have experience in their application. The chapter outlines some of the common methods used by scientists to learn about wild dingoes, some of the insights into dingo biology that have been gained through applying such methods, and the potential of some new and emerging methods to improve our knowledge of the ways of the elusive dingo.

The challenges of studying dingoes

Scientists seek to collect information about dingoes to improve our understanding of these iconic predators and to provide management advice to a broad range of clients, including government agencies, land managers, conservationists and farmers. However, the secretive nature and keen senses of dingoes make it very difficult to gather such information.

Aborigines were the first Australian people to observe that dingoes were elusive, and many of the names they gave to dingoes reflect their cryptic behaviour. For example, people from the Anbarra community in Arnhem Land may refer to the dingo as an-gugurkuja, which is derived from the masculine-noun class 'an' and the verb stem '-gugurkuja' (fearful one), alluding to the propensity of wild dingoes to run away from humans.[37] Dingoes have been a part of the daily life and mythology of many Aboriginal people for thousands of years, and their interactions led to a broad traditional understanding of dingo ecology.[37, 38] As a top predator, dingoes played an important role in regulating the ecosystems on which Aboriginal people relied, and were direct competitors for food resources.[37]

European colonists in Australia quickly learned that dingoes are experts at keeping their activities secret. Dingoes have been described with words like 'cunning', 'sneaking', 'slinking' and 'skulking',[39] and the Oxford English Dictionary[40] defines the verb *dingo* as 'to retreat, back out, act in a cowardly or treacherous manner'. The famous historian C.E.W. Bean described the dingo as 'flashing low like a streak of lightning through the grass' (Bean 1963, as cited in[39]).

Dingoes are generally most active at dawn, dusk and night, and they roam over large areas that often encompass rugged and heavily vegetated terrain. Like all *Canis* species, dingoes have highly developed senses of smell and hearing; a dingo will often detect human presence long before the human is lucky enough to catch a glimpse of it. In areas where humans persecute them, dingoes are believed to modify their behaviour, making observation of their activity even more difficult.[22, 41–44]

Of all of its senses, the dingo's highly tuned olfactory abilities pose particular problems for researchers. In general the canid sense of smell is considerably more sensitive than that of humans, and dingoes may be thousands, even millions, of times better at detecting some odours.[45] Unfortunately for researchers, many of the smells to which dingoes and other dogs are particularly sensitive are scents that are found in human body odour. For example, dogs can detect a light human fingerprint on a glass slide left outdoors for up to two weeks, and for up to six weeks if the slide is kept indoors.[46]

Ethics, legislation and dingo research

Any time that scientists interact with animals there is a risk that the animals will be harmed or experience stress, and undoubtedly the most important aspect of any type of animal research is developing ways of minimising negative effects. There are a multitude of ways in which scientists can minimise harm to the study animals, many of which have been reviewed by other authors.[47, 48] Some of the ways in which scientists approach such ethical problems will be discussed in this chapter.

At the most basic level scientists impose ethical requirements upon themselves. People who work with wildlife usually do so because they have a passion for learning about the natural world and for protecting and conserving animals. Thus, most scientists genuinely care about the well-being of their study animals and have a strong desire to treat them humanely. In addition to personal ethics, scientists who work with animals in Australia are subject to legal obligations at state, territory and federal levels.

Ethical treatment of animals in scientific research is overseen by the Australian Code for the Care and Use of Animals for Scientific Purposes 2013 (the Code), which aims to 'promote the ethical, humane and responsible care and use of animals for scientific purposes'.[48] In conjunction with the Code, research must comply with legislation pertaining to the duty of care on people in charge of animals. For example, in Queensland the duty of care is overseen by the *Animal Care and Protection Act 2001* (Qld). The Code requires that researchers obtain approval to conduct work with animals from an Animal Ethics Committee (AEC), of which there are several in Australia.

An AEC must consist of at least one qualified veterinarian, an animal researcher, an animal welfare representative and an independent person who is not associated with an establishment or involved in animal research. This mix ensures that the views of society will be broadly represented in the decision-making process. The AEC evaluates requests based on detailed information provided by the researcher, such as the justification for the work, the effect it will have on animals, why the research must involve live animals, and how the researchers will manage pain and distress in the study animals. If the AEC is satisfied that the research is ethical and the use of animals is justified, then approval will be granted. The AEC requires researchers to submit several reports on the progress of the research. In particular, annual reports and final reports must be submitted to the AEC and it must be notified promptly of any unexpected adverse events. Approval can be suspended or withdrawn at any time during the research if the AEC feels that a researcher is not meeting its requirements or if the study animals are unreasonably stressed.

Dingo scientists working within an organisation such as a government department or university are bound by additional organisational requirements, which they must satisfy before they are allowed to go anywhere near a wild dingo. Legal approval to work with wildlife must also be obtained. Applications are made to National Parks agencies outlining the way in which the work will be conducted, including how many animals will be involved, the methods that will be used and steps that will be taken to reduce the risk to the study animal, other wildlife and the public. Approvals by these agencies are generally subject to the methods having been approved by an AEC. If the animal is legally considered to be a pest and will be captured and released, researchers need additional permission to release it back into the wild.

As with any type of science, dingo researchers follow a strict procedure in order to answer questions about dingo biology. This procedure is broadly known as the scientific method, and involves a sequence of stages whereby the researcher builds upon the knowledge gained by past research to create new knowledge. A wide range of methods of collecting data can be employed by dingo researchers, broadly divided into non-invasive and invasive methods. These will be discussed in turn.

Non-invasive methods

Methods which involve little or no interaction with animals are known as non-invasive methods. For a method to be considered non-invasive, it must not cause animals to exhibit a severe stress response, and their survival or reproduction should be unaffected.[49] Non-invasive methods of collecting data can be passive, whereby a scientist collects information from samples such as faeces. Alternatively they may be active, whereby the animal interacts with some sort of sampling device that collects samples, for example saliva or hair. Non-invasive sampling devices can be placed where they will be encountered by travelling animals or baited with a motivating lure, such as odours or food substances which attract animals and encourage them to interact with the device.[50]

Social science

Social scientists can use methods such as questionnaires and interviews to collect information about people's knowledge of the biology of wild animals or the animals' relationship with people. Such knowledge can provide a great deal of insight into the biology of dingoes and the requirements for effective dingo management. Few studies have used social science as a tool to increase our understanding of dingoes. Published work has focused on dingoes' impacts on the livelihoods of pastoralists, the threat they pose to domestic animals, and perceptions of their pest status.[51–54]

Butler and colleagues[51] mailed questionnaires to 3000 landholders in the Wet Tropics of Queensland to investigate the incidence of 'wild dog' attacks on domestic dogs. They also interviewed cattle farmers and pest management officers. Twelve respondents (6%) to the mail survey reported incidents in which their dogs had been attacked or killed, and another 12 incidents were reported by farmers and pest managers. Studies which investigated dingo/wild dog perceptions among pastoralists[52, 53] and randomly selected adults in Victoria[54] reported that dingoes/wild dogs were a serious problem, and there was strong support for lethal control.

While these studies generally paint the dingo in a negative light it is important to note that the choice of terminology used by social scientists, and the demographic groups they survey, can affect the type of responses.[55] For example, the two above-mentioned studies surveyed a societal group which is most likely to come into conflict with dingoes.[52, 53] Both Johnston and Marks[54] and Butler *et al.*[51] used the term 'wild dog' in their questionnaires, which is likely to have had more negative connotations than 'dingo'.

Direct observation of behaviour

Perhaps the simplest way of recording information about dingoes is through direct observation of their behaviour. Researchers can observe dingoes with the naked eye, from hides or aircraft, or through binoculars or infrared scopes.[22, 56] Observers may stake out areas that are regularly used by dingoes, such as dens, watering points, track junctions and river crossings, and wait for a dingo to arrive, or may actively lure them to an area using carrion or other odour attractants or by mimicking their howls. Corbett, one of the foremost dingo biologists and author of numerous papers and books on dingoes, noted the effect that lures may have on dingo behaviour must be considered by researchers.[22]

Aircraft have been used to observe dingoes and their prey, especially large animals such as kangaroos, (*Macropus* spp.) and emus (*Dromaius novaehollandiae*).[8, 57] Spotlighting can be an effective way of determining information about some animals, but the secretive behaviour of dingoes generally makes it an ineffective survey method.[58, 59] Most field scientists carry a camera with which to make a permanent record of their observations. Recent advances in technology and the affordability of camera traps have enabled scientists to take photographs of wild animals without being physically present.[60]

Camera traps

Camera traps take photographs or video footage of animals when they contact some kind of trigger or pass in front of motion sensors. Camera traps can be used for myriad purposes, such as studies of animal behaviour, their activity patterns, and their distribution in time and space.[61] When used in conjunction with other sampling methods such as hair traps or GPS tracking studies, they can reveal a great deal of information about animal biology which could not be gained in other ways. There are hundreds of published studies which have used camera traps, either as a primary source of data collection or to supplement other methods[61, 62] and they are increasingly being used by scientists who study dingoes.[63–67]

Camera traps can be used to monitor animal movements and activity so that study sites can be chosen or to determine areas that should not be used. For example, during preparation for dingo trapping for a tracking study on the fringes of urban areas, I used camera traps to monitor potential trap sites. I sought to identify not only areas used by dingoes, but also areas that were used by humans, domestic animals or threatened species. In this manner I was able to choose trap sites which had the best chance of catching dingoes while minimising the chances of catching something else. Trapping an endangered animal would have brought a swift end to my research plans, and would have been an extremely unpleasant experience for all involved! By deploying odour attractants and hair traps in front of the camera traps I was able to investigate behavioural responses to different attractants while simultaneously collecting hair for DNA and stable isotope analysis. The camera trap footage also allowed description of the general morphology of the animals from which I collected samples (D. Morrant, unpublished data).

In an experimental study investigating the interactions between radio-collared dingoes, cats and foxes within a 37 km^2 fenced enclosure, Moseby *et al.*[66] placed two camera traps at a site where animal carcasses were left. The footage from these cameras enabled them to monitor the study animals and determine whether any cats or foxes that were not part of the study had gained access to the enclosure during the experiment. Camera traps have also been used to monitor freshwater crocodile (*Crocodylus johnstoni*) nests in the northern area of Lake Argyle, Western Australia, revealing that in this area dingoes are the major nest robbers,[64] and to investigate poison bait uptake by dingoes/wild dogs and other animals.[63, 68] Brook and her team[65] used camera traps to compare the activity of dingoes and feral cats on nine pairs of large rangeland properties in northern and central Australia, where one property in each pair was subject to lethal control of dingoes and one was not. They found that dingo control leads to a higher abundance of feral cats, and that the cats on non-baited properties adjusted their activity time to correspond with times when dingoes were least active.

These studies are but a handful of those that have utilised camera traps to aid in dingo research. With improvements in technology, such as camera traps which can send images via mobile or satellite networks, their importance in studies of dingo ecology will only increase.

Spoor

Spoor is a general term used to describe any sign left behind by an animal that can provide clues about behaviour and activity. Dingoes mark boundaries or places of interest within their territories using scratch marks in soil, and faeces, urine and other secretions such as those from scent glands.[22, 56] The footprints of travelling dingoes can indicate densities and patterns of land use, and the depressions where they have slept can provide information about their resting sites. Signs such as these are used to collect information about dingo distribution, abundance and activity, and in some cases as a source of biological samples.

Scats

Dingo faeces can provide a wealth of information about their ecology. Scientists refer to faeces as 'scats' possibly because it makes them feel better about roaming the countryside collecting what others might call poo, dung or other names that can't be mentioned here. Dingo scats are most easily found on roads and in areas where dingoes were known to have deposited them in the past. They are often left where they are likely to be seen by other dingoes or where the smell may be carried on the wind, such as on top of rocks, embankments, logs or clumps of grass. In some circumstances the spatial distribution of scats can help to establish patterns of habitat use by dingoes[22] although in areas where some habitats contain dense vegetation the discovery of scats may be biased towards open areas. This bias may also have implications for the results of dietary analysis because scats deposited by animals living in open areas may be better represented than those from animals residing in areas with dense vegetation.

When I conducted dingo research in the Wet Tropics, for example, I generally found scats only along vehicle tracks and in open grassland. During searches for scats I always paid particular attention to sites where I had collected scats on previous occasions, and usually found more on subsequent visits. On two occasions dingoes deposited their scats on the tracks my vehicle had left on gravel roads where it had been turned around. This may have been a way for the dingoes to communicate to other dingoes that a human had spent time at the location and that it was unsafe.

Dingo scats have been collected and analysed to glean information about dingo biology, particularly in relation to diet and more recently for information about the genetic status of dingo populations. Many studies have used the information provided by dingo scats to make inferences about their diet and the effects of their predation on native and feral fauna.[9, 20, 22, 25, 32, 41, 69–74] Mammal hairs, bone, feathers, seeds and insect exoskeletons can all be identified to provide information about the types of animals on which dingoes feed, even down to species level in some instances. Analysis of scat contents has shown that dingoes primarily prey on small to medium-sized mammals up to ~15 kg, which is the mass of the average adult dingo,[22] that they may switch primary prey species in response to changes in prey abundance and that medium-sized mammals such as rabbits (*Oryctolagus cuniculus*) and bandicoots (*Perameles* and *Isoodon* spp.) and relatively-larger macropods such as swamp wallabies (*Wallabia bicolor*), agile wallabies (*Macropus agilis*) and kangaroos (*Macropus* spp.) are often important prey species.

Tracks

Some dingo scientists have used dingo footprints (tracks) to gain valuable insights into dingo ecology. The size of dingo tracks can reveal information about the animals in the area. For example, two sets of tracks of slightly different size may indicate a male and female moving through the area, or two animals of the same sex but of different sizes. A set of large tracks with several smaller tracks probably suggests a female with pups. Although this information is useful, the inferences that can be made from it are limited and generally serve only as a basis for further investigation.

However, a type of dingo track that has proven useful to scientists is the track on a soil/sand plot. Scientists sometimes clear areas around lure stations or at regular intervals along dirt tracks. Sometimes an additional layer of soil or sand is spread across the area to make visualisation of the tracks easier. Any dingo crossing these cleared areas will leave footprints.

While luring dingoes to scent stations can provide useful information, attracting animals to the plot increases the risk of bias.[75] Allen *et al.*[43] introduced an activity index for assessing dingo populations using the mean number of dingo tracks recorded on tracking plots placed at intervals along roads to determine the relative abundance of dingo populations. No lures are used in this method, so the animals crossing the plots represent a sample that has not been artificially drawn to the area. This method has been improved on using various statistical methods and has been applied to the study of several species.[58, 75–77] For example, Edwards *et al.*[78] used a variation to investigate habitat use by dingoes and feral cats in Central Australia. They observed that dingoes showed no obvious habitat preference whereas cats preferred mulga woodland, which may have offered protection from attack from dingoes.

Behavioural response to odours

Little is known about the role of chemical signals in dingo society;[16] however, many of the lessons from studying other canids such as domestic dogs and coyotes (*Canis latrans*) can be applied to dingoes. Their sense of smell is used to detect and analyse chemical signals to provide information about their environment, and is likely to be the most refined of their senses.[45] While their acute sense of smell is clearly useful for detecting prey, it also plays an important role in many social functions and in detecting danger.[45, 79]

Dingoes and other dogs often behave in predictable ways when exposed to particular odours, and these behaviours have been the subject of much research. One odour might be more likely to motivate them to rub or roll whereas another may prompt them to urinate.[80] Behavioural responses and visitation rates to a broad range of odour attractants have been investigated for coyotes[80–89] and grey wolves (*Canis lupus*).[90] While attractant research from Australia has primarily concentrated on their use as lures for lethal control in the form of trapping, the insights have proven useful to other types of dingo research.[91–94] A behavioural response to odours that are of particular interest to many dingo scientists is rubbing or rolling.

Anyone who has spent time with a dog will know that they are drawn like magnets to vile-smelling substances, such as rotten meat or faeces – in many cases the smellier the substance the better! The reasons behind the *Canis* spp. tendency to roll in pungent substances are poorly understood, although they often roll on novel substances or substances related to food.[79, 95] Several reasons have been proposed to explain this behaviour: (1) they are attempting to increase their attractiveness to conspecifics, (2) they are attempting to familiarise themselves with new or changed odours, (3) they are attempting to conceal their own scent and (4) they feel a strong attraction or aversion to the odour.[79] Rubbing and rolling are generally unwanted behaviours in the context of wild dog capture, as they may interfere with bait ingestion and the operation of capture devices such as leg-hold traps;[83] however, in some instances, dingo researchers may want to encourage dingoes to rub or roll so that they can collect samples using a hair trap.[50]

Hair traps

Scientists have long realised the potential of mammal hair for species identification.[50] Traditionally, hair has been analysed using a microscope to look for unique patterns which can identify individual species.[96] However, since the development of DNA analysis techniques, scientists have realised that hair can be a useful source of DNA.[50] Hair can be collected from wild animals using several devices. Hair traps used for wild canids, including dingoes, generally consist of a board or post covered in a bristly (e.g. carpet) or sticky (e.g. double-sided tape) substance that traps hair. They are placed in areas where they are likely to be encountered by a dingo, and an odour that elicits rubbing or rolling is applied. This odour will not only inspire any visiting dingo to rub or roll on the hair trap, it also serves as a lure which attracts them to the trap from nearby areas. When a dingo rubs or rolls on the hair trap it leaves behind hairs that can be analysed to provide information (Fig. 7.1).

While hair traps are undoubtedly useful, there are several problems associated with them. Danielle Stephens, a geneticist who conducted PhD research on the molecular ecology of wild dogs in Australia, observed that multiple visits to a hair trap by more than one animal can lead to contamination of DNA samples, and that hair trap designs which ensure that only one animal is sampled can inspire fear in dingoes, which could reduce the number of samples that can be collected.[97, 98] In addition, the quality of the DNA collected from hair traps was generally poorer than from saliva and scats. Stephens suggested that the collection of fresh scats was

Figure 7.1: Camera trap footage of a dingo in the Wet Tropics of Australia rubbing on a hair trap that has been baited with an odour attractant.
Source: Damian Morrant.

the most practical non-invasive approach to monitoring the genetic status of dingoes and other wild dogs.

Invasive field methods

When sampling methods involve interactions that have the potential to cause stress or injury to an animal or affect their ability to reproduce, they are known as invasive sampling methods. Animals will not necessarily be adversely affected by invasive sampling but there is an acknowledgement that such methods hold inherent risks. Invasive sampling methods often involve the study animals being captured, and in some circumstances they may be killed. Methods of invasive sampling can be grouped into three broad categories.[49]

1. Non-lethal – Animals are captured or restrained and measurements or biological samples are collected, then the animal is released.
2. Lethal – Animals are euthanased in the field before or after collection of samples.
3. Post-lethal – Measurements or biological samples are taken from animals or parts of animals which have been euthanased by pest managers, hunters or museum collectors, or which have died of other causes.

It should be noted, as discussed in the section on ethics, that while invasive methods are sometimes needed to obtain appropriate data, scientists and regulatory bodies take every possible step to minimise stress or potential injury to animals. While some past researchers used lethal methods to obtain scientific data, such research has become increasingly rare. Most contemporary scientists prefer methods that do not require the death of any animal.

Bounties and other control operations

Dingo scientists sometimes use information from the carcasses or pelts of animals that have been euthanased during control operations by government agencies, landholders or pest managers. This type of sampling is an example of a post-lethal method. While these methods rely on animals being euthanased they do not involve the researcher causing distress or death to the animal, nor are the animals euthanased for the purposes of the research. Rather, scientists collect useful data from a process that would occur whether or not they were involved. Such samples offer important biological information which contributes to the management and conservation of dingoes.

Useful information can be obtained from data from bounties paid during dingo control programs. Bounty payment records have allowed inferences to be made about dingo population dynamics and their geographic distribution, and the effects of fluctuations of their populations on feral pigs.[99, 100] However, Allen and Sparkes[101] cautioned against using bounty payments as an index of dingo abundance because control effort and techniques are highly variable across time and in areas with different types of livestock production. They warned that changing trends in livestock production may have significant effects on the area in which dingoes were trapped, as well as the human effort dedicated to control, which would affect the number of scalps presented for bounties.

The carcasses and pelts from dingo control operations, including bounty programs, however, provide a more tangible source of scientific evidence. From a single carcass it is possible to collect a broad range of information about animals including genetic status (DNA and skull measurements), diet (from stomach contents and stable isotopes in tissue), disease and parasite burden, coat colour and a broad range of other morphological characteristics.

Carcasses have provided much of the information about the diseases and parasites carried and/or transmitted by wild dingoes.[102–109] For example Smout *et al.*[103] made the first observation of the parasitic nematode *Ancylostoma ceylanicum* in wild canids from carcasses and scat samples collected from dingoes in the Wet Tropics. Many of the diseases that have been identified in dingoes can be transferred to humans, livestock, domestic animals and native fauna, and their discovery has important implications for domestic and wild animal management.

The skulls and pelts of culled dingoes have been examined to investigate the genetic status of wild dingo populations. For their evaluation of genetic analyses, skull morphology and visual appearance for assessing dingo purity, Elledge and colleagues[110] augmented their sample of 60 wild dog carcasses collected in south-east Queensland with 40 skulls from western Queensland that are maintained at the Queensland Museum. Their results improved on past criteria for visually assessing the genetic purity of dingoes, and suggested that the process of hybridisation in wild dingoes could be slowed by culling obvious hybrids that had visual characteristics such as sable and patchy coat colours.

Stable isotope analysis of tissue samples

An isotope is a variety of a chemical element. Although isotopes of the same element share the same chemical properties, they have a different mass.[40] Because of this difference in mass, isotopes behave differently in many environmental and physiological processes. Stable isotopes are isotopes that do not decay (unlike radioisotopes), a property which makes them useful for scientific investigations of animal diet.

The tissues of consumers (an organism which obtains its nutrients by consuming other organisms) are built from the nutrients in their diet.[111] When herbivores eat plants, their tissues are therefore synthesised from the chemicals in the plants, including stable isotopes. If a predator eats the herbivore, the chemicals from which its body is composed are incorporated into the predator's tissue. By analysing the stable isotope composition of predator tissue, ecologists are able to determine information about characteristics such as the habitats in which they hunt, seasonal changes in prey selection and their position in the food web.[111] For example, Darimont and Reimchen[112] measured carbon and nitrogen stable isotopes in segments of grey wolf hair from British Columbia and found that wolves supplement their diet with salmon during autumn.

Metabolically inactive tissue such as hair, skin and nails reflect the diet of animals during the time of their growth, providing a snapshot of the animal's diet at that time.[110] Long hairs such as tail hairs and vibrissae (whiskers) can be cut into sections and their isotopic composition can be compared to investigate changes in feeding patterns over time.[113–115] Each segment of hair further along represents the portion of time before the previous segment.

I use stable carbon isotopes from hair and vibrissae to support investigation of the diet of dingoes in the Wet Tropics. Stable isotope analysis can use samples that were obtained invasively or non-invasively. I obtained hair and vibrissae using both types of collection. Some samples came from carcasses provided by pest managers and farmers or from animals that had been killed by motor vehicles, whereas others were obtained using hair-trapping methods.

Stable carbon isotope values in plants are determined by the physiological pathways of carbon fixation.[116] Woody vegetation, such as rainforest, predominantly employs C3 photosynthesis whereas grasses (including sugarcane) use C4 photosynthesis.[115] The important fact is that these different pathways result in different ratios of carbon isotopes in grasses and woody plants. As mentioned above, when a predator such as a dingo eats a herbivore it incorporates its stable isotopes into its own tissues. The dingoes I am studying move between forest, open grassland and sugarcane fields. The study aims to discover if they primarily hunt prey in open habitats or forested areas; this is important because the open habitats are generally disturbed and primarily home to common and abundant animals, some of which are considered to be agricultural pests, whereas the forested habitats are more natural and home to less-abundant animal species, some of which are threatened. Thus, if dingoes hunt primarily in the forest they may be a greater conservation concern than if they focus on open-dwelling species. The results of this work are yet to be published; however, there are indications of a pattern which suggests that prey are sourced primarily from one of the land use types.

Capturing dingoes

Dingoes are notoriously difficult to catch, a trait for which they are often greatly (and begrudgingly) admired.[117, 118] Several authors have detailed the extraordinary measures that are required to successfully trap dingoes, many of which involve predicting dingo behaviour and/or the use of deception.[117, 119–121] Despite several general principles to trapping, almost every trapper has their own set of methods which they think work best.

Attractive lures are needed (e.g. rotten meat, dog urine, the contents of anal glands). However, while a particular lure may work well in one location or season, it may be relatively ineffective in another because the dingoes may not associate it with something interesting or, even worse, associate it with something bad, as is the case with dingoes which have escaped from traps in the past. Some trappers use call lures in addition to the lure or bait at the trap. Call lures may be placed high in a tree so that their smell is carried on the wind, encouraging passing dingoes to venture closer to the trap site. In a similar vein carcasses may be dragged along roads, or blood trails dripped from moving vehicles. Sometimes decoys such as pieces of sheepskin are placed above the trap to distract the dingo's attention away from the trap at the crucial moment. Occasionally the trapper will need to try several variations of trap set-up before a particular dingo can be captured, and despite all the efforts some dingoes prove nearly impossible to catch (see Chapter 9).

The dingo's keen sense of smell plays an important role in bringing it to the trap but it is perhaps also the greatest enemy of the trapper. Human scent can arouse suspicion in a dingo and deter it from stepping into the trap, so dingo trappers go to considerable lengths to minimise their smell at a trap site. Such measures can include avoiding insecticides or sunscreen, wearing long sleeves, gloves and hats to cover the body, scrubbing exposed skin under running water to slough off dead skin flakes, and kneeling on mats. When setting traps in the humid wet tropics I wear a tennis-style wristband and headband to prevent drops of sweat from falling on the ground. Traps and trapping equipment are often boiled to remove odours, and may be dipped in purpose-made wax to cover smells. Often traps are dyed black during the boiling process, using a special pigment, in case an exposed piece of shiny metal should give its location away. While it is unlikely that a trap can be set without leaving any human scents behind, I imagine that the visual and olfactory message a visiting dingo gets from a carefully prepared and set trap may be something like 'A human was here but only briefly so it's not worth worrying' rather than 'DANGER – A HUMAN WAS HERE!'.

Trapping dingoes for research

Scientists sometimes need to capture dingoes in order to collect biological samples and measurements, or fit them with devices such as tracking collars or identification tags. Leg-hold trapping is the most commonly used method for these purposes. While other techniques of capture are available, such as cage traps, there is a general consensus that they increase the likelihood of bias because many dingoes are simply too clever or wary to be trapped by these methods.[22]

When used responsibly by experienced operators, leg-hold traps can generally be used without causing long-term injury. However, poorly set traps and inexperienced animal handling are a sure way to cause injury to an animal, and an animal is more likely to sustain an injury the longer it stays trapped. I will therefore not discuss the fine detail of trapping dingoes but instead focus on some additions to trapping protocols which may enable experienced trappers to improve the humaneness and target specificity of their methods, and allow non-trappers to learn some of the ways scientists make their work as humane as possible. I acknowledge that in areas where dingoes are hard to capture some of these additions could reduce the likelihood of success, and may not be adopted. Nonetheless they should be viewed as best practice.

Leg-hold traps used in research are fitted with rubber pads, which cushion the animal's leg against the steel jaws and significantly decrease the likelihood of injury and stress.[122] It is illegal to use leg-hold traps without such padding in most Australian states, and illegal to use any type of leg-hold trap in Tasmania. Despite rubber padding, however, animals may still sustain injuries from leg-hold traps and non-target species can be trapped accidentally. It is therefore important for trappers to carefully consider the inherent risks of leg-hold trapping, and to set their traps in a way that minimises them.

Potential trap sites should be monitored with camera traps before trapping to minimise the likelihood that a trap site is visited by non-target native species, humans or domestic animals. In areas where daytime activity of humans, domestic animals or other non-target animals is high, heavy concrete pavers or similar can be carefully placed over the trap in the morning to prevent non-target captures. The large flat paver causes minimal disturbance to the site and can be removed before nightfall to re-arm the trap.

Traps should be attached firmly using stakes, chains or heavy drag logs. I use a 50 cm star picket, and a chain wrapped around a sturdy tree for redundancy. Traps should be fitted with a pan-tension device, which allows the trapper to adjust the tension on the trap so that animals under a certain mass won't be caught, thus reducing non-target captures in the form of pups and other animal species.[123, 124] In addition, because a heavy mass is required to set off a trap with a pan-tension device, the animal is more likely to be caught firmly as it will have placed its foot as far as possible into the trap area. Camera trap footage shows that trapped dingoes sometimes vent their frustration on branches that have been strategically placed in the vicinity of traps, so I suggest that, where possible, these should also be used.

In areas where mobile telephone reception is available, text messaging (or SMS) trap alerts should be used, or VHF trap alerts in isolated areas. Trap alerts which work on a mobile network send a text message to the trapper the instant the trap springs. Trap alerts which emit VHF radio signals emit a constant signal which changes when an animal is trapped. These devices must be monitored with a receiver, and the signal should be checked frequently throughout the day and night. If the design of the trap alert permits, it may be possible to secure the device to the trap chain. In such cases it would be possible to radio-track the device in the unfortunate event that an animal manages to pull a trap free of the stakes or chains that secure it in place. Traps should still be checked each morning in case a trap alert fails.

Sometimes trapped dingoes may be sedated, depending on whether it is deemed necessary for the purposes of data collection, or for ethical reasons. They are usually held on a restraint board with straps fitted over their neck, chest and hindquarters to reduce the risk that they

Figure 7.2: A sedated dingo is fitted with a GPS collar for a tracking study. Note the muzzle and restraints which were used because of the risk that the animal could arouse from its sedation while samples were being collected, and injure itself or field workers.
Source: Damian Morrant.

will be injured if they struggle, or that the researchers will be bitten by a struggling animal (Fig. 7.2).

Once dingoes are caught they may be processed to collect information such as mass and body measurements, tissue and blood samples may be taken, microchips or ear-tags may be injected/attached for future identification, and if they are to be monitored for a tracking study a VHF or GPS collar will be fitted. If the animals are released, as in tracking studies, the sampling can be described as non-lethal. Past researchers sometimes used lethal sampling measures, but such methods are becoming less common.

While this section has focused on trapping dingoes, the core principles apply to most aspects of dingo research. For example, the considerations about leaving odours at a trap site also apply to camera or hair trapping because dingoes will be deterred from survey stations or may change their behaviour if they detect a human presence.

Tracking dingo movements

By determining how dingoes use the landscape we can answer many questions about the size of their ranges, the ways in which they use the habitats that are available to them, their movement and activity patterns, and how they interact with other species such as feral cats and foxes. The range of questions that can be answered with analysis of animal movement has resulted in tracking becoming a common method for wildlife research. Tracking is often conducted in conjunction with other field methods, such as dietary analysis, as the insights that can be gained from understanding an animal's use of the landscape can enhance the researcher's ability to draw conclusions. Given the value of tracking for improving our understanding of dingo biology, it seems reasonable to devote attention to explaining its use and application.

Probably the simplest form of dingo movement study involves tagging individuals and recording the distances between where they were tagged and where they were relocated. Whitehouse[125] caught dingo pups by hand at den sites and fitted them with ear-tags before release. He then recorded the locations at which the pups were subsequently seen (two animals) or recaptured using traps (12 animals). The data showed that dingo pups could move up to 34.4 km from their dens within the first 12 months of life, and that males moved further than females.

Several dingo researchers since Whitehouse[125] have used radio-tags to monitor the behaviour and physiology of wild dingoes.[22, 34, 56, 63, 126–133] Such research is known as radio-telemetry. Dingoes are captured and fitted with collars that contain tracking devices, which allow their movements to be tracked without visual observation. Through radio-telemetry dingo scientists have learned a great deal about the ranges, habitat use and movement and activity patterns of dingoes.

In the earliest examples of telemetry involving dingoes, researchers used VHF collars to investigate range and movement patterns.[22, 34, 130, 131] The signals from a VHF transmitter in a purpose-made collar are picked up by the researcher, usually at predetermined time intervals, using an antenna and receiver (Fig. 7.3). The researcher then uses a method called triangulation to determine the animal's location (Fig. 7.4). The closer the researcher is to the dingo, the more accurate the results are likely to be. This type of work is often arduous and time-intensive and usually results in relatively few locations within a day's work. Accuracy can be significantly reduced if the animal is moving when the bearings are taken.

More recently, developments in GPS and other satellite technology have enabled scientists to study dingo movements in much more detail.[56, 63, 126–129, 132, 133] GPS telemetry is the most commonly used telemetry system among contemporary dingo researchers, and enables them to gain much more information than does VHF telemetry, for a fraction of the work effort.

The GPS system was developed by the US government. GPS receivers determine their location in a similar way to the triangulation method used in VHF tracking. However, in a GPS collar it is the GPS device that calculates the animal's location, by triangulating its location on Earth relative to several satellites. A GPS receiver picks up signals from satellites that broadcast their location, status and a precise time from atomic clocks. The receiver then calculates its distance from all the satellites from which it receives a signal and uses that information, plus the location information sent from the satellite, to determine its position on the Earth's surface. This method enables relatively accurate locations, with only a few metres of error, to be determined; however, accuracy is dependent on the GPS receiver.[134]

Data collected using GPS can be recorded at short time intervals throughout the day and night. Locations are stored in a memory chip in the collar and can be retrieved by recovering

Figure 7.3: Ecologist Rob Appleby using a yagi (antenna) and a VHF receiver to locate a radio-collared dingo on Fraser Island.
Source: Bradley Smith.

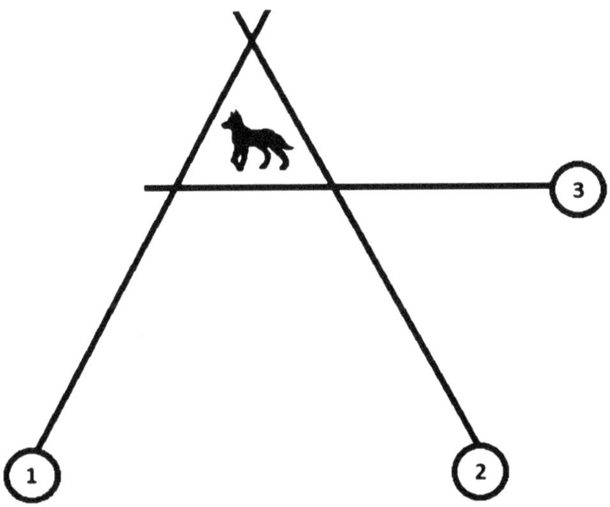

Figure 7.4: Triangulation of a radio-collared dingo. A bearing to the VHF-collared animal from point 1 (a known location) is determined by measuring with a compass the direction from which the signal originates. The researcher then moves to two other known locations (points 2 and 3) and again determines bearings. The researcher draws lines on a map along those bearings from the three locations. The area where the lines overlap represents the dingo's approximate location when the bearings were recorded.
Source: Damian Morrant.

the collar or, in some models, via transmission by SMS, UHF or satellite signals. Many GPS collars have drop-off devices with a release mechanism set for a predetermined time or when on command by the researcher via a remote signal (e.g. UHF), enabling retrieval of the collar and freeing the animal from its encumbrance.

Many questions about dingo biology can be answered using radio or GPS telemetry. Robley and colleagues[133] tracked nine wild dogs in Victoria to investigate their movements and habitat selection. They used a technique called compositional analysis[135] to compare the availability of habitats with those which were actually used. The study animals preferred habitats with an open vegetation structure, which may facilitate dingo hunting tactics.[42] Harden[130] measured the duration of active and rest periods by determining at 15-minute intervals whether dingoes were stationary or moving. Some animals were tracked almost continuously for up to 100 hours. Harden used VHF telemetry methods, which are time-intensive and laborious, so the large dataset from which these observations were made is impressive.

While tracking data are used to answer many biological questions about dingoes they are most commonly used to investigate characteristics of home range and utilisation distributions, which can in turn be used to answer questions about other characteristics such as habitat use. The chapter will now focus further on determination of home range and utilisation distributions.

Home range and utilisation distribution

A dingo's 'home range' is the area that it uses during its normal activities of hunting, mating and caring for its young.[136] A home range does not necessarily include all areas that the dingo visits, just those that it normally uses. Within a home range is a territory, which is the area that the dingo and/or dingo pack protects from intruders through aggression or through 'advertisement' using scratch marks in soil, faeces, urine and other secretions.[22, 56, 136] Territories are usually smaller than the home range and encompass areas that are relatively important to the dingo, such as den sites, water points and hunting areas. *Core areas* within territories, for example where animals regularly sleep or hunt, may be used at a higher intensity than the rest of the territory.

The utilisation distribution represents the probabilities that an animal will be within an area of its home range at any point in time.[137, 138] They are generally represented as a series of contours showing successive intensities of use; 50%, 75%, 85% and 95% contours are most commonly used in analyses (Fig. 7.5). Construction of utilisation distributions allows the researcher to visualise areas which are used most through to least intensively, and to conduct analyses using this information.

This section discusses some of the methods that can be used to estimate home range and utilisation distributions. While it will not discuss investigation of territories and core areas *per se*, the methods that are used for estimating such areas are usually extensions of those used for home range and utilisation distributions. For example, Allen *et al.*[132] used 50% kernels to represent core areas of use within dingo home ranges in their study of peri-urban dingoes in southern Queensland.

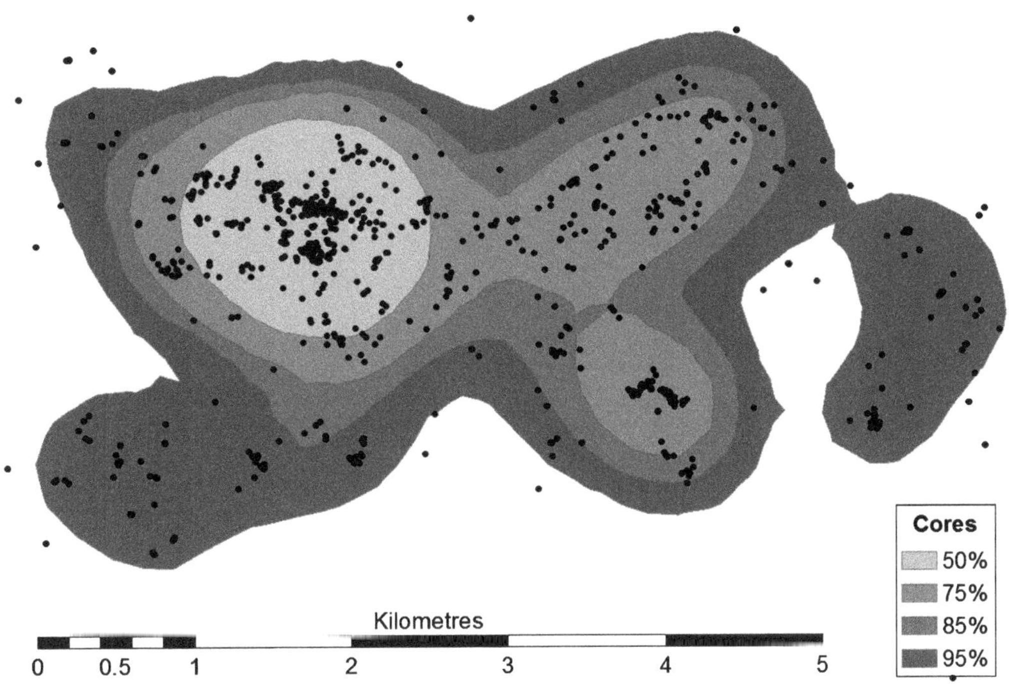

Figure 7.5: Kernel density estimate for a male dingo in the Wet Tropics of Australia showing 50%, 75%, 85% and 95% cores. Past dingo researchers used 85% or 95% cores to define home range.
Source: Damian Morrant.

Scientists estimate dingo home range and utilisation distributions in order to make inferences about the way in which dingoes use the landscape. They may be interested in preference for particular habitats for resting or hunting, dispersal of juvenile animals, likelihood of encountering bait or survey stations, interactions with livestock, or simply the size of the area needed to sustain an individual dingo or its pack. There are many different ways in which home ranges and utilisation distributions can be estimated and there is no consensus on which is best.[139, 140] There is also disagreement about methods of determining the parameters that should be used within many of these methods. In the face of so much uncertainty, selecting the best method can be very difficult.

The choice of home range estimator should depend on the biological question that is being asked and, ideally, more than one estimator should be used to determine whether the conclusions are robust.[140] As mentioned, the mathematical procedures involved in different home range estimators, and the choice of parameters, are complicated and they are worthy of a chapter (or chapters) to themselves. This chapter will provide a general overview of the two methods that have been used in the past, and discuss methods that may be more suitable in some situations.

Traditionally, minimum convex polygons[141] have been the estimator of choice for dingo research in Australia, and this method is one of the simplest HR estimators to apply.[22, 34, 63, 126, 128–130, 132, 133, 142] Minimum convex polygons (MCPs) enclose tracking locations by drawing a line that connects the outermost locations. Although the 100% MCP, which encloses all locations, is sometimes used, it is more common to see studies using 95% MCP in which the outermost 5% of locations are not included, in an attempt to exclude locations that may have been visited rarely or only once.

Density estimators produce better estimates of home range than do MCPs.[138, 143] Recently published studies investigating dingo home range have used a type of density estimator called a kernel density estimator (KDE).[126, 127, 132] One advantage of KDE over MCP is that it allows utilisation distributions to be determined, which allows more-detailed analysis of the way in which dingoes use the available habitats.

MCP and KDE have been criticised because they may be more likely than alternative estimators to omit areas that are known to be used, or include areas that are not used.[140, 144–146] There are many choices of alternative estimator; each method has strengths and weaknesses in particular situations. Many of them are better suited to detailed information about dingo movements that can be obtained using modern GPS tracking technology. They can work better than MCP/KDE when a lot of data are available, and estimates are informed by the relative position of each location. Some estimators can make use of information from data that are auto-correlated (i.e. if sequential locations are close together in time then the location of one data point is influenced by the location of the point before it) to more accurately plot utilisation distributions.

To provide an example, outlier-restricted edge polygons (OREP), movement-based kernel density estimators (MKDE) and adaptive local convex hulls (a-LoCoH) (Fig. 7.6), generally produce home ranges which more accurately reflect the actual use of the landscape by tracked animals. They do this by placing greater importance on areas which are used intensively, excluding unused areas and outlying locations using standardised procedures.[145–150] Methods such as MKDE also use the information from auto-correlated locations to inform home-range estimates. However, while these may be better suited in some situations, they may not perform as well as other methods such as the KDE if the distribution of use varies smoothly over space.[140] Thus, as mentioned above, the choice of home range estimator should be governed by the aims of the research.

To illustrate the different home range estimates that may result from different methods, Figure 7.6a–f shows the locations and HRs for a male dingo tracked in the Wet Tropics of Queensland using all the methods mentioned above. As can be seen, the choice of estimator can greatly affect results. The MCP and KDE home ranges generated for dingoes in the Wet Tropics often included deep rainforested valleys and large urban areas which were rarely visited by study animals (D. Morrant, unpublished data), and it was clear that using only these estimators would have overemphasised the use of some habitats, thereby underestimating the importance of others, particularly with MCP. Therefore, my study used MCP and kernels primarily for comparison of home-range sizes with past research and MCP to show areas, which theoretically were available to dingoes. I determined habitat preference by conducting compositional analysis of habitat use,[135] comparing the available habitats within MCP with the used habitats within areas determined using OREP, a-LoCoH and MKDE. I then compared the results from analysis using each of the three methods. Analysis that compared

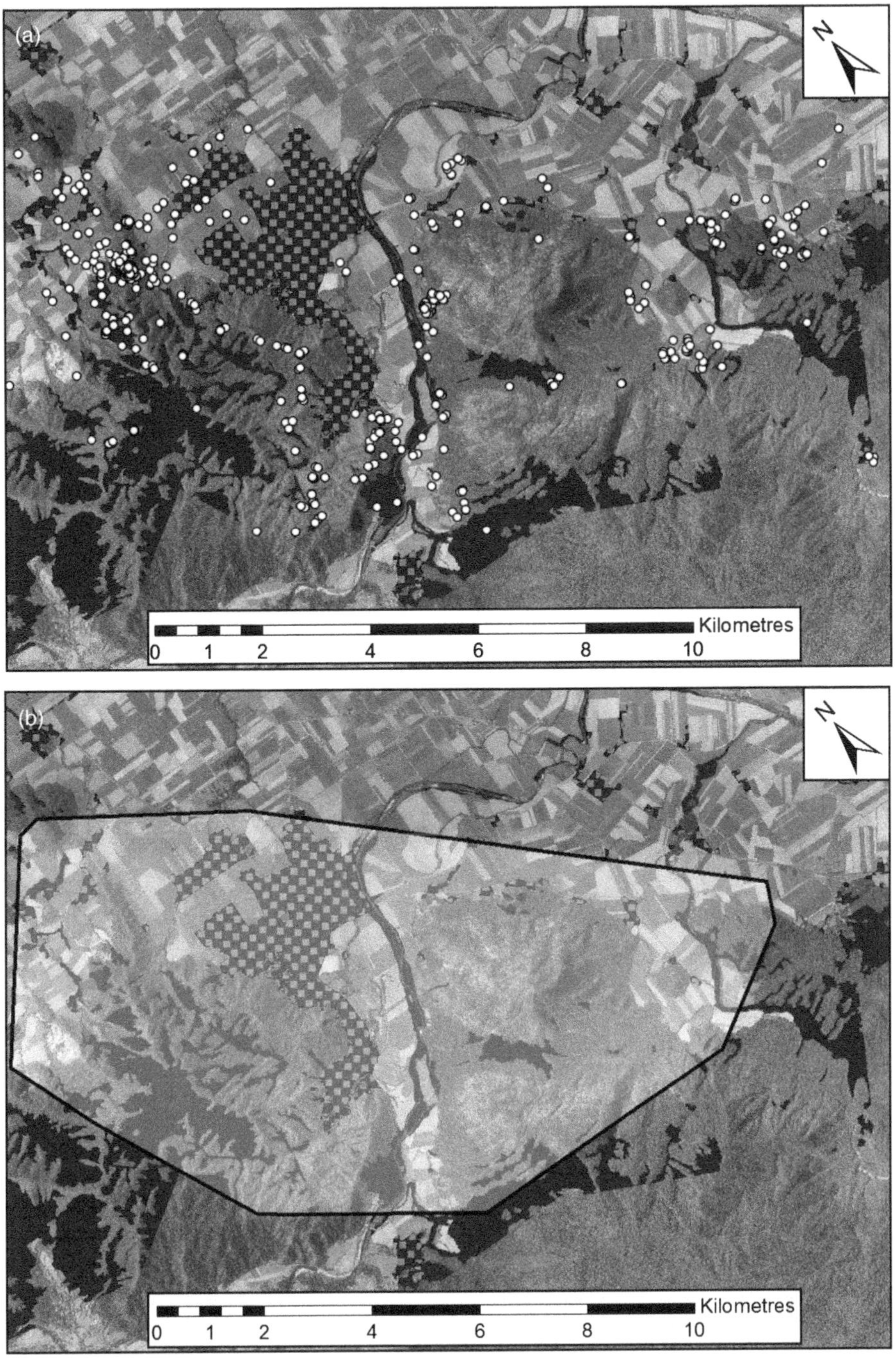

Figure 7.6: Tracking data for a male dingo in the Wet Tropics of Australia and home ranges generated using five different estimators. Tracking locations are overlaid on home ranges and are represented by black dots. Chequerboard pattern = urban areas, dark grey = rainforest. a) Tracking locations. b) 95% Minimum convex polygon. c) 85% fixed kernel density estimator. d) 95% movement-based kernel density estimator. e) 95% adaptive local convex hull. f) Outlier-restricted edge polygon.

Source: Damian Morrant.

(Continued)

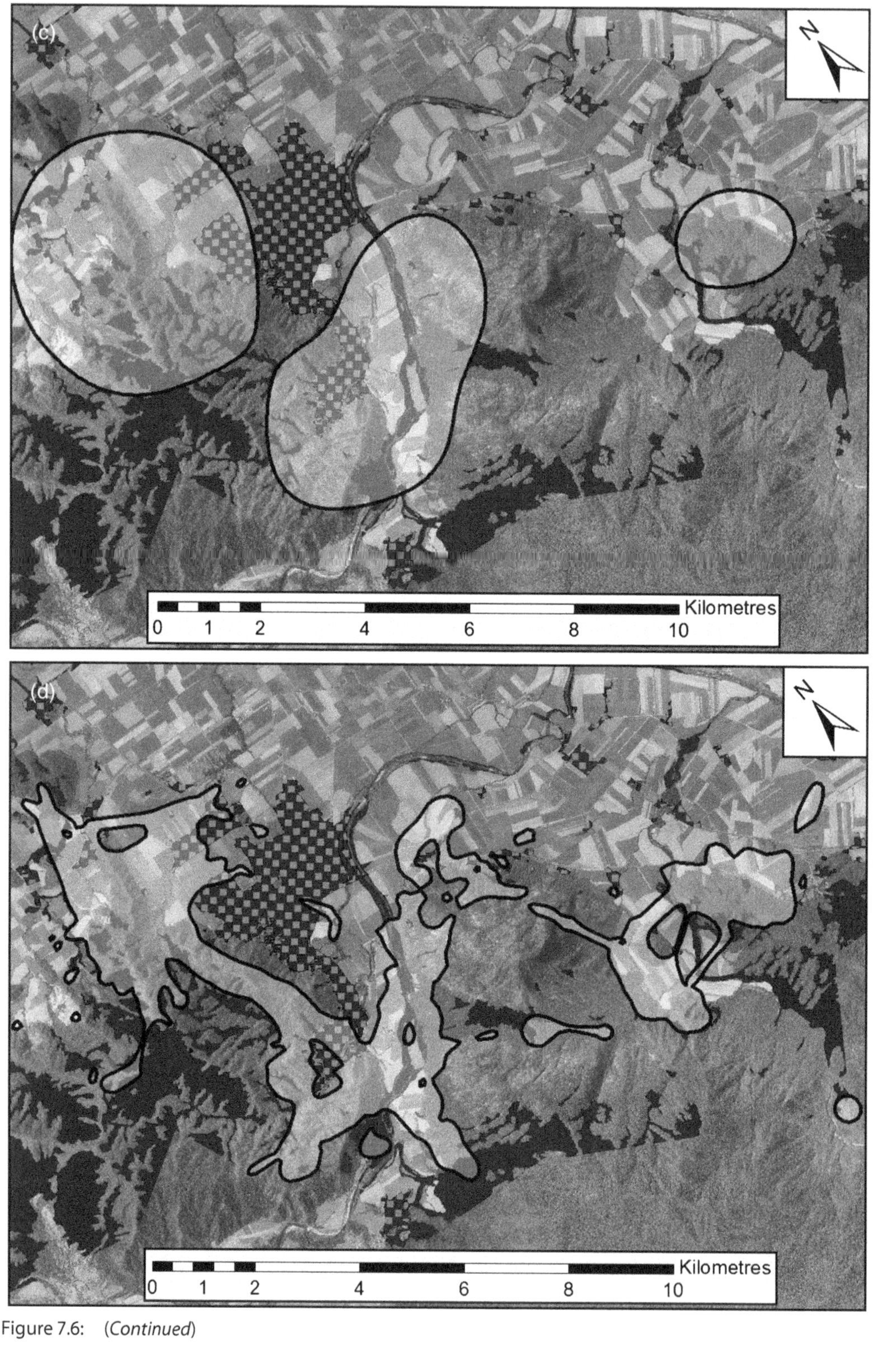

Figure 7.6: (*Continued*)

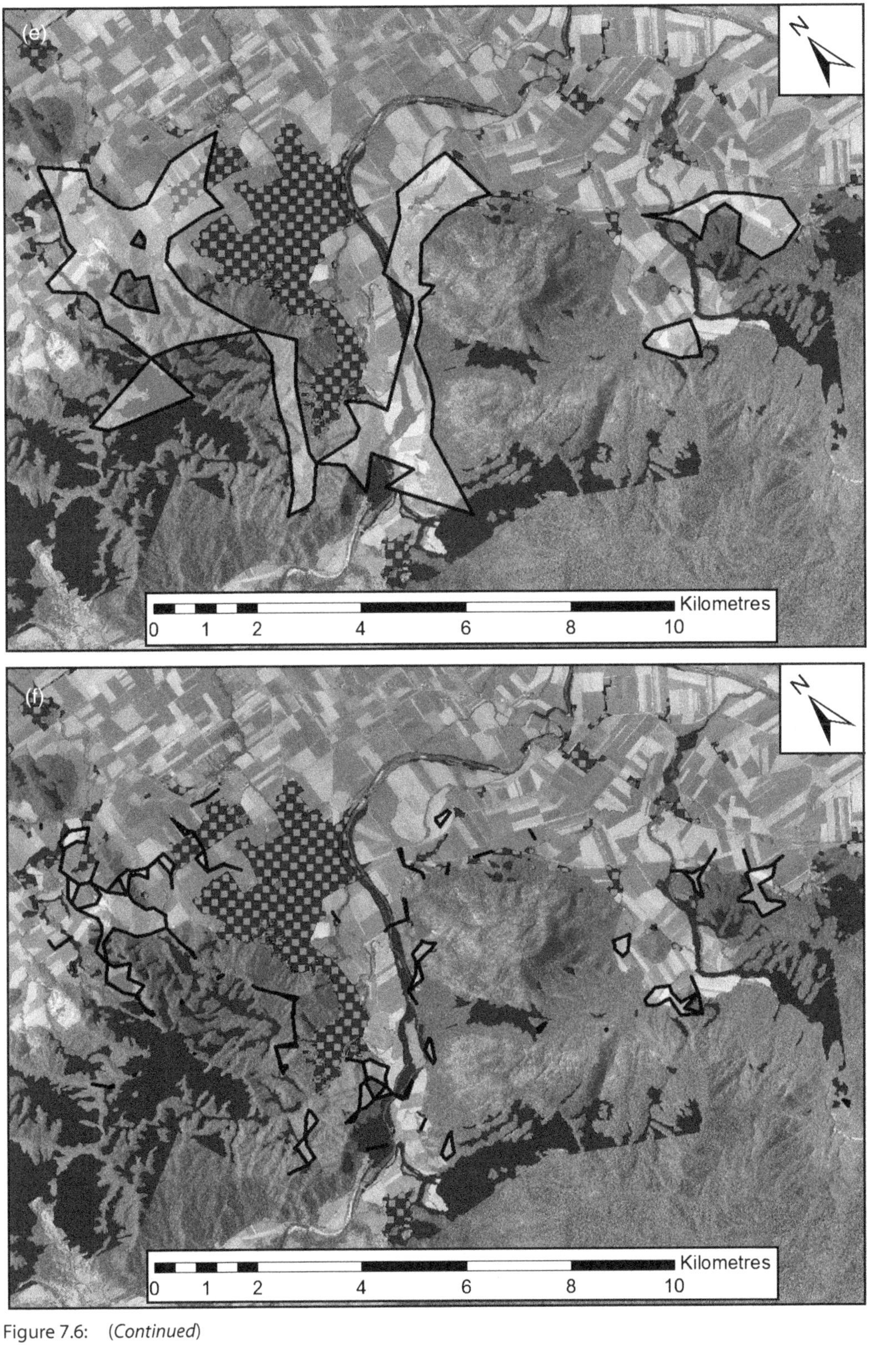

Figure 7.6: (*Continued*)

the used habitats within each of the three estimators habitat with the available habitats within MCP resulted in similar ranking of habitat preference, particularly for the most highly-ranked habitats (D. Morrant, unpublished data).

The future of dingo research

Scientists, like dingoes, are often fascinated by novel items, particularly when they come in the form of new methods or electronic gadgets which can unlock secrets that were previously out of the reach of science. In most cases, novel methods and technology used to study other species can also be applied to dingoes. The methods used to study animals will therefore constantly evolve. Zoologist Freeman from the Oxford Navigation Group in the UK, for instance, has collaborated with engineers from Microsoft Research to develop postage-stamp sized GPS tags for animals. These can communicate with aerial drones similar to those used for military applications and increasingly for conservation research. A drone can detect even a weak signal from a tag, fly towards it and download the tracking information from its memory while simultaneously recording imagery of the landscape through high-definition cameras. The tags can network with other tags in the area, which means that 'obtaining data from a single member of a tracked wolf pack could result in all the data for the individuals it has been in contact with'.[151] In light of Freeman's wolf analogy, the potential applications of such technology to dingo research seem obvious.

Cameras mounted to individual animals can provide fascinating insights into their behaviour. There can be no doubt about the usefulness of video footage filmed from a dingo's perspective. From such imagery scientists could measure all manner of dingo-related attributes such as social interactions, prey selection and hunting tactics, proportion of diet consisting of carrion, attacks on livestock, agonistic interactions with feral cats and foxes, and habitat use, to name a few. Undoubtedly, the best-known device of this type is the Crittercam®, which was conceived in 1986 by marine biologist, film-maker and inventor Greg Marshall. Crittercam® has been used by more than 30 scientific groups to study the behaviour of over 50 different species.[152] While much of the early work focused on marine animals, recently Terrestrial Crittercam® has been developed for deployment on land-dwelling animals, including carnivores such as African lions, hyenas and grizzly bears.[153, 154]

These are just two examples of new technology that could be useful for dingo research. There are myriad ways in which modern technology could be useful to researchers, and future scientific work will draw on such new methods and gadgets to enhance our understanding of dingo biology.

Conclusion

There are many ways in which scientists can learn about wild dingoes, and the ways in which new discoveries can be made are limited only by the imagination. This chapter has discussed the most common field methods, but these are by no means the only ways of collecting scientific data relating to dingoes. Measurements of biological phenomena may rely on modern equipment such as camera traps or GPS trackers, or may require nothing more than household items such as a broom for creating tracking plots. Scientists rarely employ only one method, often building a more complete picture by sourcing their information in different ways. With broad range of methods of gathering data and the potential for new discoveries using high-tech gadgets, future research holds the potential to provide fascinating insights into the lives of our iconic top predator.

Acknowledgements

I am grateful to Sarah Gill, Leila Brook, Euan Ritchie, James Butler and Bradley Smith for their helpful contributions to this chapter.

References

1. Kipling R (1912) *Just So Stories for Little Children.* Macmillan, London.
2. Allen BL, Engeman RM, Allen LR (2011) Wild dogma. II. The role and implications of wild dogma for wild dog management in Australia. *Current Zoology* **57**, 737–740.
3. Allen BL, Engeman RM, Allen LR (2011) Wild dogma: an examination of recent 'evidence' for dingo regulation of invasive mesopredator release in Australia. *Current Zoology* **57**, 568–583.
4. Letnic M, Crowther MS, Dickman CR, Ritchie EG (2011) Demonising the dingo: how much wild dogma is enough? *Current Zoology* **57**, 668–670.
5. Allen BL, Fleming PJ, Allen LR, Engeman RM, Ballard G, Leung LK-P (2013) As clear as mud: a critical review of evidence for the ecological roles of Australian dingoes. *Biological Conservation* **159**, 158–174.
6. Glen A (2012) Enough dogma: seeking the middle ground on the role of dingoes. *Current Zoology* **58**, 856–858.
7. Newsome AE (1990) The control of vertebrate pests by vertebrate predators. *Trends in Ecology and Evolution* **5**, 187–191.
8. Newsome AE, Catling PC, Cooke BD, Smyth R (2001) Two ecological universes separated by the dingo barrier fence in semi-arid Australia: interactions between landscapes, herbivory and carnivory, with and without dingoes. *Rangeland Journal* **23**, 71–98.
9. Newsome AE, Catling PC, Corbett LK (1983) The feeding ecology of the dingo. II. Dietary and numerical relationships with fluctuating prey populations in south-eastern Australia. *Australian Journal of Ecology* **8**, 345–366.
10. Newsome AE, Corbett L (1982) The identity of the dingo. II. Hybridization with domestic dogs in captivity and in the wild. *Australian Journal of Zoology* **30**, 365–374.
11. Newsome AE, Corbett LK (1985) The identity of the dingo. III. The incidence of dingoes, dogs and hybrids and their coat colours in remote and settled regions of Australia. *Australian Journal of Zoology* **33**, 363–375.
12. Newsome AE, Corbett LK, Best LW, Green B (1972) The dingo. *Australian Meat Research Committee Review* **14**, 1–11.
13. Newsome AE, Corbett LK, Carpenter SM (1980) The identity of the dingo. I. Morphological discriminants of dingo and dog skulls. *Australian Journal of Zoology* **28**, 615–625.
14. Newsome AE, Corbett LK, Catling PC, Burt RJ (1983) The feeding ecology of the dingo. I. Stomach contents from trapping in south-eastern Australia, and non-target wildlife also caught in dingo traps. *Australian Wildlife Research* **10**, 477–486.
15. Newsome AE, Parer I, Catling PC (1989) Prolonged prey suppression by carnivores: predator-removal experiments. *Oecologia* **78**, 458–467.
16. Corbett L, Newsome AE (1975) Dingo society and its maintenance: a preliminary analysis. In *The Wild Canids: Their Systematics, Behavioral Ecology and Evolution.* (Ed. MW Fox). Van Nostrand Reinhold, New York.
17. Corbett LK (1974) Dingoes, feral dogs and crossbreeds in victoria. *Australian Mammalogy* **1**, 303–304.

18. Corbett LK (1985) Morphological comparisons of Australian and Thai dingoes: a reappraisal of dingo status, distribution and ancestry. *Proceedings of the Ecological Society of Australia* **13**, 277–291.
19. Corbett LK (1988) Social dynamics of a captive dingo pack: population regulation by dominant female infanticide. *Ethology* **78**, 177–198.
20. Corbett LK (1995) Does dingo predation or buffalo competition regulate feral pig populations in the Australian wet-dry tropics? An experimental study. *Wildlife Research* **22**, 65–74.
21. Corbett LK (1995) Dingoes: expatriate wolves or native dogs? *Nature Australia* **25**, 46–55.
22. Corbett LK (2001) *The Dingo in Australia and Asia*. JB Books, Adelaide.
23. Corbett LK (2001) The conservation status of dingoes (*Canis lupus dingo*) in Australia, with particular reference to New South Wales: threats to pure dingoes and possible solutions. In *A Symposium on the Dingo*. (Eds C Dickman and D Lunney) pp. 10–19. Royal Zoological Society of New South Wales, Sydney.
24. Corbett LK, Newsome AE (1975) Dingo society and its maintenance: a preliminary analysis. In *The Wild Canids: Their Systematics, Behavioural Ecology and Evolution*. (Ed. MW Fox) pp. 369–378. Van Nostrand Reinhold, New York.
25. Corbett LK, Newsome AE (1987) The feeding ecology of the dingo. III. Dietary relationships with widely fluctuating prey populations in arid Australia: an hypothesis of alternation of predation. *Oecologia* **74**, 215–227.
26. Catling PC, Corbett LK, Newsome AE (1992) Reproduction in captive and wild dingoes (*Canis familiaris dingo*) in temperate and arid environments of Australia. *Wildlife Research* **19**, 195–209.
27. Shaughnessy PD, Newsome AE, Corbett LK (1975) An electrophoretic comparison of three blood proteins in dingoes and domestic dogs. *Australian Mammalogy* **1**, 355–360.
28. Thomson PC (1984) The use of buffer zones in dingo control. *Journal of Agriculture of Western Australia* **25**, 32–33.
29. Thomson PC (1986) The effectiveness of aerial baiting for the control of dingoes in north-western Australia. *Wildlife Research* **13**, 165–176.
30. Thomson PC (1992) The behavioural ecology of dingoes in north-western Australia. I. The Fortescue River study area and details of captured dingoes. *Wildlife Research* **19**, 509–518.
31. Thomson PC (1992) Capture of dingoes from a helicopter with tranquiliser darts loaded with ketamine hydrochloride and xylazine hydrochloride. *Wildlife Research* **19**, 601–603.
32. Thomson PC (1992) The behavioural ecology of dingoes in north-western Australia. III. Hunting and feeding behaviour, and diet. *Wildlife Research* **19**, 531–541.
33. Thomson PC (1992) The behavioural ecology of dingoes in north-western Australia. II. Activity patterns, breeding season and pup rearing. *Wildlife Research* **19**, 519–530.
34. Thomson PC (1992) The behavioural ecology of dingoes in north-western Australia. IV. Social and spatial organisation, and movements. *Wildlife Research* **19**, 543–563.
35. Thomson PC, Rose K (1992) Age determination of dingoes from characteristics of canine teeth. *Wildlife Research* **19**, 597–599.
36. Thompson DF (1984) Dingoes and sheep in pastoral areas. *Journal of Agriculture of Western Australia* **25**, 27–31.
37. Meehan B, Jones R, Vincent A (1999) Gulu-kula: dogs in Anbarra society, Arnhem Land. *Aboriginal History* **23**, 83–106.
38. Smith BP, Litchfield CA (2009) A review of the relationship between indigenous Australians, dingoes (*Canis dingo*) and domestic dogs (*Canis familiaris*). *Anthrozoos* **22**, 111–128.
39. Parker M (2007) The cunning dingo. *Society and Animals* **15**, 69–78.
40. Oxford English Dictionary (1989) *Oxford English Dictionary: The Definitive Record of the English Language*. Clarendon Press, Oxford.
41. Wallach AD, Murray BR, O'Neill AJ (2009) Can threatened species survive where the top predator is absent? *Biological Conservation* **142**, 43–52.

42. Fleming P, Corbett L, Harden R, Thomson P (2001) *Managing the Impacts of Dingoes and Other Wild Dogs*. Bureau of Rural Sciences, Canberra.
43. Allen L, Engeman R, Krupa H (1996) Evaluation of three relative abundance indices for assessing dingo populations. *Wildlife Research* **23**, 197–205.
44. Robertshaw JD, Harden RH (1986) The ecology of the dingo in north-eastern New South Wales. IV. Prey selection by dingoes, and its effect on the major prey species, the swamp wallaby, *Wallabia bicolor* (desmarest). *Wildlife Research* **13**, 141–163.
45. Lindsay SR (2000) Sensory abilities. In *Applied Dog Behavior and Training*. pp. 127–165. Iowa State University Press, Ames.
46. King JE, Becker RF, Markee J (1964) Studies on olfactory discrimination in dogs. III Ability to detect human odour trace. *Animal Behaviour* **12**, 311–315.
47. Gannon WL, Sikes RS (2007) Guidelines of the American society of Mammalogists for the use of wild mammals in research. *Journal of Mammalogy* **88**, 809–823.
48. National Health and Medical Research Council (2013) *Australian Code for the Care and Use of Animals for Scientific Purposes*. 8th edn. National Health and Medical Research Council, Canberra.
49. Pauli JN, Whiteman JP, Riley MD, Middleton AD (2010) Defining non-invasive approaches for sampling of vertebrates. *Conservation Biology* **24**, 349–352.
50. Kendall KC, McKelvey KS (2008) Hair collection. In *Non-invasive survey methods for carnivores*. (Eds RA Long, P MacKay, WJ Zielinski and JC Ray) pp. 141–182. Island Press, Washington.
51. Butler JR, Linnell JD, Morrant D, Athreya V, Lescureux N, McKeown A (2013) Dog eat dog, cat eat dog: social–ecological dimensions of dog predation by wild carnivores. In *Free-ranging Dogs and Wildlife Conservation*. (Ed. ME Gompper) pp. 117–143. Oxford University Press, Oxford.
52. Rural Management Partners (2004) *Economic Assessment of the Impact of Dingoes/wild dogs in Queensland*. Queensland Department of Natural Resources and Mines, Brisbane.
53. Hewitt L (2009) *Major Economic Costs Associated with Wild Dogs in the Queensland Grazing Industry*. AgForce, Brisbane.
54. Johnston MJ, Marks CA (1997) *Attitudinal Survey on Vertebrate Pest Management in Victoria*. Report. Victorian Department of Natural Resources and Environment, Institute of Animal Science, Frankston.
55. Krosnick JA (1999) Survey research. *Annual Review of Psychology* **50**, 537–567.
56. Purcell B (2010) *Dingo*. CSIRO Publishing, Melbourne.
57. Caughley G, Grigg GC, Caughley J, Hill GJE (1980) Does dingo predation control the densities of kangaroos and emus? *Australian Wildlife Research* **7**, 1–12.
58. Edwards MG, de Preu ND, Shakeshaft BJ, Crealy IV (2000) An evaluation of two methods of assessing feral cat and dingo abundance in Central Australia. *Wildlife Research* **27**, 143–149.
59. Read J, Eldridge S (2010) An optimised rapid detection technique for simultaneously monitoring activity of rabbits, cats, foxes and dingoes in the rangelands. *Rangeland Journal* **32**, 389–394.
60. Meek PD, Ballard AG, Banks PB, Claridge AW, Fleming PJS, Sanderson JG, Swann DE (Eds) (2014) *Camera Trapping: Wildlife Management and Research*. CSIRO Publishing, Melbourne.
61. O'Connell AF, Nichols JD, Karanth KU (2011) Introduction. In *Camera Traps in Animal Ecology: Methods and Analyses*. (Eds AF O'Connell, JD Nichols and KU Karanth). Springer, Tokyo.
62. Rowcliffe JM, Carbone C (2008) Surveys using camera traps: are we looking to a brighter future? *Animal Conservation* **11**, 185–186.
63. Robley A, Woodford L, Lee P, Kingston V, Peters W, Klippell D, Gormley A (2009) *Assessing the Effectiveness of Ground-based Baiting for the Control of Wild Dogs*. Arthur Rylah Institute for Environmental Research Technical Report Series. Victorian Department of Sustainability and Environment, Melbourne.

64. Somaweera R, Webb JK, Shine R (2011) It's a dog-eat-croc world: dingo predation on the nests of freshwater crocodiles in tropical Australia. *Ecological Research* **26**, 957–967.
65. Brook LA, Johnson CN, Ritchie EG (2012) Effects of predator control on behaviour of an apex predator and indirect consequences for mesopredator suppression. *Journal of Applied Ecology* **49**, 1278–1286.
66. Moseby KE, Neilly H, Read JL, Crisp HA (2012) Interactions between a top order predator and exotic mesopredators in the Australian rangelands. *International Journal of Ecology* **2012**, 250352.
67. Wang Y, Fisher DO (2012) Dingoes affect activity of feral cats, but do not exclude them from the habitat of an endangered macropod. *Wildlife Research* **39**, 611–620.
68. Glen AS, Dickman CR (2003) Effects of bait-station design on the uptake of baits by non-target animals during control programs for foxes and wild dogs. *Wildlife Research* **30**, 147–149.
69. Allen BL, Leung LK (2012) Assessing predation risk to threatened fauna from their prevalence in predator scats: dingoes and rodents in arid Australia. *PLoS ONE* **7**, e36426.
70. Brook LA, Kutt AS (2011) The diet of the dingo (*Canis lupus dingo*) in north-eastern Australia with comments on its conservation implications. *Rangeland Journal* **33**, 79–85.
71. Vernes K, Dennis A (2001) Mammalian diet and broad hunting strategy of the dingo (*Canis familiaris dingo*) in the wet tropical rainforests of north-eastern Australia. *Biotropica* **33**, 339–345.
72. Robertshaw JD, Harden RH (1985) The ecology of the dingo in north-eastern New South Wales. II. Diet. *Australian Wildlife Research* **12**, 39–50.
73. Robertshaw JD, Harden RH (1986) The ecology of the dingo in north-eastern New South Wales. IV. Prey selection by dingoes, and its effect on the major prey species, the swamp wallaby, *Wallabia bicolor* (Desmarest). *Australian Wildlife Research* **13**, 141–163.
74. Burnett S (1995) *The Distribution and Conservation Significance of Dingoes, Foxes and Feral Cats in the Uplands of the Wet Tropics World Heritage Area.* Report to the Wet Tropics Management Authority, Cairns.
75. Gese EM (2004) Survey and census techniques for canids. In *Canids: Foxes, Wolves, Jackals, and Dogs*. (Eds C Sillero-Zubiri, M Hoffman and DW Macdonald). IUCN World Conservation Union/SSC Canid Specialist Group, Gland, Switzerland/Cambridge, UK.
76. Edwards GP, De Preu N, Crealy IV, Shakeshaft BJ (2002) Habitat selection by feral cats and dingoes in a semi-arid woodland environment in central Australia. *Austral Ecology* **27**, 26–31.
77. Engeman RM, Allen L, Zerbe GO (1998) Variance estimate for the allen activity index. *Wildlife Research* **25**, 643–648.
78. Edwards GP, de Preu ND, Crealy IV, Shakeshaft BJ (2002) Habitat selection by feral cats and dingoes in semi-arid woodland environment in Central Australia. *Austral Ecology* **27**, 26–31.
79. Harrington FH, Asa CS (2003) Wolf communication. In *Wolves: Behaviour, Ecology, and Conservation*. (Eds LD Mech and L Boitani) pp. 66–103. University of Chicago Press, Chicago.
80. Kimball BA, Mason JR, Blom FS, Johnston JJ, Zemlicka DE (2000) Development and testing of seven new synthetic coyote attractants. *Journal of Agricultural and Food Chemistry* **48**, 1892–1897.
81. Bullard RW (1982) Wild canid associations with fermentation products. *Industrial and Engineering Chemistry Product Research and Development* **21**, 646–655.
82. Bullard RW, Turkowski FJ, Kilburn SR (1983) Responses of free-ranging coyotes to lures and their modifications. *Journal of Chemical Ecology* **9**, 877–888.
83. Kimball BA, Johnston JJ, Mason JR, Zemlicka DE, Blom FS (2000) Development of chemical coyote attractants for wildlife management applications. In *Proceedings of the 19th Vertebrate Pest Conference*. Davis: University of California.
84. Ebbert S, Fagre D (1987) Importance of attractant qualities for improving a new coyote delivery system. In *Great Plains Wildlife Damage Control Workshop Proceedings*. Paper 60. http://digitalcommons.unl.edu/gpwdcwp/60.

85. Phillips R, Blom F, Engeman R (1990) Responses of captive coyotes to chemical attractants. In *Proceedings of the 14th Vertebrate Pest Conference.* Sacramento.
86. Graves GE, Boddicker ML (1987) Field evaluation of olfactory attractants and strategies used to capture depredating coyotes. In *Great Plains Wildlife Damage Control Workshop Proceedings.* Paper 67. http://digitalcommons.unl.edu/gpwdcwp/67.
87. Roughton R (1982) A synthetic alternative to fermented egg as a canid attractant. *Journal of Wildlife Management* **46**, 230–234.
88. Turkowski FJ, Popelka ML, Bullard RW (1983) Efficacy of odor lures and baits for coyotes. *Wildlife Society Bulletin* **11**, 136–145.
89. Martin D, Fagre DB (1988) Field evaluation of a synthetic coyote attractant. *Wildlife Society Bulletin* **16**, 390–396.
90. Goodmann PA (1978) *Scent Rolling in Grey Wolves* (*Canis lupus*). Purdue University, Lafayette.
91. Jolly SE, Jolly LM (1992) Pen and field tests of odour attractants for the dingo. *Journal of Wildlife Management* **56**, 452–456.
92. Jolly SE, Jolly LM (1992) Environmental influences on the ability of captive dingoes to find meat baits. *Journal of Wildlife Management* **56**, 448–452.
93. Hunt RJ, Dall DJ, Lapidge SJ (2007) Effect of a synthetic lure on site visitation and bait uptake by foxes (*Vulpes vulpes*) and wild dogs (*Canis lupus dingo, Canis lupus familiaris*). *Wildlife Research* **34**, 461–466.
94. Mitchell J, Kelly A (1992) Evaluating odour attractants for control of wild dogs. *Wildlife Research* **19**, 211–219.
95. Sillero-Zubiri C, Macdonald DW (1998) Scent-marking and territorial behaviour of Ethiopian wolves *Canis simensis*. *Journal of Zoology* **245**, 351–361.
96. Brunner H, Coman B (1974) *The Identification of Mammalian Hair.* Inkata Press, Melbourne.
97. Pauli JN, Hamilton MB, Crain EB, Buskirk SW (2008) A single-sampling hair trap for mesocarnivores. *Journal of Wildlife Management* **72**, 1650–1652.
98. Stephens D (2011) The molecular ecology of Australian wild dogs: hybridisation, gene flow and genetic structure at multiple geographic scales. PhD thesis. University of Western Australia, Perth.
99. Harden RH, Robertshaw JD (1987) Ecology of the dingo in north-eastern New South Wales. V. Human predation on the dingo. *Australian Zoologist* **24**, 65–72.
100. Woodall PF (1983) Distribution and population dynamics of dingoes (*Canis familiaris*) and feral pigs (*Sus scrofa*) in Queensland, 1945–1976. *Journal of Applied Ecology* **20**, 85–95.
101. Allen L, Sparkes EC (2001) The effect of dingo control on sheep and beef cattle in Queensland. *Journal of Applied Ecology* **38**, 76–87.
102. Durie PH, Riek RF (1952) The role of the dingo and wallaby in the infestation of cattle with hydatids (*Echinococcus granulosus* (Batch, 1978) Rudolphi, 1805). *Australian Veterinary Journal* **28**, 249–254.
103. Smout FA, Thompson RC, Skerratt LF (2013) First report of *Ancylostoma ceylanicum* in wild canids. *International Journal for Parasitology: Parasites and Wildlife* **2**, 173–177.
104. Coman BJ (1972) Helminth parasites of the dingo and feral dog in Victoria, with some notes on the diet of the host. *Australian Veterinary Journal* **48**, 456–461.
105. Johnson AM, Phillips P, Jenkins D (1990) Prevalence of *Toxoplasma gondii* antibodies in dingoes. *Journal of Wildlife Diseases* **26**, 383–386.
106. Brown B, Copeman DB (2003) Zoonotic importance of parasites in wild dogs caught in the vicinity of Townsville. *Australian Veterinary Journal* **81**, 700–702.
107. Barber JS, Gasser RB, Ellis J, Reichel MP, McMillan D, Trees AJ (1997) Prevalence of antibodies to *Neospora caninum* in different canid populations. *Journal of Parasitology* **83**, 1056–1058.
108. Starr TW, Mulley RC (1988) *Dirofilaria immitis* in the dingo (*Canis familiaris dingo*) in a tropical region of the Northern Territory, Australia. *Journal of Wildlife Diseases* **24**, 164–165.

109. Dunsmore JD, Burt RJ (1972) *Filaroides osleri* in dingoes in south-eastern Australia. *Australian Veterinary Journal* **48**, 548–551.
110. Elledge AE, Allen LR, Carlsson B-L, Wilton AN, Leung LKP (2008) An evaluation of genetic analyses, skull morphology and visual appearance for assessing dingo purity: implications for dingo conservation. *Wildlife Research* **35**, 812–820.
111. Crawford K, McDonald RA, Bearhop S (2008) Applications of stable isotope techniques to the ecology of mammals. *Mammal Review* **38**, 87–107.
112. Darimont C, Reimchen T (2002) Intra-hair stable isotope analysis implies seasonal shift to salmon in gray wolf diet. *Canadian Journal of Zoology* **80**, 1638–1642.
113. Hobson KA, Schell DM, Renouf D, Noseworthy E (1996) Stable carbon and nitrogen isotopic fractionation between diet and tissues of captive seals: implications for dietary reconstructions involving marine mammals. *Canadian Journal of Fisheries and Aquatic Sciences* **53**, 528–533.
114. Cerling TE, Wittemyer G, Rasmussen HB, Vollrath F, Cerling CE, Robinson TJ, Douglas-Hamilton I (2006) Stable isotopes in elephant hair document migration patterns and diet changes. *Proceedings of the National Academy of Sciences of the United States of America* **103**, 371–373.
115. Wurster CM, Robertson J, Westcott DA, Dryden B, Zazzo A, Bird MI (2012) Utilization of sugarcane habitat by feral pig (*Sus scrofa*) in northern tropical Queensland: evidence from the stable isotope composition of hair. *PLoS ONE* **7**, e43538.
116. Ehleringer JR, Cerling T (2002) C3 and c4 photosynthesis. In *Encyclopedia of Global Environmental Change*. (Eds HA Mooney and JG Gandadell) pp. 186–190. John Wiley and Sons, Chichester.
117. Hudson L (1974) *Dingoes Don't Bark*. Rigby, Adelaide.
118. Rolls EC (1993) *From Forest to Sea: Australia's Changing Environment*. University of Queensland Press, Brisbane.
119. Holden P (1991) *Along the Dingo Fence*. Hodder and Stoughton, Sydney.
120. Bacon JS (1960) *The Australian Dingo: King of the Bush*. Glasson and Oliver, Sydney.
121. Murray AG (1972) Trapping dingoes: a dying art. *Walkabout* **26**, 32–34.
122. Fleming PJS, Allen LR, Berghout MJ, Meek PD, Pavlov PM, Stevens P, Strong K, Thompson JA, Thomson PC (1998) The performance of wild-canid traps in Australia: efficiency, selectivity and trap-related injuries. *Wildlife Research* **25**, 327–338.
123. Turkowski FJ, Armistead AR, Linhart SB (1984) Selectivity and effectiveness of pan tension devices for coyote foothold traps. *Journal of Wildlife Management* **48**, 700–708.
124. Phillips RL, Gruver KS (1996) Performance of the paws-i-trip™ pan tension device on 3 types of traps. *Wildlife Society Bulletin* **24**, 119–122.
125. Whitehouse SJO (1977) Movements of dingoes in western Australia. *Journal of Wildlife Management* **41**, 575–576.
126. Newsome TM, Ballard GA, Dickman CR, Fleming PJS, Howden C (2013) Anthropogenic resource subsidies determine space use by Australian arid zone dingoes: an improved resource selection modelling approach. *PLoS ONE* **8**, e63931.
127. Newsome TM, Stephens D, Ballard G-A, Dickman CR, Fleming PJS (2013) Genetic profile of dingoes (*Canis lupus dingo*) and free-roaming domestic dogs (*C. l. familiaris*) in the Tanami Desert, Australia. *Wildlife Research* **40**, 196–206.
128. Claridge AW, Mills DJ, Hunt R, Jenkins DJ, Bean J (2009) Satellite tracking of wild dogs in south-eastern mainland Australian forests: implications for management of a problematic top-order carnivore. *Forest Ecology and Management* **258**, 814–822.
129. Eldridge SR, Shakeshaft BJ, Nano TJ (2002) The impact of wild dog control on cattle, native and introduced herbivores and introduced predators in central Australia. Unpublished report. Bureau of Rural Sciences, Canberra.
130. Harden RH (1985) The ecology of the dingo in north-eastern New South Wales. I. Movements and home range. *Australian Wildlife Research* **12**, 25–37.
131. McIlroy JC, Cooper RJ, Gifford EJ, Green BF, Newgrain KW (1986) The effect on wild dogs, *Canis. f. familiaris*, of 1080-poisoning campaigns in Kosciusko National Park, NSW. *Wildlife Research* **13**, 535–544.

132. Allen BL, Goullet M, Allen LR, Lisle A, Leung LKP (2013) Dingoes at the doorstep: preliminary data on the ecology of dingoes in urban areas. *Landscape and Urban Planning* **119**, 131–135.
133. Robley A, Gormley A, Forsyth DM, Wilton AN, Stephens D (2010) Movements and habitat selection by wild dogs in eastern Victoria. *Australian Mammalogy* **32**, 23–32.
134. National Coordination Office for Space-Based Positioning, Navigation, and Timing (2014) *How GPS Works.* National Coordination Office for Space-Based Positioning, Navigation, and Timing, Washington.
135. Aebischer NA, Robertson PA, Kenward RE (1993) Compositional analysis of habitat use from animal radio-tracking data. *Ecology* **74**, 1313–1325.
136. Burt WH (1943) Territoriality and home range concepts as applied to mammals. *Journal of Mammalogy* **24**, 346–352.
137. Hayne DW (1949) Calculation of size of home range. *Journal of Mammalogy* **30**, 1–18.
138. Worton BJ (1989) Kernel methods for estimating the utilization distribution in home-range studies. *Ecology* **70**, 164–168.
139. Laver PN, Kelly MJ (2008) A critical review of home range studies. *Journal of Wildlife Management* **72**, 290–298.
140. Fieberg J, Börger L (2012) Could you please phrase 'home range' as a question? *Journal of Mammalogy* **93**, 890–902.
141. Mohr CO (1947) Table of equivalent populations of North American small mammals. *American Midland Naturalist* **37**, 223–249.
142. Newsome TM, Ballard GA, Dickman CR, Fleming PJ, van de Ven R (2013) Home range, activity and sociality of a top predator, the dingo: a test of the resource dispersion hypothesis. *Ecography* **36**, 914–925.
143. Kenward RE (2001) *A Manual for Wildlife Radio Tagging.* Academic Press, London.
144. White GC, Garrot RA (1990) *Analysis of Wildlife Radio-tracking Data.* Academic Press, San Diego.
145. Getz WM, Fortmann-Roe S, Cross PC, Lyons AJ, Ryan SJ, Wilmers CC (2007) Locoh: non-parameteric kernel methods for constructing home ranges and utilization distributions. *PLoS ONE* **2**, e207.
146. Getz WM, Wilmers CC (2004) A local nearest-neighbor convex-hull construction of home ranges and utilization distributions. *Ecography* **27**, 489–505.
147. Kenward RE, Casey NM, Walls SS (2009) *Ranges 8: For the Analysis of Location and Tracking Data.* Anatrack, London.
148. Benhamou S, Cornélis D (2010) Incorporating movement behavior and barriers to improve kernel home range space use estimates. *Journal of Wildlife Management* **74**, 1353–1360.
149. Benhamou S (2011) Dynamic approach to space and habitat use based on biased random bridges. *PLoS ONE* **6**, e14592.
150. Benhamou S, Riotte-Lambert L (2012) Beyond the utilization distribution: identifying home range areas that are intensively exploited or repeatedly visited. *Ecological Modelling* **227**, 112–116.
151. Marks P (2013) Animals tracked with tiny tags summon their own drones. *New Scientist* **2938**, 19 http://www.newscientist.com/article/mg22029383.800-animals-tracked-with-tiny-tags-summon-their-own-drones.html#.VJ3hVsAw
152. National Geographic (2014) *Greg Marshall: Vice President, Remote Imaging.* National Geographic Society, Washington.
153. National Geographic (2014) *Crittercam FAQs.* National Geographic Society, Washington.
154. Moll RJ, Millspaugh JJ, Beringer J, Sartwell J, He Z (2007) A new 'view' of ecology and conservation through animal-borne video systems. *Trends in Ecology and Evolution* **22**, 660–668.

8
An ecological view of the dingo

Chris Johnson

The dingo, it is often said, is the apex predator in Australian ecosystems. What this means is that the dingo is the continent's largest predator (on land, at least) and has no predator of its own. Of course, humans regularly kill dingoes and have done so for as long as dingoes have lived in Australia (see Chapter 5). This makes our own species the true apex predator in Australia. However, we want to understand what role the dingo plays as a predator in situations where we choose to leave it alone. This is a crucial question for wildlife management and biodiversity conservation in Australia. Dingoes prey on livestock as well as wildlife. In response to this threat, lethal control of dingo populations – mostly by broad-scale distribution of meat baits laced with 1080 poison – is practised throughout the vast pastoral lands of Australia.[1] Does this disrupt important ecological services that dingoes would otherwise provide as apex predators?

The impact of apex predators on ecosystems is currently a lively topic in ecological science, for two reasons. First, there is growing evidence that the effects of apex predators can be very strong, to the extent that their removal triggers cascading effects through entire food webs resulting in net loss of species diversity and degradation of ecological function.[2, 3] Apex predators are able to have such strong effects because, by definition, their abundance is not controlled by the impacts of predators at higher levels in the food web. Therefore, population size of apex predators is free to rise to the point where it is limited by the availability of prey, so that apex predators have large impacts on their prey and can cause large reductions in their abundance. If those prey species are large generalist herbivores (as the prey of large mammalian predators often are), population control by predation regulates their rate of consumption of plant material and protects vegetation from overutilisation. Vegetation condition can therefore be better and plant diversity higher in landscapes with apex predators; in turn, diverse and healthy vegetation communities furnish habitat for many species of smaller animals.

Apex predators also have large impacts on populations of smaller predators (mesopredators). This can be a result of apex predators killing mesopredators for food. In addition, apex predators often indulge in what can only be described as malicious killing of mesopredators.[4] Lynx, for example, frequently kill red foxes and leave their bodies uneaten.[5] Understandably, mesopredators fear apex predators and may avoid hunting at times and in places where they face a high risk of an unpleasant encounter with a large and lethal enemy. In this way, apex predators not only directly reduce numbers of mesopredators, they also limit their hunting opportunities by instilling fear and wariness. Because mesopredators (species like foxes, small cats, weasels, stoats and so on) are typically versatile predators capable of reaching high population densities, limitation of their abundance and behaviour by apex predators is capable of reducing the total intensity of predation on smaller prey species, and indirectly protecting some prey species that would be vulnerable to overhunting.[6]

Removal of apex predators can therefore have two consequences for ecosystems: outbreaks of herbivores, and mesopredator release – the increase in abundance and/or activity of smaller predators. The combined effects of these changes may include degradation or destruction of sensitive vegetation communities, and overhunting of small prey, potentially leading to extinction of some populations and even species. These kinds of effects have been documented in cases such as the extirpation of wolves in North America,[7–10] the disappearance of all large predators from habitat patches following their isolation on islands in an artificial lake in Venezuela[11] and the extirpation of big cats in parts of Africa.[12]

The second reason for the interest in ecological effects of apex predators is that their populations are in decline worldwide, mostly because of the effects of habitat loss and multiple causes of mortality due to humans. These causes of mortality include direct persecution intended to reduce predation on livestock.[3] In Australia, persecution of dingoes is often ineffective but it can succeed in reducing dingo abundance if it is consistently applied in a coordinated fashion over large areas.[1, 13–15] Over large parts of south-eastern and south-western Australia dingoes have been effectively eliminated by heavy control, backed up by barrier fencing to prevent reinvasion

Apex predators on other continents are large animals like wolves and big cats. These animals are bigger than dingoes, and they are more likely to concentrate their hunting on large herbivores. It is therefore not obvious that dingoes in Australia should have the same kinds of effects as observed for apex predators elsewhere in the world.[16] I explore the evidence on this question later in this chapter. But first, it is important to consider another question regarding the dingo's place in the Australian environment. Before the arrival of the dingo Australia's large mammalian carnivores were marsupials. Did the dingo have anything to do with the extinction of marsupial carnivores, and does the dingo now make a good fit to the ecological space that they once occupied?

The dingo versus marsupial carnivores

Large carnivorous marsupials were a prominent part of the fauna of Australia from at least 20 million years ago until ~50 000 years ago, when humans occupied the continent.[17] Probably the most formidable of these species was the marsupial lion (*Thylacoleo carnifex*). The marsupial lion was a powerfully built predator with a highly specialised dentition: most of its teeth were reduced or absent, but the third premolars filled most of the jaw and worked as long cutting blades, and one set of incisor teeth were elongated to form sharp points that must have been used for stabbing. To complete its weaponry, the large thumbs on the marsupial lion's forepaws were each equipped with a long curved claw to pierce and grasp large prey. Large individuals of the marsupial lion may have reached 150 kg, close to the mass of an African lioness. The species was widespread across Australia until ~40 000 years ago when it disappeared quite abruptly, a casualty of the larger event known as Australia's Pleistocene megafaunal extinction, which removed all the largest vertebrates of Australia. What caused the megafaunal extinction is controversial, but recent evidence supports human impact, probably via hunting, as the predominant cause.[18–20] The next largest carnivore of the time was the giant musky rat-kangaroo (*Propleopus oscillans*). Interestingly, Ride *et al.*[21] interpreted the dentition of *P. oscillans* as being distinctly canid-like, and suggested that it was an opportunistic predator analogous to dogs. It was also extinct by 40 000 years ago.

The thylacine (*Thylacinus cynocephalus*) (Fig. 8.1) and devil (*Sarcophilus*) were also widespread over continental Australia during the Pleistocene. Both survived the megafaunal extinction but disappeared from the mainland at approximately the time dingoes arrived, to survive only in Tasmania, which dingoes did not reach (Chapter 3). This coincidence in time and space of dingo arrival with disappearance of the two marsupial carnivores suggests that the dingo caused their extinction. Corbett[22] argued that because of a more complex social organisation which allowed cooperative hunting, the dingo may have fared better in conditions of resource shortage than the thylacine, and so outcompeted it. Furthermore, the skeletal morphology of the thylacine suggests that it was a slower runner than the dingo.[23] It is also possible that direct killing of thylacines and devils by dingoes could have helped the marsupials on their way to extinction; here also, the cooperative behaviour of dingoes might have given them an edge in direct conflict with thylacines and devils. Thylacines in Tasmania were larger than most dogs and were evidently able to dominate them,[24] but Letnic *et al.*[25] recently presented evidence that while mainland thylacines were similarly large when humans first

Figure 8.1: Thylacines housed at the National Zoological Park, Washington about 1903. The thylacine is thought to have become extinct on the Australian mainland roughly 5000 years ago, and in Tasmania in 1936.
Source: Smithsonian Institution Archives, Image SIA2007–0070.

came to Australia ~50 000 years ago, they subsequently underwent a reduction in size and by the Holocene were about the same size as dingoes, or even slightly smaller. Another possible advantage that dingoes might have enjoyed could have been their association with people: possibly, the fact that some dingoes spent part of their lives with people and eventually moved back into the wild (see Chapter 4) provided a demographic buffer for the wild dingo population that might have helped to sustain it while the fully wild thylacine and devil died out.[19]

So, there is persuasive, if circumstantial, evidence that dingoes caused the extinction of thylacines and devils on mainland Australia, perhaps through a combination of ecological competition and direct aggression. But there is an alternative hypothesis: from ~5000 years ago the human population of mainland Australia appears to have increased dramatically as part of the social and demographic change referred to by archaeologists as 'intensification', and this increase was associated with improvements in hunting technology.[26–28] These changes were widespread across mainland Australia, but there is little or no evidence for them in Tasmania. We know from rock art and archaeological evidence that people hunted both thylacines and devils.[19] Perhaps increased hunting pressure due to intensification, in combination with reduction of shared prey populations as a further consequence of the increasing human population, caused extinction of the thylacine and devil? Wild dingoes would have been subject to the same pressures, but their intermittent association with people might have afforded them some protection. To complicate matters, it is quite plausible that human impact and competition with dingoes worked together to push thylacines and devils to extinction on the mainland. There is the additional factor of an increasingly variable and somewhat drier climate over the last few thousand years due to intensification of the El Niño-Southern Oscillation system.[29, 30]

At the moment, we have no archaeological or fossil evidence to show which of these mechanisms was most important in driving the thylacine and devil extinct on the mainland. There is a tantalising hint from depictions in rock art that in the north of Australia thylacines disappeared before dingoes arrived.[19] Walsh[31] divided the rock paintings of the Kimberley region into distinct stylistic periods in chronological sequence. Thylacines appear frequently in the earlier styles and make their final appearance in paintings from Walsh's 'clothes peg figure' period. The ages of these paintings are mostly unknown,[32] but it is possible that the clothes peg figures were painted more than 20 000 years ago.[33] Then, there is (apparently) a long span of paintings named by Walsh the 'clawed hand' period, which contains depictions of neither thylacines nor dingoes. Dingoes do not appear until the 'Wandjina' paintings, which were mostly painted within the last 3000 years.[33] This history suggests that thylacines disappeared for reasons other than pressure from dingoes. Possibly, further analysis with more accurate dating of rock art could help resolve whether thylacines and dingoes interacted in northern Australia and, in conjunction with further dates on subfossils, reconstruct the pattern of decline of the thylacine.

Recently, Prowse *et al.* took an alternative approach.[34, 35] They built a comprehensive mathematical model of the interactions among predators (people, dingoes, thylacines, devils) and their prey (represented by kangaroos) in prehistoric Australia. The model included components describing the effects of rainfall on vegetation, allowing them to simulate different El

Niño regimes, the effects of vegetation on kangaroos and the direct and indirect effects of predators on kangaroos and one another. They ran the model from before arrival of the dingo until just before European impact, allowing the human population to increase at a rate inferred from the archaeological record, and they watched what happened through thousands of simulations that included all plausible combinations of input variables.

A model like this cannot tell us what actually happened in the past. But it can show which of many possible histories is most plausible, given present knowledge, and we can experiment with model simulations to test which processes might have been most influential in arriving at the final result. The answer in this case was surprisingly clear: the most influential factor leading to thylacine and devil extinction was human population increase, and it had its strongest effect by reducing abundance of prey for devils and thylacines. The other factors – direct and indirect impacts of dingoes, climate variability, and hunting of thylacines and devils by people – all increased the likelihood of extinction, but were far less significant.

This modelling establishes the possibility that the thylacine and devil went extinct for reasons mostly unrelated to the dingo, which stepped into the ecological space that was vacated with their disappearance. If so, to what extent was this simply a changing of the guard, in which the newcomer replaced the ecological roles that had been performed by the earlier species without significant change or disruption to that role?

There are striking similarities between the thylacine and dingo, and some rather more subtle differences. The thylacine was a classically wolfish animal, with teeth that were canid-like and therefore similar to those of the dingo, except that the thylacine's dentition was slightly more specialised for carnivory. Because the thylacine was probably a relatively slow-moving predator, it may have hunted prey of its own size or smaller that it caught by ambush rather than by long fast pursuits.[23] The prey are most likely to have been animals like medium-sized kangaroos (probably including females and subadults from large species), small macropods, possums and so on. The thylacine's skull was more susceptible to the stresses that would be imposed by a large struggling prey animal than was the skull of the dingo, suggesting that while there was probably considerable overlap between the two species in their use of small to medium prey, the dingo was more capable of handling prey animals larger than itself, which would generally require social hunting for their capture[36] (see Plate 14). The dingo is a versatile hunter, capable of switching from solitary hunting of small prey to social hunting of large prey depending on prey abundance.[22, 37] It was probably more versatile in this respect than the relatively specialised thylacine.

The dingo (unlike the thylacine) is a scavenger as well as a predator, but as a scavenger it is outclassed by the devil. With its robust skull, massive jaws and bone-crushing teeth, the devil can consume all parts of a carcass, whether taken as prey or found as carrion.[23, 38] So, it is quite possible that the efficiency with which the scavenging function was executed declined with the loss of the devil and its replacement by the dingo. Otherwise, it does seem that the dingo, as a generalist predator and scavenger, occupied essentially the same ecological niche that had previously been the combined domain of the more specialised thylacine and devil, and it probably took over hunting of a broadly similar spectrum of prey species that may even have included more large animals than were taken by the thylacine.

Another way to look at this problem is to ask whether there is evidence that the replacement of the thylacine and devil by the dingo upset the ecological balance of Australia in some way. For example, given the evidence noted above that the dingo is a more effective pursuit predator than the thylacine was – on account of its speed, stamina and ability to mount cooperative chases of fast-moving prey – its arrival in Australia may have put increased pressure on some prey which were (mostly) fast enough to escape thylacines but no match for the dingo. A significant shift like this might be apparent in the extinction of some prey species when the dingo took over as the dominant non-human predator on mainland Australia.

Here the evidence is clear: with one ambiguous exception, we know of no extinctions of prey species coinciding with the arrival of the dingo. The exception is the Tasmanian native hen (*Gallinula mortierii*), a 2 kg flightless bird that occurred from southern Queensland to Tasmania during the Pleistocene[39] but is now restricted to Tasmania. The most recent record of the native hen on mainland Australia dates to 4670 years ago.[40] Being quite large and flightless, Tasmanian native hens are vulnerable to mammalian predators. Baird[40] argued that because the disappearance of this species from mainland Australia approximately coincided with the arrival of the dingo, and because it persisted in Tasmania, dingoes could have caused their extinction. However, increased hunting pressure from humans provides a plausible alternative in this case as it does for the thylacine and devil. There is an especially close match between changes in the human population and the extinction of the Tasmanian native hen from mainland Australia. The hen lives close to permanent water but nests on dry ground – both the eggs and incubating adults are therefore easy to harvest. The sparse fossil record for the species suggests that by the Holocene it was restricted to wetland habitats in southern Victoria. We have a detailed archaeological record for this region and period, and it shows that people increased their use of such habitats and probably began to occupy them year-round rather than seasonally as they had previously done, at about the same time that the hen population went extinct.[41]

On the other hand, it seems that all other prey species survived the onslaught of the dingo in the mid-Holocene. This is in stark contrast to the recent impact of introduced placental predators. The red fox and feral cat have been the main contributors to extinction of more than 20 species of Australian mammals over the last 200 years[19, 42] and there is evidence that feral cats are driving current declines of small mammals across northern Australia.[43, 44] This contrast provides powerful evidence that, while the extinction of the thylacine and devil may be deeply regrettable events, the replacement of those carnivores by the dingo did not herald a significant disruption to the ecosystems of mainland Australia.

Do dingoes limit populations of prey?

The dingo invasion may not have caused an ecological upheaval in prehistoric Australia, but it is quite clear that dingoes have large effects on the contemporary ecology of Australia.[45] Dingoes occur in a huge range of habitats, from tropical rainforests to deserts and alpine grasslands. While they tend to hunt medium- to large-sized species as their preferred prey when such species are available, they can switch to preying on small mammals, reptiles and even invertebrates and vegetation when larger prey are scarce; they also feed readily on carrion and hunt livestock, especially sheep and goats.[1, 45–49] This ecological versatility allows dingoes to survive in most Australian ecosystems, where they are capable of affecting populations of many other species.

Perhaps the most consistent ecological effect of dingoes is to limit population growth of the larger animals that are often their preferred prey. This became clear soon after European settlement in Australia, when destruction of dingo populations was quickly followed by rapid increase of kangaroos.[50] A good example comes from the experience of the pioneer grazier Albert Wright, who took up grazing land in the Dawson Valley of central Queensland in the 1860s.[51] Initially both dingoes and kangaroos were uncommon in the region, apparently because of hunting by Aborigines. Dingoes quickly increased following the disruption and depletion of the Aboriginal population. This alarmed the new settlers, who saw the threat of predation on their sheep flocks increase. They responded by organising a mass poisoning campaign, which was effective: 'The strychnine campaign had depleted the dingoes; few now howled round his [Albert Wright's] night camps.' This success against the dingo population was followed by a massive 'plague of marsupials': 'As he rode he noted that thousands of wallabies and kangaroos were invading the open country – his night camps were nearly bare of grass even after the long wet season'.[51] The settlers responded to this new threat with heavy persecution of macropods as well as dingoes, striking an unstable balance in which continuous suppression of both wild herbivores and predators was applied as a precarious defence of the pastures and security of domestic grazing animals. This is still the situation throughout most of the pastoral regions of Australia.

The effect of dingoes on abundance of kangaroo and emu populations has been confirmed by more recent research. Early surveys of kangaroos on either side of the dingo fence showed repeatedly that kangaroos were far more abundant 'inside' the fence, where heavy suppression reduces or eliminates dingo populations, than in the outside country where dingo control is sporadic and less effective.[52, 53] Attributing this contrast in kangaroo abundance to dingo predation is not straightforward, because there are other differences across the dingo fence: mostly, sheep are grazed inside the fence and cattle outside, and the inside landscapes also tend to be more productive and better supplied with water.[54, 55] Subsequent studies of the dingo fence have taken in a larger range of land types and used designs that controlled for potentially confounding environmental differences.[15, 56, 57] These provide compelling evidence that the abundance of dingoes is the factor responsible for contrasting kangaroo abundance on either side of the fence. In a recent (as yet unpublished) study Euan Ritchie and I worked on large cattle stations in north Queensland, central Australia and Western Australia, all far north of the dingo fence, that do not control dingoes and compared them with nearby properties that do control dingoes. We found that dingo control produced variable but mostly significant effects on dingo activity, which on average was halved in the presence of control (see also[13]). This difference produced a disproportionate effect on kangaroo abundance, which on average was seven times higher where dingoes were controlled.

Two other studies confirmed the strength of the effect of dingoes on abundance of kangaroos, and added evidence that dingo predation not only reduces the average size of kangaroo and emu populations but regulates population size in those species: that is, under the influence of dingoes the abundance of prey populations is more stable than when dingoes are removed. Pople *et al.*[57] used aerial survey data to study trends in kangaroo and emu abundances on either side of the dingo fence in South Australia over a 15-year period. Outside the fence, kangaroo and emu populations were consistently low, and showed little or no increase in response to good rainfall. Inside the fence, and without dingoes, numbers of kangaroos and emus were much higher and they increased rapidly to very high levels in good seasons then crashed in poor seasons. Letnic and Crowther[58] compared estimates of kangaroo abundance over a series of locations sampling a gradient in average rainfall, contrasting sites on either side of the fence at each location. In the absence of dingoes, abundance of kangaroos increased steeply with average rainfall, but in their presence the response to rainfall was much less, and kangaroo abundance was more similar among sites as well as being lower.

These are significant observations for management of rangeland vegetation. The condition of vegetation is likely to be better, and conflicts with the grazing industry lessened, if kangaroo populations are held at low numbers and are prevented from fluctuating greatly in response to variations in rainfall. If dingoes accomplish this, the result should be better and more consistent vegetation condition with more complete coverage of the ground by grasses and other herbs in areas with higher dingo activity (other things being equal).[15, 59] It is quite likely that dingoes also reduce populations of medium-sized ungulates such as feral goats, feral pigs and deer; there is anecdotal evidence for this[1] but effects of dingoes on wild ungulates are much less studied than on kangaroos and emus.

Most studies of the effects of dingoes on large herbivores have been done in the semi-arid and arid rangelands of the drier interior of Australia. We know less about interactions of dingoes and their prey in relatively high-rainfall forest habitats, but the evidence available suggests that the effects of dingoes on forest macropods are similar to those in the rangelands. Colman *et al.*[14] surveyed dingo activity in forest conservation reserves in eastern New South Wales, comparing reserves in which dingoes were not controlled to nearby and environmentally similar reserves in which poison baiting was used to suppress dingo populations. Counts of macropods and the density of macropod dung (an index of intensity of herbivory by macropods) were lower where dingo populations were not subject to poison baiting, than in matched sites with baiting. Also, the structural complexity of vegetation between ground level and a height of 200 cm was greater where dingo populations were uncontrolled. This suggests that when macropods are abundant (partly due to rarity of dingoes), they impose heavy grazing and browsing pressure which keeps the herb and shrub layer simple and open. An earlier study examined prey selection by dingoes in forests of north-eastern New South Wales, and demonstrated a strong preference for a mid-sized macropod, the swamp wallaby (*Wallabia bicolor*).[60] Abundance of swamp wallabies was lower where dingoes were abundant, and breeding was also disrupted: at a site with low dingo predation the wallabies bred predominantly in late summer, but where predation was high breeding extended throughout the year. Probably this was because harassment of wallabies by dingoes caused high mortality of pouch-young, leading to disruption of synchrony of breeding as some females re-bred to produce replacement young outside the typical breeding season.[60]

So, the evidence that dingoes reduce populations of large native herbivores is strong, and it seems likely that the effects of dingoes are important across a broad range of environments, from deserts to forests. The evidence noted above for regulation of population size of herbivores is particularly significant: it suggests that predation by dingoes is unlikely to push abundance of those species all the way down to zero, and hence cause local or complete extinction. The studies which reported large reductions of macropod and emu abundance under dingo predation did not find that those prey species disappeared; rather, they persisted at low to moderate abundance. At these intermediate levels of abundance, the productivity of populations is generally high because of weak intraspecific competition for resources. That is, fertility and the growth and survival of young should be strong, because of high per capita availability of food and habitat. Prey populations with these characteristics should be resistant to further declines in abundance, making extinction unlikely provided that the predator population is reasonably stable.

Dingoes prey on a variety of small-bodied species as well as kangaroos, emus and introduced ungulates. If predation by dingoes affects the abundance of larger prey, it might have similarly important effects on small-bodied species like reptiles, small mammals and ground-dwelling birds. Does predation by dingoes threaten the survival of any such species? Corbett[22] argued that early in the history of European settlement of Australia dingoes increased in abundance as a result of reduced Aboriginal predation, along with other developments such as the introduction of additional prey in the form of domestic livestock and the creation of artificial water points for those newcomers. He went on to speculate that increased dingo predation was partly responsible for the wave of extinctions of medium-sized native mammals that followed European settlement of Australia. There is some evidence that the abundance of dingoes did increase soon after European settlement – as in the experience of Albert Wright, recounted above – although we know surprisingly little about this. On the other hand, there is no compelling evidence that this was connected with rapid decline leading to extinction of any prey species. Where we have good evidence on the causes of decline and extinction of small and medium-sized mammals in recent Australian history, they have been most closely linked to the arrival and impact of smaller invasive predators, especially the red fox.[19, 61, 62] Foxes continue to threaten many native species today, and it is becoming clear that feral cats are also a significant threat.[44, 63]

Even allowing that dingoes might not have caused dangerous declines of prey species in the past, several authors have argued that predation by dingoes is currently a significant threat to many species of Australian wildlife, so that if (for example) dingoes were able to return to parts of New South Wales from which they have been extirpated, extinctions of native species might follow.[16, 64] On the other hand, there are several reasons to expect that predation by dingoes should have relatively low impact on smaller native species, certainly compared with the effects of predation by foxes and feral cats.

First, dingoes tend naturally to occur at low population densities, on the order of only one or two individuals for every 10 square km.[1, 65] Small marsupials and native Australian rodents typically live at population densities that are hundreds or thousands of times greater than this; in most cases, the intensities of predation that dingo populations are capable of imposing are not likely to trouble them greatly. In contrast, red foxes live at densities 10 or 20 times higher than that of a typical dingo population[65] and densities of feral cats are similar or higher still.[66, 67] These two predators are therefore more capable of killing small-bodied prey at unsustainable rates.

Second, small predators like red foxes and feral cats use their habitat in a more 'fine-grained' way than dingoes. Brook[68] recently used GPS tracking to examine space use by individual dingoes and feral cats in the same landscape, a semi-arid savanna in the Kimberley region of north-western Australia. She found dingoes used large ranges but that the distribution of activity within the range was highly skewed, so that the range of each dingo contained large spaces that were rarely visited by its owner and several hotspots of activity that were frequently used. Feral cats used smaller ranges and spread their activity more evenly through the range, using every part of the range at higher and more uniform intensity. The result of this more intensive use of habitat is that a population of feral cats throws a more complete blanket of predation over the landscape, leaving no areas where some prey can escape high predation, while the behaviour of dingoes is more likely to leave such gaps. Also, because cats are smaller and more agile than dingoes they can enter smaller spaces and climb rock piles and trees, making it harder for prey such as small mammals to find refuges from the threat of predation. Third, as noted above, while dingoes hunt a wide range of prey, small and large, they tend to prefer medium-sized and large mammals. This means that they have relatively low impact on small mammals, reptiles and birds. Foxes and cats on the other hand prefer such small-bodied prey, especially small mammals.[47, 48, 69, 70]

None of this means that dingo predation can never threaten the survival of populations of prey. One recent case where dingoes have been identified as a serious threat is the northern hairy-nosed wombat (*Lasiorhinus krefftii*). There is only one surviving population of this species in the wild and it contains not much above 100 individuals, in Epping Forest National Park in central Queensland. This population is slowly increasing from a low point of as few as 20 or 30 animals,[71, 72] but it suffered a significant setback when dingoes invaded its range and killed as many as nine individuals;[73] Epping Forest National Park is now surrounded by a dingo-proof fence. The northern hairy-nosed wombat was once more widespread and there is no evidence that its initial decline to the edge of extinction was caused by predation (the decline was probably driven mainly by overgrazing of its habitat by livestock).[72] There are other cases in which dingoes may threaten the survival of prey species,[16] but they are similar to the northern hairy-nosed wombat in that the animals in question appear to have become rare for reasons other than predation by dingoes and are now highly vulnerable to any additional stressor. Another situation in which dingoes potentially threaten the survival of prey populations could occur if the population density of dingoes is unnaturally high. For example, Newsome *et al.*[74] studied dingoes that raided a refuse tip around isolated townships in the Tanami Desert in Central Australia, and showed that access to this food subsidy boosted local density and resulted in increased predation on small mammals. Finally, it is possible that dingoes could have particularly significant impacts on native predators of similar size to foxes and cats, given the evidence (reviewed below) on the strong effects of dingoes on foxes and cats. The only native species on mainland Australia that falls into this category is the spotted-tailed quoll (*Dasyurus maculatus*). This species has declined severely and is now rare on the mainland, like all species of quolls, due largely to impacts of foxes and cats. The effects on it of dingoes are mostly unknown.[75]

To summarise, contemporary evidence suggests that predation by dingoes does not pose a general threat to the viability of species of native wildlife in Australia, and is likely to do so only under exceptional circumstances. This is consistent with the fact that dingoes coexisted for several thousand years with many species of native wildlife, including species that have recently gone extinct or are now threatened with extinction. To the extent that predation caused these recent declines and extinctions, the culprits are the red fox and feral cat.[19] Compared to these recent arrivals, the dingo is a benign presence in Australian ecosystems. All the same, it could be argued that predation by dingoes must make matters worse for species threatened by foxes and cats, even if by itself it would not drive them extinct, because it adds to the total predator-caused mortality of populations of these animals. But as the next section will show, this seems not to be the case.

Dingoes, mesopredators and prey

In the early 1980s I was doing research on a cattle property named Wallaby Creek in north-eastern New South Wales, a place with an exceptionally diverse mammal fauna.[76] Among its other values, Wallaby Creek was one of the few sites in New South Wales where the rufous bettong (*Aepyprymnus rufescens*), a rat-kangaroo vulnerable to predation by the red fox, remained abundant. The property owners – two exceptionally knowledgeable naturalists – maintained that this was because they did not poison dingoes, and the dingoes 'hunted' the foxes. In four years of fieldwork, during which I regularly saw dingoes and often wandered out at night for the pleasure of watching rufous bettongs, I did not see a single fox.

That observation points to an extremely strong effect of dingoes on foxes. Control by predation alone does not seem sufficient to explain this. Dingoes do occasionally eat foxes, as well as cats, but foxes and cats never make more than a small contribution to their diet.[48, 77] However, dingoes also kill foxes and cats without eating them, as shown in an experimental study in the arid north of South Australia. Moseby *et al.*[78] established a large predator-proof enclosure (the 'dingo paddock'), into which they released a pair of adult dingoes.[78] At ~37 km^2, the dingo paddock was large enough to provide territory and prey for the dingoes at a density typical of the area, without food supplementation, and the habitat was natural. They released seven foxes and six cats into the enclosure. These animals and the two dingoes were fitted with GPS collars so that their movements and spatial interactions could be recorded. The dingoes killed all the foxes within a month; the cats also died over a three-month period, and at least three of these deaths were due to attacks by dingoes. In all these cases, the bodies of the foxes and cats were left uneaten where they had been killed (although they did not use the carcasses as food, the dingoes seemed fascinated by them and returned frequently to view them).

This study was based on a small number of animals, and the fact that dingoes eradicated foxes and cats inside the dingo paddock but have clearly not done so in other parts of the arid zone where foxes and cats persist in the presence of dingoes suggests that it overestimated the rates at which dingoes kill foxes and cats. Nonetheless, the study provided clear evidence that dingoes have a propensity to attack and kill foxes and cats. If that is generally true, we should expect two consequences: (1) the abundance of foxes or cats, or both, should be inversely related to the abundance of dingoes and (2) foxes and cats should be more wary in the presence of dingoes, and use behaviours that minimise their risk of encountering dingoes. This risk-minimisation could be accomplished either by avoiding habitats or locations where dingoes are most likely to occur, or by reducing activity at times when dingoes are most active.

The first prediction has been tested in several studies that surveyed the abundance of foxes and cats in sites with contrasting dingo abundance – except that so far these studies have not determined the actual numbers of predators living in the areas studied, relying instead on indirect indicators of abundance such as counts of tracks on prepared sand-plots, counts of scats, or rates of detection on remote cameras. This is a significant limitation (discussed further below) but the studies nonetheless provided a series of cases in which indices of abundance of foxes or cats were lower where abundance of dingoes was (evidently) higher. These cases include dingoes and foxes across the Tanami Desert,[79] dingoes and foxes in conservation reserves with and without dingo poisoning in the forests of eastern New South Wales,[14, 80] dingoes and foxes across the dingo fence in South Australia, New South Wales and Queensland,[15] dingoes and cats on paired cattle stations in north and central Australia with and without poison baiting of dingoes,[13] and dingoes and cats in conservation reserves managed by the Australian Wildlife Conservancy in northern Australia.[81] In some cases in which dingoes coexist with both foxes and cats, indices of abundance of foxes were lower where abundance of dingoes was higher, but abundance of cats was apparently unaffected.[14, 15] This could indicate that dingoes have weaker effects on cats than on foxes. Another possibility is that cats benefit from reduction of fox abundance due to dingoes, and this offsets the direct effects on them of dingoes. This is plausible, given that we have evidence that foxes suppress cats in the same way that it appears dingoes suppress foxes. As a result, fox control (in the absence of dingoes) leads to increased abundance of cats.[63, 82, 83]

We know less about whether the presence of dingoes forces changes in the behaviour of foxes or cats. Some observations of foxes avoiding locations regularly used by dingoes,[84, 85] and rapidly occupying sites following removal of dingoes,[86] suggest the existence of a sophisticated strategy of behavioural avoidance of dingoes by foxes. The most dramatic instance of this was when poison baiting was used to reduce numbers of dingoes in the vicinity of one of the last two small populations of the rufous hare-wallaby (*Lagorchestes hirsutus*) surviving in the Tanami Desert. Foxes had been absent from the site, but a week or two after the dingoes were poisoned foxes appeared and killed all the resident wallabies.[87] The local fox increase was too rapid to be due to population growth, and must have been caused by a small number of foxes (perhaps just a single animal) taking the opportunity to move into the area to hunt hare-wallabies.

There is also good evidence that cats modify their behaviour to minimise encounters with dingoes. The study by Brook *et al.*,[13] which showed that cats were more abundant on cattle stations where dingoes were controlled, also found a shift in timing of activity of cats. In the absence of control, dingoes tended to show crepuscular behaviour, being most active around dawn and dusk, probably because these are the best times of day to hunt large macropods (timing of activity was measured by detections on remote cameras). Under these conditions, cats (which are strongly nocturnal) were most active in the middle of the night. On properties where dingoes were baited, the dingoes reduced their activity at dusk, and concentrated their activity in the hours before dawn. This could be because poison baiting is often accompanied by opportunistic shooting and, to minimise their risk of being shot at, dingoes shifted their activity to times when shooters were least active. The shift by dingoes from dusk and to pre-dawn activity was accompanied by a sharp increase in activity of cats around dusk. This was probably beneficial to cats, because the nocturnal small mammals that cats like to hunt also have a peak in activity around dusk and in the early evening. By focusing their hunting at this time cats may well increase their hunting success.

Cats avoid dingoes in space as well as in time, even though the two species overlap in their use of habitat. Brook's[68] simultaneous GPS tracking, revealing the fine-scaled movements of cats and dingoes in the same landscape, showed that cats avoid coming within several hundred metres of dingoes and at any moment are more likely to be a kilometre away from a dingo than would be expected if the movements of cats and dingoes were independent. To maintain this separation, cats must frequently move from one place to another depending on the position of the nearest dingo, a task that would be more challenging when the local abundance of dingoes is high, or if dingoes sometimes pursue cats as Moseby *et al.*'s[78] study suggested.

If dingoes do reduce the abundance of foxes and cats, this might be beneficial for species vulnerable to predation by those mesopredators. Less obviously, changes in behaviour of foxes and cats, forced by their need to avoid dingoes, could reduce their efficiency as predators. This is because we expect predators to hunt in the places and at the times where they have the greatest likelihood of finding and killing their preferred prey. If foxes and cats are obliged to compromise their decisions about where and when to hunt in order to reduce risk to themselves, the effect would be to move their hunting behaviour away from its optimum. The result, in theory, should be a lower hunting success and therefore reduced impact on prey. The changes in behaviour of cats in the presence of dingoes illustrate how this might work. Cats shifted their activity times away from the peak in activity of small mammals, and the extra movements forced on them by the need to remain a safe distance from dingoes presumably often disrupted and frustrated their hunting behaviour – although it has to be said that the effects of these behaviour changes on hunting success are not known, and measuring them in the wild will be extremely difficult.

Whatever the mechanisms involved, there are strong positive correlations between the presence or abundance of dingoes and the abundance and diversity of species threatened by foxes and cats. These have been observed at the fine spatial scales illustrated by the anecdotes of the rufous bettong and rufous hare-wallaby, as well as at whole-landscape and continental scales. At the landscape scale, bilbies (*Macrotis lagotis*) are most likely to persist in those parts of the Tanami Desert which are occupied by dingoes, probably due to the fact that activity of foxes is lower where dingoes are present.[79] Broad surveys of sites in arid Australia have found repeatedly that activity of small or medium-sized mammals is positively related to indices of abundance of dingoes,[15, 59, 88] and surveys of forest sites with contrasting dingo abundance reveal a very similar pattern.[14] Letnic and co-workers examined in detail the effects of contrasting activity of dingoes on an endangered arid-zone rodent, the dusky hopping mouse (*Notomys fuscus*).[89, 90] They showed that abundance of dusky hopping mice was negatively related to activity of foxes, and positively to dingoes. There was also a positive relationship of hopping mouse abundance to mean annual rainfall that was independent of the effects of predators or of other species such as livestock, rabbits and kangaroos. The magnitude of the positive effect of dingoes on hopping mice was similar to that of rainfall; this is an impressive finding, given the clear importance of rainfall in the ecology of rodents in arid Australia. Contrasts in the local activity of dingoes and foxes also affected behaviour of hopping mice. Where activity of dingoes was high and foxes low, hopping mice spent more time feeding on experimental food patches than they did under the opposite conditions. This suggests that foxes had two distinct types of negative effect on hopping mouse populations: they killed and ate hopping mice (the occurrence of hopping mice in the diet of foxes was greater than in that of dingoes) and their presence also reduced food intake by hopping mice, which were reluctant to spend long periods feeding where the risk of being found by a fox was high.

Finally, at the scale of the entire continent, there is a strong positive relationship between the overlap of the geographic ranges of marsupial species with regions where dingoes are abundant, and the extent to which those species have maintained their original geographic range.[91] In other words, range declines of marsupial species (including declines to extinction) have occurred mostly where dingoes have been removed or are naturally rare. Geographic overlap with dingoes is a stronger predictor than other factors potentially associated with decline of marsupial species, such as overlap with sheep or rabbits.

Trophic cascades

A trophic cascade is a series of changes in the abundances of species, originating from a change in a single species (or guild of ecologically analogous species) and spreading through the network of interactions that link species together into an ecosystem. As shown at the beginning of this chapter, the removal or suppression of apex predators can trigger trophic cascades because apex predators not only interact strongly with herbivores, they have the capacity (and often, it seems, the motivation) to affect smaller predators as well, while herbivores and small predators have large effects on still other species. There is a great deal that we do not yet know about the role of the dingo in Australian ecology, but the last two sections make it clear that there are grounds to expect that removal or restoration of dingoes could trigger trophic cascades involving many other species.

However, the potential complexity of the influences of dingoes on other species makes it difficult to interpret the ecological pathways involved. For example, if we find that high activity of dingoes is correlated with healthy small-mammal communities in a certain ecosystem – leaving aside for a moment the problem of whether we can infer that the two things are causally related – is this because of the effect of dingoes on mesopredators? Or because dingoes control kangaroo populations and thereby protect the food resources of small mammals from the impacts of over-grazing? Or perhaps because the more complex vegetation that forms when large herbivores are limited by predation provides refuges that allow small mammals to better evade their predators, regardless of any direct effect of dingoes on foxes and cats?

Figure 8.2 illustrates a preliminary analysis of this problem. Colman *et al.*[14] used a statistical procedure called structural equation modelling to identify the interaction pathways that best account for the differences in ecological variables that they recorded in forest sites with and without dingo control in eastern New South Wales.[14] This showed, for example, that the activity of small mammals was affected about equally strongly by density of ground vegetation (a positive effect), activity of foxes (negative) and activity of dingoes (positive). Vegetation density in turn was positively affected by the grazing and browsing activity of macropods, which was reduced by high dingo activity. There was also a strong effect of fire history on vegetation density, so fire history indirectly affected small-mammal activity independently of predator activity. Poison baiting was linked to reduced dingo activity and increased fox activity.

These pathways are probably best regarded as hypotheses that are supported by current evidence, rather than as firm conclusions. This is because the number of sites surveyed in this study (14) was small relative to the complexity of the interactions interpreted by the statistical modelling. So, as Colman *et al.*[14] pointed out, the statistical power of the analysis to uncover all significant pathways was low (note also that for the sake of clarity Fig. 8.2 does not show all pathways modelled by Colman *et al.*, in particular relationships involving possums and gliders are excluded). Furthermore, establishing the directions of cause and effect of the pathways depicted in Fig. 8.2, as distinct from circumstantial association, will ultimately require some form of experimentation. Ideally, an experimental approach would encompass independent manipulation of each of the proposed causal factors in the diagram – dingo, fox and macropod abundance and vegetation density – to test their effects on small-mammal abundance and other variables of interest, while controlling for other causal factors and important extrinsic drivers like fire history. This would entail several lifetimes of work by a

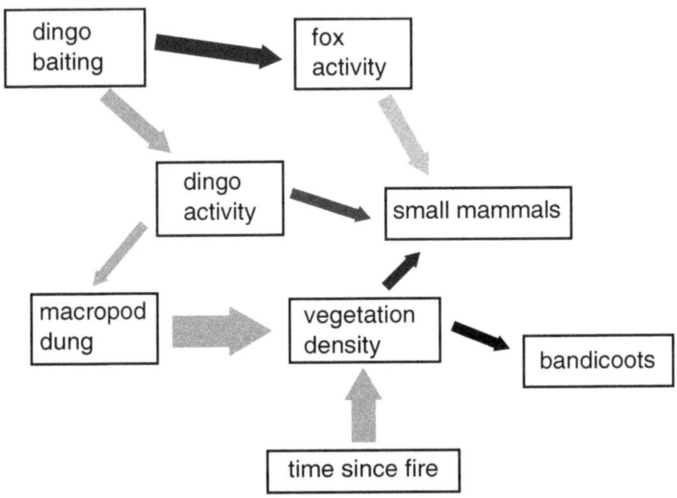

Figure 8.2: Ecological relationships among mammals, vegetation structure and fire in sclerophyll forest of eastern Australia.[14] The width of each arrow is correlated with the strength of the effect. Positive relationships, in which an increase in the magnitude of one variable is associated with an increase in the other, are indicated by black arrows, negative relationships by grey arrows.

small army of ecologists. While we wait for that to happen, we can at least conclude that there is a reasonably high likelihood that a simple management decision – whether or not to lay poison baits for dingoes – is likely to have far-reaching effects that lead ultimately to changes in vegetation structure and diversity of smaller animals.

Colman *et al.*'s[14] study and the others reviewed in this chapter do not demonstrate that dingo predation is the only or even necessarily the strongest causal factor influencing the myriad processes and interactions that make up a functioning ecosystem. For example, the link between macropod activity and vegetation structure revealed by Colman *et al.*'s analysis was considerably stronger than the hypothesised effect of dingoes on macropod activity. Therefore, the indirect effect of dingo abundance on vegetation structure will be diluted and sometimes overridden by other factors, the most obvious of which in this case is fire. In other words, the effects of changes in dingo abundance on ecosystem processes are likely to be complex and difficult to predict. Nonetheless, dingoes as apex predators have an especially significant role because of the potential for their effects to spread so far through the ecosystem, by virtue of the many strong links that connect them, directly and indirectly, to so many other species. Perhaps the only other factor that has such far-reaching importance in Australian terrestrial ecology is fire. To an ecologist, therefore, playing with dingoes is like playing with fire.

Controversies and uncertainties

The idea that dingoes provide ecological benefits – or, more to the point, that control of dingo populations can incur environmental costs – is controversial. To a remarkable extent, the controversy has focused on the suggestion that dingoes are capable of reducing the abundance or modifying the behaviour of foxes and cats to a degree that provides net benefit for populations of small native vertebrates that are threatened by foxes and cats. In particular, there have been strong criticisms of the design and methodology of the ecological studies that have produced evidence that higher abundance of dingoes is associated with lower activity of foxes and cats. Allen *et al.* took particular aim at studies that used sand-plot tracking to measure activity of dingoes, foxes and cats in landscapes with differing populations of dingoes.[92, 93]

Probably the most significant criticism is that these studies (and many others described in this chapter) were non-experimental. Rather than doing controlled manipulations of dingo populations, they relied on pre-existing contrasts in dingo abundance (especially those produced by long-term control) and examined correlates of those contrasts to infer likely effects of dingo abundance. In the absence of controlled manipulations, such studies do not demonstrate cause and effect. Two published studies set out to experimentally manipulate abundance of dingoes and monitor the effect of the change on other species of wildlife (especially foxes and cats). Eldridge *et al.*[94] conducted poison baiting on half of each of three cattle stations in central Australia, leaving the other half of each station as an unbaited control. Unfortunately, their baiting programs produced very little change in dingo activity, perhaps because the study was conducted during a run of unusually good seasons when abundance of all species of wildlife remained high. Allen *et al.*[95] implemented a similar experimental design on cattle stations in South Australia and Queensland, and claimed that their result demonstrated that lethal control of dingoes had no significant effects on other predators. However, in this study it was not clear whether baiting reduced dingo activity, and some of the statistical analysis presented by the authors suggest that it did not.[96] So far, the only firm conclusion that we can draw from these experimental studies is that short-term programs of poison baiting often fail to produce reductions in dingo abundance. Experimental tests of the effects of dingoes on other species are still badly needed, but it is clear that conducting such experiments is a very difficult challenge.

There is a larger problem in testing the ecological effects of dingoes using experiments: the effects are likely to emerge over a wide range of spatial and temporal scales, from the local and immediate outcomes of interactions between individual animals through the consequences of those interactions for demography of populations and ultimately to emerging properties of changed interactions among species which affect processes at landscape and larger scales and over many generations of predator and prey populations. If we want to understand how dingoes fit into Australian ecology, we need studies that address all these spatial and temporal scales, but there is no feasible experimental design that can address them all adequately. For some questions, such as what effects dingoes and other predators have had on prey in the past and how they influence the structure of prey assemblages over large regions, experiments are not possible. Therefore, to require experimentation as the sole criterion for reliability of inference in this case is unrealistic. On the contrary, we need to use a broad range of research strategies, of which field experimentation is one kind. Viewed in that light, the studies reviewed in this chapter amount to useful progress towards an understanding of the ecological role of dingoes in Australia.

Published surveys of the relationship of dingo to fox and cat populations have also been criticised because they used indices of abundance, rather than measures of absolute population size, to describe differences in population sizes of dingoes, foxes and cats.[97] Indirect measures of abundance were used because the effort required to capture and mark individual animals of any of these species, and so infer absolute population size, is so great that it is impossible to accomplish over the number of independent sites needed to demonstrate statistical relationships among abundances of the different species. Hayward and Marlow[97] criticised the use of abundance indices (e.g. track counts on sand plots) because they are potentially subject to biases unrelated to the true population size of the animals responsible for them. This can be true, although it should be noted that many of the studies reviewed in this chapter used designs that minimised the likely impacts of these biases on their results. In any case, Hayward and Marlow's conclusion that because there are uncertainties in the interpretation of abundance of dingoes, foxes and cats we have no basis to change the management of dingoes, is faulty because it ignores other evidence relevant to ecological management of dingoes, such as their role in regulating populations of large herbivores.

Another source of uncertainty is due to the recurring problem of deciding what a dingo is. While it is possible to define the characteristics of a 'true' dingo in principle and show how dingoes differ from domestic dog breeds,[98] the ease with which dingoes and domestic dogs are able to interbreed means that in many parts of Australia dingo–dog hybrids are common in wild populations. Hybridisation has been especially pervasive in south-eastern Australia while in parts of central and northern Australia dingo populations show little evidence of genetic admixture with domestic dog breeds.[99, 100] How important are these differences, and is there any evidence that 'pure' dingoes are better (or worse) for the environment than hybrids? There is still no clear answer to this question. Most of the studies described in this chapter have not attempted to distinguish genetically pure dingoes from hybrids, because it is not practical to do so. However, it is notable that there is no clear indication that the ecological effects of dingoes are fundamentally different in south-eastern Australia, where hybrids are frequent, than in central and northern Australia where they are rare. One common feature is that all studies were conducted in places where dingo populations were stable, with little indication of strong demographic effects from domestic dogs establishing in the wild at high rates. In other words, they were in places where feral dogs, showing characteristics of distinct domestic breeds and indicative of recent establishment in the wild, were rare. In these situations, the animals display many characteristics that are close to the dingo type, whether they are in fact genetically pure dingoes or have mixed ancestry. Animals like this are sometimes referred to as 'dingo-like canids'. As far as we can tell, the ecological traits of these animals are not distinguishable from those of 'true' dingoes.

Conclusion

Australia is unusual among the continents in having few large predators, and it seems particularly strange that the largest living mammalian predator is a relatively recent arrival which, compared with large predators elsewhere, is of only moderate size. One interpretation of this puzzling fact is that predation is not a significant force in Australian ecology. However, a look into the past shows a very different picture. Until recently Australia had a diverse suite of large carnivorous marsupials, which paralleled the predator assemblages of other continents. The comparative absence of large predators is a recent anomaly, produced by the extinction of species like the marsupial lion, thylacine and others. These extinctions caused a drastic simplification of the ecological structure of wildlife communities in Australia. Seen in this light, the dingo is one of the most significant species of wildlife in Australia: it replaces at least some of the ecological function of those extinct large carnivores and, by exerting control of the key processes of herbivory and predation, it provides a stabilising influence which confers a measure of resilience on native wildlife communities. For this reason, if for no others, we should value the dingo and protect it as part of the natural fabric of Australian ecology.

References

1. Fleming P, Corbett L, Harden R, Thomson P (2001) *Managing the Impacts of Dingoes and Other Wild Dogs*. Bureau of Rural Sciences, Canberra.
2. Estes JA, Terborgh J, Brashares JS, Power ME, Berger J, Bond WJ, Carpenter SR, Essington TE, Holt RD, Jackson JBC, Marquis RJ, Oksanen L, Oksanen T, Paine RT, Pikitch EK, Ripple WJ, Sandin SA, Scheffer M, Schoener TW, Shurin JB, Sinclair ARE, Soule ME, Virtanen R, Wardle DA (2011) Trophic downgrading of planet Earth. *Science* **333**, 301–306.
3. Ripple WJ, Estes JA, Beschta RL, Wilmers CC, Ritchie EG, Hebblewhite M, Berger J, Elmhagen B, Letnic M, Nelson MP, Schmitz OJ, Smith DW, Wallach AD, Wirsing AJ (2014) Status and ecological effects of the world's largest carnivores. *Science* **343**, 1241484-1–1241484-11.
4. Palomares F, Caro TM (1999) Interspecific killing among mammalian carnivores. *American Naturalist* **153**, 492–508.

5. Helldin JO, Liberg O, Gloersen G (2006) Lynx (*Lynx lynx*) killing red foxes (*Vulpes vulpes*) in boreal sweden: frequency and population effects. *Journal of Zoology* **270**, 657–663.
6. Crooks KR, Soule ME (1999) Mesopredator release and avifaunal extinctions in a fragmented system. *Nature* **400**, 563–566.
7. Berger J, Stacey PB, Bellis L, Johnson MP (2001) A mammalian predator–prey imbalance: grizzly bear and wolf extinction affect avian neotropical migrants. *Ecological Applications* **11**, 947–960.
8. Ripple WJ, Beschta RL (2008) Trophic cascades involving cougar, mule deer, and black oaks in Yosemite National Park. *Biological Conservation* **141**, 1249–1256.
9. Ripple WJ, Beschta RL (2012) Trophic cascades in Yellowstone: the first 15 years after wolf reintroduction. *Biological Conservation* **145**, 205–213.
10. Ripple WJ, Beschta RL, Fortin JK, Robbins CT (2014) Trophic cascades from wolves to grizzly bears in Yellowstone. *Journal of Animal Ecology* **83**, 223–233.
11. Terborgh J, Lopez L, Nunez PV, Rao M, Shahabuddin G, Orihuela G, Riveros M, Ascanio R, Adler GH, Lambert TD, Balbas L (2001) Ecological meltdown in predator-free forest fragments. *Science* **294**, 1923–1926.
12. Brashares JS, Prugh LR, Stoner CJ, Epps CW (2010) Ecological and conservation implications of mesopredator release. In *Trophic Cascades: Predators, Prey and the Changing Dynamics of Nature*. (Eds J Terbough and JA Estes) pp. 221–240. Island Press, Washington.
13. Brook LA, Johnson CN, Ritchie EG (2012) Effects of predator control on behaviour of an apex predator and indirect consequences for mesopredator suppression. *Journal of Applied Ecology* **49**, 1278–1286.
14. Colman NJ, Gordon CE, Crowther MS, Letnic M (2014) Lethal control of an apex predator has unintended cascading effects on forest mammal assemblages. *Proceedings of the Royal Society B: Biological Sciences* **281**.
15. Letnic M, Koch F, Gordon C, Crowther MS, Dickman CR (2009) Keystone effects of an alien top-predator stem extinctions of native mammals. *Proceedings. Biological Sciences* **276**, 3249–3256.
16. Fleming PJS, Allen BL, Ballard G-A (2012) Seven considerations about dingoes as biodiversity engineers: the socio-ecological niches of dogs in Australia. *Australian Mammalogy* **34**, 119–131.
17. Johnson CN (2014) The rise and fall of large marsupial carnivores. In *Carnivores of Australia*. (Eds AS Glen and C Dickman). CSIRO Publishing, Melbourne.
18. Rule S, Brook BW, Haberle SG, Turney CSM, Kershaw AP, Johnson CN (2012) The aftermath of megafaunal extinction: ecosystem transformation in Pleistocene Australia. *Science* **335**, 1483–1486.
19. Johnson C (2006) *Australia's Mammal Extinctions: A 50,000 Year History*. Cambridge University Press, Melbourne.
20. Bird MI, Hutley LB, Lawes MJ, Lloyd JON, Luly JG, Ridd PV, Roberts RG, Ulm S, Wurster CM (2013) Humans, megafauna and environmental change in tropical Australia. *Journal of Quaternary Science* **28**, 439–452.
21. Ride WDL, Pridmore PA, Barwick RE, Wells RT, Heady RD (1997) Towards a biology of *Propleopus Oscillans* (*Marsupialia: Propleopinae, Hypsiprymnodontidae*). *Proceedings of the Linnean Society of New South Wales* **117**, 243–328.
22. Corbett L (1995) *The Dingo in Australia and Asia*. University of New South Wales Press, Sydney.
23. Jones ME (2003) Convergence in ecomorphology and guild structure among marsupial and placental carnivores. In *Predators with Pouches: The Bology of Carnivorous Marsupials*. (Eds ME Jones, C Dickman and M Archer) pp. 285–296. CSIRO Publishing, Melbourne.
24. Paddle RN (2000) *The Last Tasmanian Tiger*. Cambridge University Press, Cambridge.
25. Letnic M, Fillios M, Crowther MS (2012) Could direct killing by larger dingoes have caused the extinction of the thylacine from mainland australia? *PLoS ONE* **7**, e34877.
26. Lourandos H (1997) *Continent of Hunter-gatherers: New Perspectives in Australian Prehistory*. Cambridge University Press, Cambridge.

27. Johnson CN, Brook BW (2011) Reconstructing the dynamics of ancient human populations from radiocarbon dates: 10 000 years of population growth in Australia. *Proceedings. Biological Sciences* **278**, 3748–3754.
28. Williams AN (2013) A new population curve for prehistoric Australia. *Proceedings of the Royal Society B: Biological Sciences* **280**.
29. Brown OJF ((2006)) Tasmanian devil (*Sarcophilus harrisii*) extinction on the Australian mainland in the mid-Holocene: multicausality and ENSO intensification. *Alcheringa* **1**(special issue), 39–48.
30. Donders TH, Wagner-Cremer F, Visscher H (2008) Integration of proxy data and model scenarios for the mid-Holocene onset of modern ENSO variability. *Quaternary Science Reviews* **27**, 571–579.
31. Walsh G (2000) *Bradshaw art of the Kimberley*. Takarakka Nowan Kas Publications, Brisbane.
32. Aubert M (2012) A review of rock art dating in the Kimberley, Western Australia. *Journal of Archaeological Science* **39**, 573–577.
33. Morwood MJ (2002) *Visions from the Past: The Archaeology of Australian Aboriginal Art*. Allen and Unwin, Sydney.
34. Prowse TAA, Johnson CN, Bradshaw CJA, Brook BW (2014) An ecological regime shift resulting from disrupted predator–prey interactions in Holocene Australia. *Ecology* **95**, 693–702.
35. Roberts RG (2014) A pardon for the dingo. *Science* **343**, 142–143.
36. Wroe S, Clausen P, McHenry C, Moreno K, Cunningham E (2007) Computer simulation of feeding behaviour in the thylacine and dingo as a novel test for convergence and niche overlap. *Proceedings. Biological Sciences* **274**, 2819–2828.
37. Thomson PC (1992) The behavioral ecology of dingoes in north-western Australia. IV. Social and spatial organisation and movement. *Wildlife Research* **19**, 543–563.
38. Jones ME (2008) Tasmanian devil. In *The Mammals of Australia*. (Eds SV Dyck and R Strahan) pp. 78–80. Reed New Holland, Sydney.
39. Baird RF (1984) The Pleistocene distribution of the Tasmanian native-hen *Gallinula mortierii mortierii*. *Emu* **84**, 119–123.
40. Baird RF (1991) The dingo as a possible factor in the disappearance of *Gallinula mortierii* from the Australian mainland. *Emu* **91**, 121–122.
41. Lourandos H (1983) Intensification: a late Pleistocene-Holocene archaeological sequence from south-western Victoria. *Archaeology in Oceania* **18**, 81–94.
42. McKenzie NL, Burbidge AA, Baynes A, Brereton RN, Dickman CR, Gordon G, Gibson LA, Menkhorst PW, Robinson AC, Williams MR, Woinarski JCZ (2007) Analysis of factors implicated in the recent decline of Australia's mammal fauna. *Journal of Biogeography* **34**, 597–611.
43. Woinarski JCZ, Legge S, Fitzsimons JA, Traill BJ, Burbidge AA, Fisher A, Firth RSC, Gordon IJ, Griffiths AD, Johnson CN, McKenzie NL, Palmer C, Radford I, Rankmore B, Ritchie EG, Ward S, Ziembicki M (2011) The disappearing mammal fauna of northern Australia: context, cause, and response. *Conservation Letters* **4**, 192–201.
44. Fisher DO, Johnson CN, Lawes MJ, Fritz SA, McCallum H, Blomberg SP, VanDerWal J, Abbott B, Frank A, Legge S, Letnic M, Thomas CR, Fisher A, Gordon IJ, Kutt A (2014) The current decline of tropical marsupials in Australia: is history repeating? *Global Ecology and Biogeography* **23**, 181–190.
45. Letnic M, Ritchie EG, Dickman CR (2012) Top predators as biodiversity regulators: the dingo *Canis lupus dingo* as a case study. *Biological Reviews of the Cambridge Philosophical Society* **87**, 390–413.
46. Allen BL, Leung LKP (2012) Assessing predation risk to threatened fauna from their prevalence in predator scats: dingoes and rodents in arid Australia. *PLoS ONE* **7**, e36426.
47. Claridge AW, Mills DJ, Barry SC (2010) Prevalence of threatened native species in canid scats from coastal and near-coastal landscapes in south-eastern Australia. *Australian Mammalogy* **32**, 117–126.

48. Cupples JB, Crowther MS, Story G, Letnic M (2011) Dietary overlap and prey selectivity among sympatric carnivores: could dingoes suppress foxes through competition for prey? *Journal of Mammalogy* **92**, 590–600.
49. Newsome TH, Ballard G-A, Crowther M, Fleming PJS, Dickman CR (2014) Dietary niche overlap of free-roaming dingoes and domestic dogs: the role of human-provided food. *Journal of Mammalogy*. In press.
50. Rolls EC (1969) *They All Ran Wild*. Angus and Robertson, Sydney.
51. Wright J (1981) *The Cry for the Dead*. Oxford University Press, Melbourne.
52. Caughley G, Grigg GC, Caughley J, Hill GJE (1980) Does dingo predation control the densities of kangaroos and emus? *Australian Wildlife Research* **7**, 1–12.
53. Newsome AE (2001) The biology and ecology of the dingo. In *A Symposium on the Dingo*. (Eds CR Dickman and D Lunney) pp. 20–33. Royal Zoological Society of New South Wales, Sydney.
54. Newsome AE, Catling PC, Cooke BD, Smyth R (2001) Two ecological universes separated by the dingo barrier fence in semi-arid Australia: interaction between landscapes, herbivory and carnivory, with and without dingoes. *Rangeland Journal* **23**, 71–98.
55. Dawson TJ (1995) *Kangaroos*. University of New South Wales Press, Sydney.
56. Grice D, Caughley G, Short J (1985) Density and distribution of emus. *Australian Wildlife Research* **12**, 69–73.
57. Pople AR, Grigg GC, Cairns SC, Beard LA, Alexander P (2000) Trends in numbers of kangaroos and emus on either side of the South Australian dingo fence: evidence for predator regulation. *Wildlife Research* **27**, 269–276.
58. Letnic M, Crowther MS (2013) Patterns in the abundance of kangaroo populations in arid Australia are consistent with the exploitation ecosystems hypothesis. *Oikos* **122**, 761–769.
59. Wallach AD, Johnson CN, Ritchie EG, O'Neill AJ (2010) Predator control promotes invasive dominated ecological states. *Ecology Letters* **13**, 1008–1018.
60. Robertshaw JD, Harden RH (1986) The ecology of the dingo in north-eastern New South Wales. IV. Prey selection by dingoes, and its effect on the major prey species, the swamp wallaby, *Wallabia bicolor* (Desmarest). *Australian Wildlife Research* **13**, 141–163.
61. Short J (1998) The extinction of rat-kangaroos (*Marsupialia: Potoroidae*) in New South Wales, Australia. *Biological Conservation* **86**, 365–377.
62. Short J (2004) Mammal declines in southern Western Australia: perspectives from Shortridge's collections of mammals in 1904–07. *Australian Zoologist* **32**, 605–628.
63. de Tores PJ, Marlow N (2012) The relative merits of predator-exclusion fencing and repeated fox baiting for protection of native fauna: five case studies from Western Australia. In *Fencing for Conservation: Restriction of Evolutionary Potential or a Riposte to Threatening Processes?* (Eds MJ Somers and MW Hayward) pp. 21–42. Springer, New York.
64. Allen BL, Fleming PJS (2012) Reintroducing the dingo: the risk of dingo predation to threatened vertebrates of western New South Wales. *Wildlife Research* **39**, 35–50.
65. Johnson CN, Ritchie EG (2013) The dingo and biodiversity conservation: response to Fleming *et al.* (2012). *Australian Mammalogy* **35**, 8–14.
66. Dickman CR (1996) *Overview of the Impacts of Feral Cats on Australian Native Fauna*. Australian Nature Conservation Agency, Canberra.
67. Denny EA, Dickman CR (2010) *Review of Cat Ecology and Management Strategies in Australia*. Invasive Animals Cooperative Research Centre, University of Canberra.
68. Brook LA (2014) Predator guild interactions in northern Australia: behaviour and ecology of an apex predator, the dingo *Canis lupus dingo*, and an introduced mesopredator, the feral cat *Felis catus*. PhD thesis. James Cook University, Townsville.
69. Kutt A (2012) Feral cat (*Felis catus*) prey size and selectivity in north-eastern Australia: implications for mammal conservation. *Journal of Zoology* **287**, 292–300.
70. Brook LA, Kutt AS (2011) The diet of the dingo (*Canis lupus dingo*) in north-eastern Australia with comments on its conservation implications. *Rangeland Journal* **33**, 79–85.

71. Gordon G, Riney T, Toop J, Lawrie BC, Godwin MD (1985) Observations on the Queensland hairy-nosed wombat (*Lasiorhinus latifrons* (Owen)). *Biological Conservation* **33**, 165–195.
72. Crossman DG, Johnson CN, Horsup AB (1994) Trends in the population of the northern hairy-nosed wombat, *Lasiorhinus krefftii* in Epping Forest National Park, central Queensland. *Pacific Conservation Biology* **1**, 141–149.
73. Banks SC, Horsup A, Wilton AN, Taylor AC (2003) Genetic marker investigation of the source and impact of predation on a highly endangered species. *Molecular Ecology* **12**, 1663–1667.
74. Newsome T, Ballard G-A, Fleming PS, Ven R, Story G, Dickman C (2014) Human-resource subsidies alter the dietary preferences of a mammalian top predator. *Oecologia* **175**, 139–150.
75. Jones ME, Oakwood M, Belcher CA, Morris K, Murray A, Woolley PA, Firestone KB, Johnson B, Burnett S (2003) Carnivore concerns: problems, issues and solutions for conserving Australasia's marsupial carnivores. In *Predators with Pouches*. (Eds ME Jones, C Dickman and M Archer) pp. 422–434. CSIRO Publishing, Melbourne.
76. Jarman PJ, Johnson CN, Southwell CJ, Stuart-Dick R (1987) Macropod studies at Wallaby Creek. I. The area and animals. *Wildlife Research* **14**, 1–14.
77. Marsack P, Campbell G (1990) Feeding behaviour and diet of dingoes in the Nullarbor region of Western Australia. *Australian Wildlife Research* **17**, 349–358.
78. Moseby KE, Neilly H, Read JL, Crisp H (2012) Interactions between a top order predator and exotic mesopredators in the Australian rangelands. *International Journal of Ecology* **2012**, 250–352.
79. Southgate R, Paltridge R, Masters P, Carthew S (2007) Bilby distribution and fire: a test of alternative models of habitat suitability in the Tanami Desert, Australia. *Ecography* **30**, 759–776.
80. Johnson CN, VanDerWal J (2009) Evidence that dingoes limit abundance of a mesopredator in eastern Australian forests. *Journal of Applied Ecology* **46**, 641–646.
81. Kennedy M, Phillips BL, Legge S, Murphy SA, Faulkner RA (2011) Do dingoes suppress the activity of feral cats in northern Australia? *Austral Ecology* **37**, 134–139.
82. Risbey DA, Calver MC, Short J, Bradley JS, Wright IW (2000) The impact of cats and foxes on the small vertebrate fauna of Herisson Prong, Western Australia. II. A field experiment. *Wildlife Research* **27**, 223–235.
83. Christensen P, Burrows N (1995) Project desert dreaming: experimental reintroduction of mammals to the Gibson Desert, Western Australia. In *Reintroduction Biology of Australian and New Zealand Fauna*. (Ed. M Serena). Surrey Beatty and Sons, Sydney.
84. Mitchell BD, Banks PB (2005) Do wild dogs exclude foxes? Evidence for competition from dietary and spatial overlaps. *Austral Ecology* **30**, 581–591.
85. Brawata RL, Neeman T (2011) Is water the key? Dingo management, intraguild interactions and predator distribution around water points in arid Australia. *Wildlife Research* **38**, 426–436.
86. O'Neill A (2002) *Living with the Dingo*. Envirobooks, Sydney.
87. Lundie-Jenkins G, Corbett LK, Phillips CM (1993) Ecology of the rufous hare-wallaby, *Lagorchestes hirsutus* Gould (*Marsupialia: Macropodidae*), in the Tanami Desert, Northern Territory. III. Interactions with introduced mammal species. *Wildlife Research* **20**, 495–511.
88. Letnic M, Koch F (2010) Are dingoes a trophic regulator in arid Australia? A comparison of mammal communities on either side of the dingo fence. *Austral Ecology* **35**, 167–175.
89. Letnic M, Crowther MS, Koch F (2009) Does a top-predator provide an endangered rodent with refuge from an invasive mesopredator? *Animal Conservation* **12**, 302–312.
90. Letnic M, Dworjanyn SA (2011) Does a top predator reduce the predatory impact of an invasive mesopredator on an endangered rodent? *Ecography* **34**, 827–835.
91. Johnson CN, Isaac JL, Fisher DO (2007) Rarity of a top predator triggers continent-wide collapse of mammal prey: dingoes and marsupials in Australia. *Proceedings. Biological Sciences* **274**, 341–346.
92. Allen BL, Engeman RM, Allen LR (2011) Wild dogma: an examination of recent 'evidence' for dingo regulation of invasive mesopredator release in Australia. *Current Zoology* **57**, 568–583.

93. Allen BL, Fleming PJS, Allen LR, Engeman RM, Ballard G, Leung LKP (2013) As clear as mud: a critical review of evidence for the ecological roles of Australian dingoes. *Biological Conservation* **159**, 158–174.
94. Eldridge SR, Shakeshaft BJ, Nano TJ (2002) *The Impact of Wild Dog Control on Cattle, Native and Introduced Herbivores and Introduced Predators in Central Australia: Final Report to the Bureau of Rural Sciences.* Parks and Wildlife Commission of the Northern Territory, Alice Springs.
95. Allen BL, Allen LR, Engeman RM, Leung LK-P (2013) Intraguild relationships between sympatric predators exposed to lethal control: predator manipulation experiments. *Frontiers in Zoology* **10**, 39.
96. Johnson C, Crowther M, Dickman C, Letnic M, Newsome T, Nimmo D, Ritchie E, Wallach A (2014) Experiments in no-impact control of dingoes: comment on Allen *et al.* 2013. *Frontiers in Zoology* **11**, 17.
97. Hayward MW, Marlow N (2014) Will dingoes really conserve wildlife and can our methods tell? *Journal of Applied Ecology*. In press.
98. Crowther MS, Fillios M, Colman N, Letnic M (2014) An updated description of the Australian dingo (*Canis dingo* Meyer, 1793). *Journal of Zoology*. In press.
99. Newsome TM, Stephens D, Ballard G-A, Dickman CR, Fleming PJS (2013) Genetic profile of dingoes (*Canis lupus dingo*) and free-roaming domestic dogs (*C. L. familiaris*) in the Tanami Desert, Australia. *Wildlife Research* **40**, 196–206.
100. Stephens D (2011) The molecular ecology of Australian wild dogs: hybridisation, gene flow and genetic structure at multiple geographic scales. PhD thesis. University of Western Australia, Perth.

9

Dingo intelligence: a dingo's brain is sharper than its teeth

Bradley Smith

Through stories told over the last few centuries, dingoes have become notorious for their intelligence. These stories are told from a variety of perspectives, including those who have lived with them in domestic situations and those that have actively tried to exterminate them. All come to the same conclusion – that dingoes are too smart for their own good. Robert Kaleski, a prominent figure in the development of many Australian dog breeds, could see the benefits of introducing dingo genes into his working dogs. To achieve this he kept a small colony of dingoes. In *Australian Barkers and Biters*[1] he wrote: 'I have had some marvellously intelligent dingoes. The last one I taught to read and write; but when I caught him practicing my signature in his spare time, I thought it best to get rid of him.' This quote, although a bit tongue-in-cheek, suggests the dingo possesses a superior intelligence, but is there any scientific evidence to support this? This chapter explores the scientific research relating to dingo intelligence (cognition) according to the findings from various captive and field-based observational studies. As the chapter will show, dingoes are becoming just as famous for solving problems as they are for causing them! Throughout the chapter, dingoes will be compared to other canids, namely wolves and domestic dogs, to provide some context to these skills and why and how they have developed. Which is smarter – dogs or dingoes? In this chapter we find out.

Cunning or clever?

Australian farmers and dingoes share a long and troubled history. The bad blood between livestock farmers and wild carnivores has existed ever since humans began keeping domestic animals. Famers yearn for a world free of any force (flora, fauna or climate) that may impact their livelihood, and try to remove or control these forces (see Chapter 5). The fact that dingoes are so resilient to control measures adds fuel to the fire, and has left them with a less-than-flattering reputation. Like their wolf and coyote relatives, dingoes are frequently described as cunning, treacherous, cowardly, promiscuous, vicious and devious creatures. The regard for dingoes is so low, for example, that calling a man a dingo is considered an expression of extreme contempt suggestive of cowardice and treachery.[2, 3] On several occasions, Australian politicians have referred to other members of Federal Parliament as being a dingo. In 1973, for instance, politician Doug Anthony declared of the then Prime Minister Gough Whitlam, 'All I can say of him in the Australian language is that he is a dingo'.[4]

But is this reputation justified? One could well argue that dingoes are simply victims of their own success. Terms such as cunning, treacherous and devious, when you think about their true meaning, describe the actions of an intelligent animal. After all, these labels are rarely applied to sheep, cattle or kangaroos, which are quite passive herbivores. The dictionary definition of cunning describes the possession of skills in deception, subtlety and ingenuity, all of which are intelligent attributes that are required for a wild predator to survive and successfully reproduce. Parker[5, 6] conducted a study on the position of the dingo in Australian discourse (by analysing a lengthy list of Australian colonial texts), and argued that resourceful, quick-witted, pragmatic and intelligent are far more appropriate labels for dingoes according to how their behaviour has been described in the literature. Parker made some interesting statements regarding why such negative images of the dingo were held. The first, a rather obvious one, is that the dingo poses a threat to a farmer's bottom line (profitability). The second one is more than a financial threat – it hurt their ego. Parker believed that farmers became increasingly frustrated by the dingo's incredible success in resisting their continued eradication attempts. She described how dingoes regularly outsmarted farmers, who did not appreciate looking intellectually inferior to animals. Every so often, individual dingoes were particularly difficult to capture. In these situations, the more 'cunning the dingo', the greater the 'prestige bestowed on the successful trapper' that was clever enough to catch them (p. 40).[5, 6]

Early Australian literature is filled with stories, poems and anecdotes featuring dingoes, the majority authored by farmers and doggers (a nickname given to those commissioned by government or land owners to exterminate problem dingoes) writing about their adventures trying to hunt and trap them. It is interesting to note that most of these texts suggest doggers hold a healthy respect for the dingo, and in some ways admired them. Bacon,[7] a New South Wales sheep farmer who spent his life trapping dingoes, is a good example of such contradiction. In his book, Bacon alternated between praising the dingo for its remarkable skills in killing native prey, and describing it as cunning and savage when those same skills were applied to killing sheep. Had these authors not had a vested interest in seeing the dingo disappear, it is likely they too would better appreciate the dingo's true intelligence that is so evident in the tales they tell. These stories, however, can still be a useful source of information about dingo behaviour, particularly their ability to avoid human capture.

Few dingo hunters (past or present) will say that dingoes are easy to trap. In fact, their ability to evade capture is a trait for which they are often greatly admired.[3, 8] It is here that dingoes can be easily distinguished from domestic dogs gone wild – the latter are far easier to capture and kill.[7] Several authors have detailed the extraordinary lengths that are required to successfully trap dingoes, many of which involve predicting dingo behaviour, and/or the use of deception.[3, 7, 9, 10] For example, the trap must be positioned delicately and located in a position likely to be visited by a dingo (i.e. based on past behaviour of specific dingoes, or where dingoes typically travel, such as around logs and water crossings). Attractive baits are required (e.g. canine urine, semen, anal glands); some also reported success imitating a dingo howl (often described as 'howling up' dingoes).[7, 10, 11] Baits or decoys must be positioned to distract the dingo's attention away from the trap (e.g. decoy traps, or hanging items in trees that divert their attention from the ground). Rolls[8] provided an account where a senior dingo hunter, Bill Baldwin, was outwitted by a dingo bitch for several months:

> No matter how carefully he set a trap she always located it – sometimes, when uncertain about exactly where a trap was, she seemed to make a point of finding it, pushing her forepaws carefully through the sand until they touched metal (p. 51).

Bacon,[7] another experienced dingo hunter, openly admitted that he was often outsmarted by dingoes, and so made a considerable effort to learn their habits. Whenever he arranged dingo drives (hunts), 'the dingoes are cunning enough to double back without giving the shooters a chance' (pp. 26–27), and when he put bells on his sheep the dingoes 'kill the bell wether first' (p. 13). He charged the top wires of a fence with electricity but 'the batteries kept running down' and the dingoes 'went along the fence until they found a suitable place to cross' (Supplement 2). He built the fence higher, only to take a 'photo of a dingo jumping over a 12 foot high fence' (p. 4).

Traditional methods of dingo control such as trapping, ground baiting (e.g. with strychnine) and shooting have changed (although they are still employed). Much focus is now on large-scale control not on individual properties, although this still occurs. Aerial baiting using sodium monofluoroacetate (a poison more commonly known as '1080') has become a popular method of pest control in Australia as a larger area can be targeted, it involves less labour than traditional techniques and has shown to be moderately successful.[12–16] However, there is some indication that baiting is not as effective as initially thought, as baits lose toxicity over time,[17] many non-target species take the baits[17–19] and the baits seem to affect only young and inexperienced (less than 24 months old) or solitary dingoes, suggesting that 'the poorer hunting success of young or lone dingoes predisposes them to taking baits' (p. 173).[16] Dingoes may have learned to avoid foreign food items (e.g. Eldridge *et al.*[20] found non-refined meat baits more effective than highly processed meat baits) or simply have a preference for natural prey, particularly in areas of high prey density.[17] Bacon[7] believed that the chances of killing a dingo decreased with age and experience:

> When young dingoes reach a certain age they are apt to learn quickly and are much harder to get. Then at a later stage, after they have been caught in a bad trap and got away, or after having taken a badly laid bait and recovered, they become known as veteran dingoes – the most difficult to destroy (p. 10).

Killing adult dingoes appears counterproductive as existing groups become fractured into smaller groupings, each with a breeding female no longer pressured not to reproduce.[21] Areas that have received large-scale baiting campaigns are soon reinvaded by new dingoes.[20] This can lead to increased attacks on livestock as the new colonisers tend to be young inexperienced dingoes. These animals are more likely to target easy prey (livestock) than wild prey that is difficult to catch.[22]

Beyond descriptions of behaviour exhibited in avoiding capture and destruction, dingoes are seen to be extremely opportunistic and resourceful. For example, on Fraser Island, a World Heritage-listed haven for dingoes visited by thousands of tourists each year, dingoes have been observed to open tents, bags, plastic containers and eskies (ice-boxes) in order to locate food.[23] On Fraser Island the dingoes quickly learned to avoid park rangers at the sight or sound of their vehicles and uniforms. I have seen dingoes constantly navigate past the many human-made obstacles used to keep them from human settlements (e.g. fences, electrified modified cattle grids at entrances to resorts and townships), and wait patiently for fish scraps near fishers who are cleaning their day's catch. On one occasion, dingo ecologist Rob Appleby, who conducted years of research on the island's dingoes, observed an adult female dingo travelling along the beach watching lines of fishers. There is some indication that the dingo travelled from fisher to fisher seemingly 'checking' the lines to see if anyone had caught anything. At one point the dingo made a direct line for a fisher who appeared to have caught a fish. It was not successful, but I have heard many stories where dingoes have stolen fish directly from the line. In similar circumstances, professional hunter Adam O'Neill[24] described a time when he was hunting for rabbits in Strzelecki Desert and was followed by dingoes for days. The dingoes, keeping their distance, picked up any rabbits he rejected or left behind, and quickly learned to take the rabbits from the spotlit area as soon as he had shot them.

While there are countless anecdotes that paint the dingo as a highly intelligent animal, is there any evidence to support this? The remainder of this chapter will discuss the scientific evidence relating to the intelligence (cognitive ability) of the dingo.

Intelligence

Before exploring the science, we must first define the concept of intelligence, or behaviour that is considered to involve higher-order functioning. The way dingoes process and respond to the information in their environment relies heavily on both instinct and intelligence. Intelligence refers to the ways in which animals take in information through their senses, how they process and retain this information, and what they act on. It encompasses abilities such as perception, learning, memory, tool use, emotion, reasoning and problem-solving. It is important to distinguish between behaviour that is an expression of innate biological factors such as simple reflexes or associations (instincts) and behaviours we term higher-order, which are more complex or flexible thought processes, such as insight, foresight, planning and reasoning. Although traditionally animals have been thought to be purely instinctual, actually it is their cognitive processes that influence their behaviour.

Every species on the planet is faced with different problems to solve, and therefore in order to survive and reproduce they only need to be as intelligent as their environment and challenges require. This is why, for example, predators such as dingoes have different, often higher-level cognitive capabilities than do herbivores, such as kangaroos. In general, the more challenging the situation in which prey have to be caught, the more likely a predator will have to use its wits to do so. Intelligence is one of the reasons why a medium-sized canid can kill prey animals that are often three or more times its size. Further differences exist between domesticated animals and their wild equivalents because we have deliberately dumbed them down. All this suggests that two different species (particularly herbivores and carnivores) can't be fairly compared on the same task or ability, or one speciesbe considered smarter than another – they all live in different niches. We can however, safely compare members of the Family Canidae, and doing so creates interesting results.

On the hunt

The best examples of higher-order behaviours in natural environments for wild canids in general are seen during social/pack hunting, which requires cooperation, coordinated strategies and cognitive mapping.[25, 26] For example, in Asia, wild dholes (*Cuon alpines*) have been observed to hunt cooperatively, with some individuals lying in ambush while others drive prey towards them.[27] Although there are some reports that dingoes working in groups are more successful than solitary dingoes in hunting large prey such as kangaroos and calves,[28] there is no direct evidence to suggest dingoes hunt cooperatively in groups. Dingo hunting behaviour depends greatly on local environments and prey availability (see Chapter 2).

Cooperative hunting is much more prominent in canids that take down large prey. Yet, despite the numerous hunts observed for several wild canids in the scientific literature, a detailed analysis of the cognitive processes involved in various hunting strategies has yet to be undertaken. Part of the reason is that it is extremely difficult to observe such behaviour in the wild. A well-known wolf researcher, David Mech,[29] made the best attempt. He observed a pack of seven wild adult arctic wolves during two instances where they displayed behaviour suggestive of complex cognitive processes during hunts of muskoxen (*Ovibos moschatus*) on Ellesmere Island (Canada). The wolves demonstrated what Mech deemed 'foresight (behaving appropriately for dealing with a future event), understanding (comprehending complex relationships), and planning (deciding to behave in a way that considers information relevant to a perceived outcome)' (p. 146).[29] In the first example, the wolf pack split up and approached the muskoxen from two directions. Had they succeeded in forcing the muskoxen to run, the waiting wolves might have successfully intercepted or ambushed their prey. In the second example, the wolves positioned themselves strategically out of sight of the muskoxen, backtracking and repositioning and waiting in ambush for about three hours, occasionally popping up and looking at the muskoxen. Mech[29] concluded that a group of wolves with considerable experience hunting together, that have learned each other's usual hunting approaches under various conditions, could easily employ highly specialised strategies which require considerable intelligence in order to be properly executed. Note that not everyone agrees that group hunting is a sign of higher order intelligence. Packard,[30] for example, argued that many such behaviours exhibited by wild wolves could be explained by relatively low-level learning processes commonly exhibited by many taxa, such as trial and error learning or observational learning.

Studying dingo cognition in captivity

Since conducting observations of a wild canid is plagued with difficulty, a lot of dingo activity takes place in private. Captivity offers an opportunity to unlock some of these secrets (also see Chapter 11). While free-ranging dingoes must seek prey, rear families and generally stay out of trouble, captive dingoes have more opportunity and energy to explore and manipulate objects within their environment. Captive subjects are also far easier to observe and film. Although some of the problems encountered in captivity may be different from those that occur in the wild and reports are often anecdotal, instances of high-order behaviour can provide valuable insight into the behavioural potential of species that are endangered or difficult to observe under field conditions, and thus neglected or underrepresented in the scientific literature. (Anecdotal accounts are often of single events or individuals. Although they provide useful information and insight, behaviours are not considered common for the species unless replicated or the individual is rigorously tested.) Accounts of high-order behaviour in wild canids (wolves) are limited to a handful of studies investigating development of flexible hunting skills, the modification of behaviour contingent on the action of their group mates,[31–34] standardised problem-solving tests conducted in captivity[35–43] and anecdotal descriptions of unusual problem-solving strategies.[26, 44, 45]

The majority of research on dingo cognition and behaviour (including all the cognitive experiments described in this chapter) has been done at the Dingo Discovery and Research Centre, a private sanctuary in Toolern Vale, Victoria. The sanctuary houses ~30 DNA-tested pure dingoes and has an active breeding program. The dingoes are housed in breeding pairs but have the opportunity to exercise and interact in social groups in large running yards. More information about the sanctuary is provided in Chapter 10. Captive dingoes are the obvious choice for cognitive experiments because it is next to impossible to conduct many of the cognitive and behavioural-based experiments on wild dingoes. Experiments require access to large numbers of animals, knowledge of their history and, practically, the ability to handle them and place them on a leash, conducting repeated trials in controlled conditions. Most importantly, the dingoes need to be comfortable around humans and strange new objects. There are some limitations to focusing purely on captive populations, but in general such findings are still applicable to wild dingoes, and in many cases showcase their behavioural potential. Unfortunately captive dingoes are not well studied either, and the majority of our knowledge of dingo cognition (including all the scientifically published experiments and observations described here) comes from studies that I have conducted or been involved in. It makes sense, then, that the remainder of this chapter is written from my personal perspective.

Did anyone call a locksmith?

Like wolves and coyotes, dingoes kept in sanctuaries and/or private households are infamous for escaping and opening locks (see mentions of such behaviour from people who have kept captive dingoes).[1, 3, 46–48] However, examples featuring dingoes and other canids, although compelling and potentially valuable, are anecdotal and fail to provide an avenue for independent evaluation through, for example, video recordings of behavioural observations. Until now, that is.

During research work at the sanctuary, I observed an adult male dingo that was a master at opening gates and locks, a veritable dingo Harry Houdini (these observations and accompanying video footage have been published).[49] Due to Teddy, staff had been forced to find new ways to latch gates. Teddy had discovered that he could open gates by manipulating the latch with his nose. To document this behaviour, I removed the padlock on a gate and set up a video camera overlooking the gate to his pen. I created an incentive for escape by moving Ayjay (a female adult dingo, Teddy's usual enclosure mate) into the adjacent enclosure. Opening the latch on the other side of the gate required Teddy to place his head inside a small opening above the latch (the space allocated for human hands) then push upwards on the latch with his nose (Fig. 9.1). While the latch was being held up, Teddy applied pressure onto the gate with his body, pushing it outwards in order to swing it open. The pressure on the gate had to occur at the same time he withdrew his head or the gate would swing shut again. Teddy was able to open the gate and escape on his fifth attempt. When his mate Ayjay was placed back

Figure 9.1: Teddy opening a gate by placing his head through a small opening and lifting the latch with his nose. Image taken from video footage.
Source: Bradley Smith.

in the same enclosure, Teddy made no attempt to escape. When I removed her again, it took Teddy only 13 seconds to open the gate so they could be together.[49] I soon realised that most dingoes have the potential to escape their surroundings or get around obstacles, whether in a captive zoo setting or in areas where humans are trying to keep them out (e.g. campsites or rubbish tips) if they are motivated enough. In captivity at least, they mostly choose to stay inside the fences because the sanctuary or enclosure has become their 'territory'; there is familiarity, food and family, and there would be little advantage in leaving.

A dingo that sets his own table

The ability to open gates and manipulate locks, although demonstrating some degree of cognitive processing, is nothing new; it has been seen in a variety of species including cows and domestic dogs. There are many explanations for this skill, with most evidence suggesting that animals can learn very complex behaviours by trial and error learning, or can be trained using instrumental learning techniques. For example, Morgan[50] describes in detail how he taught his pet dog to open a garden latch with his head. It took the dog approximately three weeks to master this task and to understand that lifting the latch was a means to get outside and onto the road. Epstein[51] trained a pigeon to move a wooden block towards a banana that had been placed out of reach, allowing the pigeon to step onto the block and retrieve the banana.

Are dingoes capable of exhibiting more than trial and error learning? Another observation I conducted at the dingo sanctuary suggests they are. During one of my many visits I was introduced to an interesting dingo named Sterling, who helped take the dingo's reputation beyond gate-opening to be the first member of the Family Canidae to exhibit tool use.[49] Tool use, often considered a hallmark of intelligence and reported in only a few species, can be defined as 'the manipulation of an inanimate object, not internally manufactured, with the effect of improving the animal's efficiency in altering the position or form of some separate object' (p. 464).[52] Tool-using animals are those that establish the effective orientation of an object to alter some condition and attain an incentive.[53] How can dingoes use tools, given their lack of limbs with grasping capabilities such as those of primates? Tool use is not restricted to use of limbs; some species, canines included, can manipulate objects with their mouths. For example, bottlenose dolphins (*Tursiops* sp.) have been observed to carry sponges during foraging but it is unknown how they actually use them[54] and New Caledonian crows manufacture hook-tools with their beaks to aid prey capture.[55]

Sanctuary staff informed me that the name-tags used to identify each dingo in a particular enclosure, positioned ~1.7 m above the ground on the external side of the entrance, kept disappearing. This was strange because the tags were at a height that the dingoes couldn't reach by standing up on their back legs, and what staff member would remove them? In an attempt to determine how the tags were vanishing, I helped the staff position a video camera in front of the enclosure housing three dingoes. A small plastic envelope containing dry dog food was placed just inside the entrance in a position similar to that of the missing name-tags. We then left the scene, since the dingoes had ignored both the name-tag and envelope of food when we were present. The video footage showed that after several unsuccessful attempts at jumping for the envelope, Sterling solved the task by moving then jumping up onto a trestle table inside the enclosure. This allowed him to gain the additional height necessary to reach the food item. To move the table, Sterling clamped his mouth onto the strut between its legs. He then walked backwards, dragging the table approximately 2 m, until it appeared that either his back leg or tail touched the enclosure mesh. He then jumped onto the table, but as he was still at least a body-length away from the envelope he had to span the gap between the table and the enclosure mesh by propping his front paws onto the mesh, gradually moving them towards the envelope. At full stretch, he reached the envelope on his second attempt (Fig. 9.2). Importantly, Sterling was never purposely trained or encouraged to exhibit this (or similar) behaviour. The behaviour therefore appeared to initially occur spontaneously (i.e. without caretaker training) and without an obvious incentive such as a food reward (i.e. food was used only in the video example and not in cases where name-tags had been removed).

Sterling didn't stop there. According to staff, he had manipulated objects within his environment from an early age. Prior to the table-pulling behaviour, in a different enclosure he was seen rolling a barrel (100 cm long, 60 cm in diameter) using both front paws, towards the enclosure fence. He was observed jumping onto the barrel then escaping over the fence (he jumped from the barrel and clambered up and over). This escape occurred during breeding season, and allowed Sterling to approach a dingo bitch located in another area of the sanctuary.

Figure 9.2: (a) Sterling at full stretch attempting to reach the food reward, positioned at the top right of the enclosure. (b) In the process of dragging the table into position. (c) Reaching the reward. Images taken from video footage.

Source: Bradley Smith.

(Continued)

Figure 9.2: (*Continued*)

In a follow-up observation, Sterling was moved to a new enclosure without tables or barrels but containing a plastic commercial dog kennel. This new enclosure was enclosed by 1.2 m high corrugated iron sheets designed to provide a wind break and prevent fighting between dingoes in adjacent pens. During regular maintenance activities, sanctuary staff had observed that the plastic kennel in Sterling's enclosure was often in a different position from where it had been previously. I installed a hidden infrared camera to film Sterling's enclosure during the night. The footage showed Sterling dragging the dog kennel with his mouth in a similar manner to that described in the table-moving example (Fig. 9.3). He dragged the kennel on two separate occasions to a different location in the enclosure, then jumped on top of it. The kennel appeared to afford him the height necessary to view dingoes in adjoining enclosures; it served as a look-out.

Teddy and Sterling provide examples of dingoes manipulating objects in a goal-oriented manner. Although the case of Teddy does not necessarily meet the definitions of tool use, Sterling's behaviour certainly does. No other dingoes at the sanctuary have exhibited similar behaviour, so it cannot be assumed that these abilities exist in all individuals of this species or across scenarios or settings for the same individual. That is, Sterling may be the exception and not the rule. This seems to be the case with most of the research on wild canids. There is a large degree of individual variation in captive wolves tested in various problem-solving tasks.[56]

Figure 9.3: (a) Sterling dragging the kennel. (b) Using the kennel as a look-out over the 1.2 m corrugated iron sheets which separate each enclosure. Images taken from video footage.

Source: Bradley Smith.

In both cases involving Sterling, an object was dragged with the mouth and used to gain the height necessary to achieve a goal (escape, view, attainment of out-of-reach objects). For Sterling, neither the plastic tags nor better views would normally be considered a primary reinforcer (e.g. food), but dragging the table or kennel afforded him greater control over his immediate environment. Analogous problems exist in the wild. For example, wild dingoes on Fraser Island have been observed to use fallen trees and logs as apparent look-outs to scan immediate surrounds that are obscured by dense understorey or long grass (although this obviously involves less complexity than the case of Sterling, who made his own look-out).

Reactions to the death of a pup

Until recently, higher-order behaviours such as tool use were thought to be uniquely human. Another example that fits into the higher-order category of behaviours is mourning the passing of relatives and close companions. Although it can be problematic to attribute human emotions (e.g. grief) to non-human animals, there is a growing body of evidence that indicates the capacity for grief has evolutionary roots. How a species reacts to the death of a conspecific (offspring, sibling, relative, member of the group) can be indicative of many complex functions of thought and feelings. Most non-human animals appear to show little interest in the remains of the dead. However, a small list of primates, elephants and dolphins exhibit behaviour that suggests they show specific interest in and concern, and render assistance to sick, injured, distressed or dead members of their species. It is still hotly debated among animal behaviourists whether these animals are truly demonstrating grief or mourning, or whether we are being anthropomorphic (imposing human emotions onto animals).

In 1817, during an expedition into the interior of New South Wales, John Oxley[57] kept a rare written account of a dingo reacting to the death of another dingo:

> About a week ago we killed a native dog, and threw his body on a small bush: in returning past the same spot today, we found the body removed three or four yards from the bush, and the female in a dying state lying close behind it; she had apparently been there from the day the dog was killed, being so weakened and emaciated as to be unable to move on our approach (p. 110).

In support of this anecdotal account, and in another world first for canids, we provide a documented case of a dingo exhibiting complex responses to the death of a conspecific. While studying the behaviour of dingoes on Fraser Island, dingo ecologist Rob Appleby observed and filmed the responses of a wild adult female dingo (the mother) and four littermates to the death of a pup.[58] Appleby witnessed the death of the pup, which occurred ~20 min after he began his observations. He could not intervene, due to both research and animal ethics guidelines. Also, he was some distance from his vehicle and it would have taken a long time to take the pup to the car then to the nearest ranger's office (not outfitted with veterinary equipment anyway). Over the next days he observed what was assumed to be at least four instances of the deceased pup being transported by the mother,[58] (Fig. 9.4) including one occasion where the mother was directly observed and filmed moving the pup.

Figure 9.4: (a) Dingo pups investigating their deceased sibling. (b) The mother picking up the deceased pup before moving it. Images taken from video footage.

Source: Rob Appleby.

It is no real surprise that dingoes show complex reactions to dead conspecifics – like all the species that have been seen to exhibit such responses, dingoes are capable of high-order cognitive ability and live in complex societies. For example, they have relatively prolonged periods of parental care and juvenile development, strong mother–infant bonding and sociality more generally. Appleby and colleagues (including myself), however, avoid making any real assertion that dingoes mourn the dead. Instead we choose a more evolutionary or adaptive interpretation, which is equally as fascinating. We argue that the mother was adopting a 'wait and see' approach to her deceased pup, which still appeared physically healthy and intact (although the cause of death couldn't be properly determined, some of the visible symptoms suggested envenomation). Given that there is considerable investment in parental care, if parents were to abandon or consume infants that exhibit ambiguous perceptual cues (information) about their state too early, fitness may be needlessly reduced. That is, if a mother was uncertain about whether her pup was dead, it would be against her interests to give up before she became certain. Provided the fitness benefits of enduring responses are not outweighed by fitness costs such as a misallocation of care (e.g. protecting dead individuals to the detriment of living offspring), then a wait and see strategy may be an adaptive one.

Cognition under the microscope

The last few examples described opportunistic observations of wild and captive dingoes performing various behaviours that show higher-order information processing. Another way to look at intelligence is to set up more structured problem-solving tasks or situations then analyse how the subjects perform. By exposing various species or subspecies to the same or similar situations, we are able to directly compare their cognitive abilities. Then, if we look at the differences between wild canids (e.g. wolves and dingoes) and domestic dogs, we are able to speculate about the effect of domestication on the cognitive abilities of dogs (assuming that wolves are the ancestors of domestic dogs). The following sections describe a series of cognitive experiments conducted with wolves, dingoes and domestic dogs. They discuss the various effects that domestication has had on cognition, and subsequent differences between wild and domestic dogs. After all, results or findings of dingo performance tell us little unless we compare them to other canids. The dingo is a great model for investigating the influence of domestication on the canine mind because it is considered by many to be a 'proto dog', i.e. it is thought to be one of the earliest forms of domestic dog. Not a domestic dog as we know them today, but one representing an early transition from the more-wolf-like common ancestor, to dog (see Chapter 2).

Nothing gets between a dingo and its dinner

Obstacles and problems related to navigation are commonly encountered in natural conditions. For example, a dingo might be required to navigate around large trees or bushes when the prey it is chasing goes out of view. The ability to take shortcuts and detours has adaptive significance; animals that can move quickly and efficiently when hunting prey or avoiding predators are more likely to survive and reproduce. A classic experiment called the 'detour' or 'barrier' task has been designed to replicate such scenarios and to assess spatial problem-solving. The detour task is quite simple in design: it requires the subject to travel around a transparent barrier to obtain a reward located on the other side. Typically, a transparent V-shaped barrier is used to separate the subject and a target object or reward (e.g. food or favoured object) which the subject can see but not reach. To solve the problem, an indirect route (detour)[59–61] must be chosen. That is, the animal must move away and out of sight of the reward, and run directly around the end of the barrier.[40, 62] Performance is rated according to the time taken (latency) to make the detour from the start location to the goal location, and errors made during the trial (e.g. stopping, reversing direction, hesitating). Genuine achievement is indicated by a continuous run, without hesitation, to the objective.[61]

The task, at least the concept, sounds relatively easy and something that dogs should have no trouble solving. After all, all they need to do is walk around a 3 m long barrier. It is surprising that studies have shown that dogs generally perform poorly on this task, and fail to significantly improve performance after repeated trials or generalise problem-solving strategies when conditions are reversed. About 40% of dogs fail to solve the task within 60 seconds on the first trial. For the remaining dogs, the latency to reach the target without demonstration is consistently reported as 20–40 seconds.[63–66] The observation of a demonstrator (human or dog) conducting a detour led to immediate improvement in the detouring behaviour. Dogs were able to learn by observing a conspecific perform a detour. Pongrácz[66] found that 40% of dogs abandoned the direction they chose in the first trial and followed the demonstrated direction (opposite side). Thus dogs showed a preference for the more conservative and socially learned behaviour.[65]

Wolves have been shown to be more proficient at this task than their domestic counterparts. Frank and Frank[40, 67] tested four six-week-old wolf puppies, and found that wolves were much more successful at the task than their domestic counterparts. On a U-shaped barrier, wolves made significantly fewer errors than the dogs of similar ages tested in the well-known studies of dog development conducted by Scott and Fuller.[62] Some caution is needed when interpreting these results, as only a small number of wolves were tested (an increased sample is likely to have produced less variability and therefore increased the significance of the interspecies difference), the dog sample were raised with a wolf mother (may have led to developmental differences from dog-raised pups), and the barrier was opaque not transparent. There are clear differences in cognitive development between dogs and wolves at six weeks of age, that were not accounted for. As yet, these are the only wolf–dog comparison studies conducted with a detour type task. Such studies need to be replicated with different wolf or wild canid samples.

The research led me to wonder how captive dingoes would respond to the detour barrier, and whether they would outperform domestic dogs. I tested 20 captive dingoes in a grassed outdoor enclosure using a V-shaped fence.[68] Dingoes were randomly allocated to one of four experimental groups: 'inward detour' was a scenario where the dingo started outside the fence and had to detour inside (Figs 9.5, 9.6); 'outward detour' was where the dingo started inside the fence and had to detour outside; 'demonstration' was when the dingo was able to watch a human detour the fence; 'door' condition was where a small opening in the fence gave the dingo quick access to the reward so that it did not have to detour. Four trials were given, then the condition was reversed to see if the dingo could adapt its problem-solving strategy to a

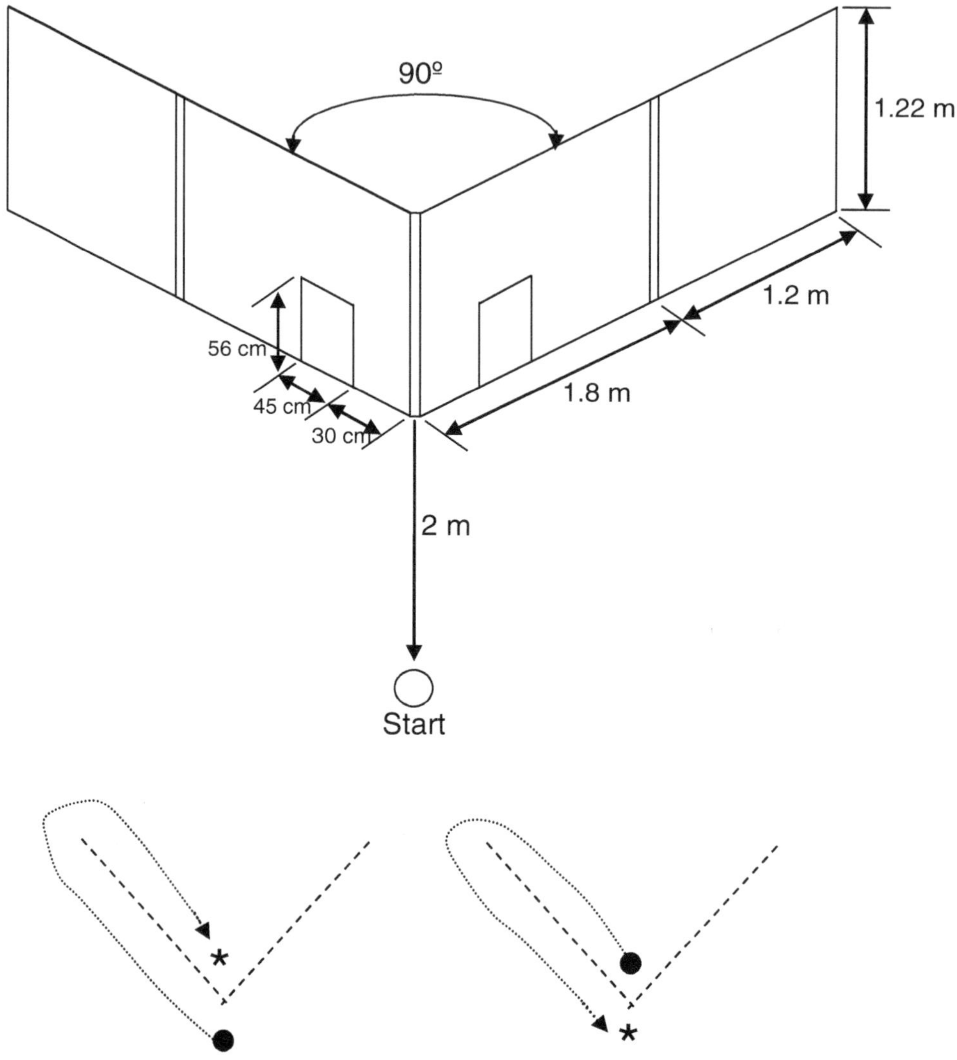

Figure 9.5: V-shaped fence design. The path of the dingo making (a) an inward detour (starting from the outside going inwards) and (b) an outward detour (starting from the inside going outwards). = dog's path. = starting position. * = location of reward/target.

Source: Smith and Litchfield.[68]

slightly different problem. In the 'door' condition, the opening was open for the first four trials but closed on the fifth, so the dingo had to detour around the fence. This was particularly difficult because a previously successful strategy was no longer available and the dingo had to think very quickly about how to solve it.

The dingoes did not seem to have any trouble with the task and successfully completed all forms of the detour task from the first trial within the 60 second time limit. Compared to domestic dogs in detour experiments conducted by other researchers, dingoes on average

Figure 9.6: An adult dingo making an inward detour on his first trial (1–4): approaching the barrier, then immediately navigating around it (most dingoes ran straight around the barrier without first approaching it).
Source: Bradley Smith.

(Continued)

Figure 9.6: *(Continued)*

completed the detour task more quickly (under 12 seconds), showed no signs of frustration or confusion (e.g. pawing at the fence, digging or barking) and made few errors (less than 1 on average). All dingoes successfully completed the task on the first trial, and could successfully generalise their problem-solving strategies when conditions were reversed (Figs 9.7, 9.8).

Apart from showing how well dingoes can solve spatial problems, the results suggest that there is a considerable difference in performance between dingoes and domestic dogs, as originally proposed by Frank.[40, 69] That is, wild canids (wolves and dingoes) raised in captivity are more adept at non-social problem-solving tasks than domestic dogs. Domestication is likely to contribute to this difference in problem-solving abilities and/or strategies between wild and domestic species of canid. The wolf evolved in a natural setting, and only individuals who learned quickly from their actions (or those of others) were likely to survive and reproduce. Frank suggested that the natural environment promotes the evolution of complex cognitive systems, whereas this selection pressure may be relaxed in an environment where domestic dogs and humans interact constantly. Humans serve as a constant buffer between the dog and the environment, mediating most significant environmental feedback.[26] Thus, over time, dogs have become reliant on humans for problem-solving, using them to solve their problems or becoming dependent on human cues (e.g. by encouraging glances or gestures).[63] It is unclear whether domestic dogs have altogether lost the ability to work independently to solve problems, or whether they are simply hindered or handicapped by their close relationship with and dependence on humans.[70]

I don't need your help!

The detour task showed that dogs and wild canids differ in their ability to solve spatial problems (problems that have no social elements). While domestic dogs might have lost some

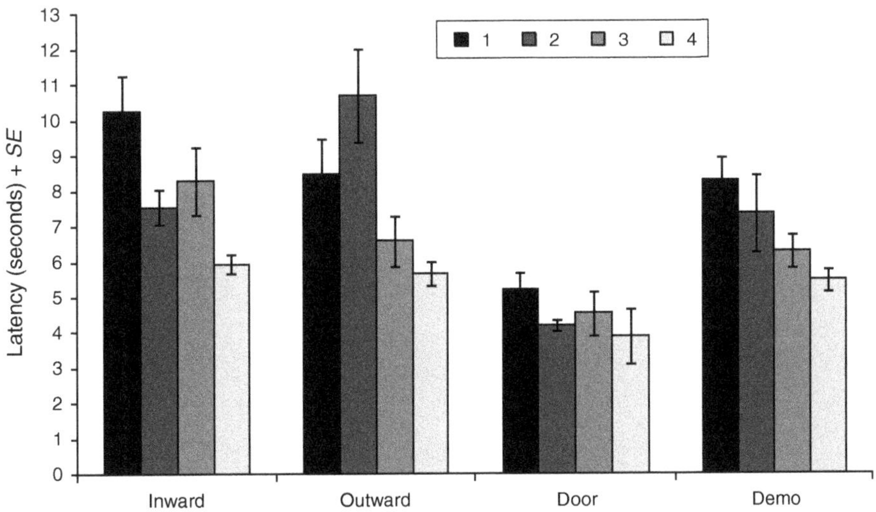

Figure 9.7: The mean latency in seconds (± s.e.m.) to complete the detour during trials 1–4 in the inward doors closed ($N = 8$), outward doors closed ($N = 4$), inward doors open ($N = 4$) and demonstration ($N = 4$) detour conditions.

Source: Smith and Litchfield.[68]

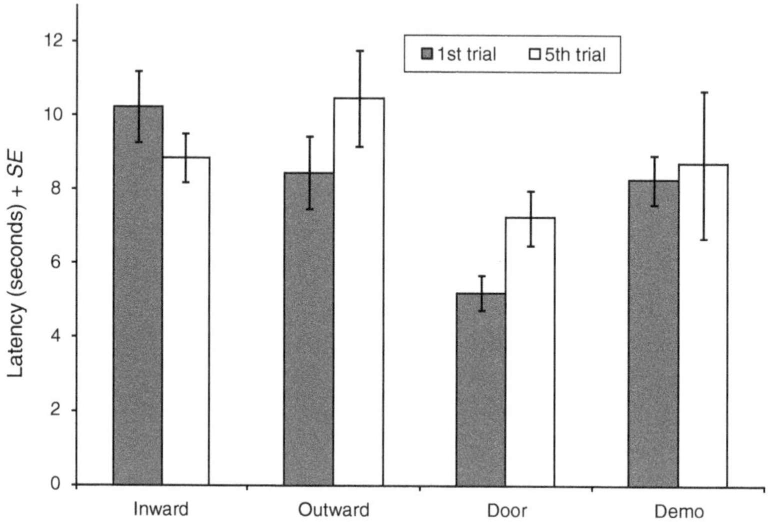

Figure 9.8: The latency to complete the detour in the inward (doors closed), outward (doors closed), inward (door open) and demonstration detour conditions for trials 1 and 5. The first four trials for every condition were the same, whereas the fifth trial was reversed.
Source: Smith and Litchfield[66]

cognitive abilities needed in the wild, they have developed specialised cognitive abilities relevant to the human-oriented environment.[71] Therefore they adopt distinctly different problem-solving strategies during times of conflict and uncertainty. These differences are most evident when dogs and wolves are confronted with a difficult or unsolvable problem in the presence of a human. In general, wolves prefer to work independently, attempt several strategies and show persistence, whereas domestic dogs often give up easily and gaze or look back towards a nearby human for assistance. For example, after presenting wolves and dogs with a puzzle box task, Frank and Frank[42] noted that wolves generally attacked the problem until they had solved it or the trial ended. In contrast, dogs investigated the puzzle box and, after discovering the food was not easily accessible, 'performed a variety of solicitation and begging gestures' towards the nearby experimenter (p. 271). Dogs may have adjusted their problem-solving strategies in such a way that they use humans as a form of social 'tool' and rely on human assistance when confronted with a problem-solving situation.[56, 72] Consider a dog bringing a leash to its owner in order to persuade the owner to open an external door and/or take them for a walk, which may be an example of social tool use. The dog is using an object to manipulate the behaviour of a human within the context of a social interaction.

In 2003, Hungarian scientist Adam Miklósi[56] and his colleagues made an important discovery relating to the problem-solving strategies adopted by dogs. They presented dogs and wolves with a simple task which involved pulling on a rope to get a reward that was otherwise out of reach. After the dogs had learned to get the reward at the end of the rope, the researchers set up the task so it was no longer solvable. That is, they fixed the rope such that it could not be pulled free yet appeared the same as before. They were particularly interested in seeing how dogs and wolves behaved in the presence of a human who was standing behind them. Dogs looked or gazed towards their owner/caregiver, while the socialised wolves appeared to ignore their owner/caregiver. During the study, seven of the nine dogs looked back at the human after trying to obtain the food reward for only about 1 minute (median. Only two of the seven wolves looked back at all, instead attempting to solve the task on their own or simply giving up.

From these experiments, Miklósi *et al.*[56] concluded that one of the key behavioural differences between dog and wolf behaviour was the dog's tendency to look at a human's face and make eye contact. Even after intensive socialisation, wolves appear less likely to use human communicative signals or engage in communicative interactions with humans.[73] Conversely, domestic dogs appear genetically predisposed to refer to humans for cues. Such referential behaviour leads to initialising and maintaining communicative interaction with humans, with dogs and humans capable of complex forms of communication as a result. Miklósi suggested that the results may shed light on the process of domestication of the dog (the transition from wolves to domestic dogs). For example, it may be that humans selected only wolves that displayed abilities to communicate or use human-given cues,[56] or that reading human cues is merely a by-product of domestication and spending time with humans over many generations.[74] I was curious to understand how dingoes would behave in this task because it would help shed light on the comparisons between wild and domestic canids, and help me establish to what extent, if any, dingoes may have been domesticated in their past.[75]

Each of the 12 dingoes I tested were positioned 1.5 m in front of a wire mesh cage and held on a leash by a familiar caretaker who was located 1 m behind them (starting position). To solve this task, which they did easily (Fig. 9.11), each dingo had to learn to pull a piece of rope towards itself in order to reach a piece of meat attached to the end of the rope; the original positions of the rope and meat were out of reach and housed within a metal dog crate. For the first five trials the dingo walked up to the cage and pulled the rope to get the reward (a piece of chicken wing; Fig. 9.9). On the sixth occasion, the trial was 'blocked' (the rope was attached to the cage by threading a separate piece of rope through the original rope threads and tying this piece to the base of the cage; Fig. 9.10). The rope could no longer be pulled out of the cage to retrieve the meat, which was not obvious by visual inspection. The first dingo tested on the blocked trial (unsolvable version of task) would not give up trying to get the rope, and eventually pulled the rope so forcefully that the cage moved. I ensured the cage was weighted down for the next dingo.

The results of the blocked trial depend greatly on the definition of looking back that is used.[75] In this chapter we argue that 'looking back' during the rope-pulling task involved the dingo standing/sitting/lying in a stationary position (there may be some movement) and, after first looking at or orienting its head/nose towards the task, the dingo then lifts or turns its head/nose and orients it towards the human experimenter's/caretaker's (holding leash, positioned behind the dingo) head/face for at least half a second. Using these strict criteria, only three of the 12 dingoes looked back during the experiment (see Fig. 9.12 for a visual example of a look back). The average duration of these gazes was 0.78 seconds (range 0.4–1.5 seconds), with only one dingo looking back at the human for longer than 1 second. It can be suggested that dingoes do not seem to look back at humans for assistance during times of uncertainty or when confronted with a difficult problem. Instead they behave in a similar way to wolves.[56]

Figure 9.9: A dingo during training condition (trials 1–5) having successfully retrieved the meat.
Source: Bradley Smith.

Unfortunately there were two problems with this study that I think are important to discuss. First, it was evident that problem-solving strategies between dogs and dingoes may be fundamentally different, and the rope task does not allow for the different strategies. It would be unlikely that wolves would spend the full 2 minutes continuously pulling or digging at the rope without trying any alternative strategies, since it would be a waste of energy to continue utilising an ineffective strategy, but they were not really allowed the opportunity. A look back for wild canids is most likely to occur after all strategies have been exhausted. Prior experience of the animals may have also influenced their behaviour during the task. That is, the rearing environments of dogs and wolves influence their problem-solving strategies. For example, some dogs may have extensive experience with owners solving problems for them, whereas wolves have more experience solving problems on their own with little or no assistance from humans. Therefore dogs may be reinforced during difficult problems when looking at their owner, whereas wolves learn that persistence generally pays off. This is indicative of artificial selection on dogs and the absence of such selection pressure in wolves, which retain certain independence in problem-solving. Second, the use of a leash with wild animals may change the task from its intended purpose to an escape task. Wild animals, particularly canids, do not react positively to being restrained (e.g. placed on a leash). Glances or looks back made by dingoes in this experiment occurred primarily when struggling to get off the lead, perhaps a reflection of an innate escape response to restraint.[30] Frank and Frank[76] tested wolves on

Figure 9.10: A dingo attempting to retrieve the meat during the blocked condition (trial 6).
Source: Bradley Smith.

various cognitive tasks and found that when restrained by a leash, wolves responded in an emotional way that seemed to interfere with their performance on standardised tests. When using a restraint, it was difficult to understand the communicative behaviour of the dingoes. When an animal was exposed to an aversive stimulus like a leash, a lot of appetitive behaviours (e.g. searching for food) may be inhibited, and behaviours more reflective of other motivations (e.g. to escape) may be exhibited. The dingo was likely to have been focused on the leash (or escaping from it), and the unavailable reward was irrelevant in this context. Even though these might be problems with this experimental methodology and we can generally be sure that wild canids don't refer back to nearby humans during difficult or confusing situations, the experiments still highlighted differences in the behaviour of dingoes and dogs, most likely the result of their very different life histories.

Do dingoes get the point?

The previous experiment suggested that wild and domestic canids communicate differently with humans in problem-solving situations. A similar finding has been reported on the ability of dogs and wolves to follow or understand referential gestures (e.g. pointing) given by humans. I wanted to establish whether dingoes could understand such human social cues.

Most highly social species must utilise information from both their immediate physical environment and conspecifics. Acquiring information from group mates regarding food, prey

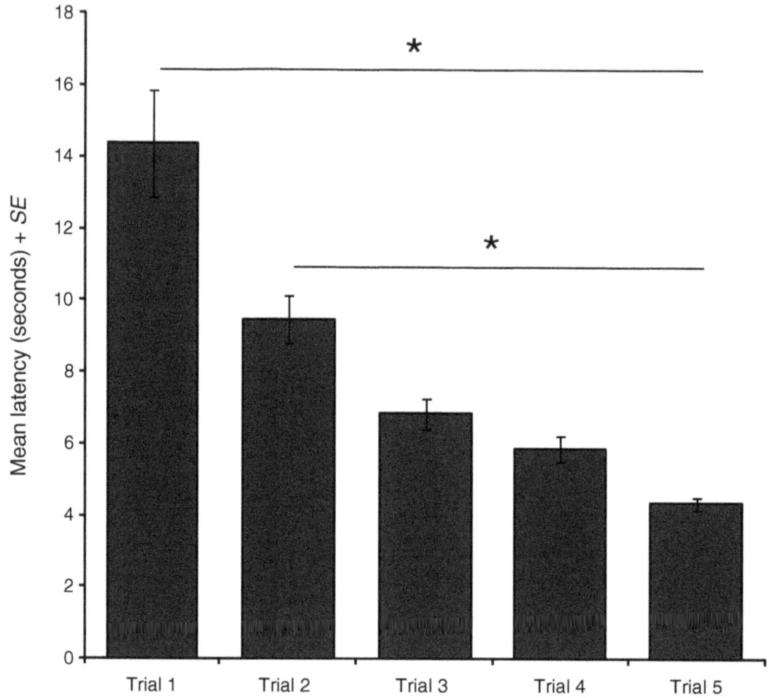

Figure 9.11: Mean latency in seconds (± s.e.m.) to obtain the food during trials 1–5. * = significantly different at $p = <0.05$.
Source: Smith and Litchfield.[75]

and predators is achieved through interpretation of their communicative gestures. This includes the ability to predict future actions of conspecifics during agonistic (aggressive or threatening), affiliative and predatory interactions. Social cues may be exchanged when a group member looks towards an object, place or individual accompanied by head or body orientation, vocalisation or movement towards a target location.[72, 77–79] For example, wolves can indicate the position of prey by freezing and orienting their whole body towards the prey.[80]As highly social species, both dogs and wolves communicate with conspecifics using signals which combine the position of the body, head and tail, facial gestures, head gestures and eye movement, and via vocalisations such as barking, howling and whimpering.[32, 81, 82]

What is interesting about dogs is that they are not only able to understand communicative signals of their own species but those of humans (many other species do not). The ability of a non-human animal to comprehend the social communication of other members of their species or of other species can be tested using an experimental procedure called the object–choice paradigm. This can easily be done at home (if your dog is not good at following your gestures, you can train it to improve). You hide a reward (e.g. food) in one of two distinct locations (e.g. two buckets), then give the dog a signal of directional attention (e.g. looking at, pointing at, tapping on, placing a marker on/next to) indicating the location of the reward. The dog must approach the correct location to successfully pass the task, signifying an understanding of the communicative nature of the experimenter's gesture.[35, 77, 83]

Figure 9.12: A dingo during the blocked condition 'looking back' at the experimenter. Image taken from video footage.
Source: Bradley Smith.

As alluded to earlier, domestic dogs are remarkably talented at reading the body language and attentional states of humans and from a very young age are able to follow a variety of human communicative gestures and cues on the object–choice task.[84–88] Such gestures and cues include variations of pointing, nodding, bowing, tapping, standing behind, physical markers and when the experimenter is located behind the incorrect container but is pointing at the correct container.[35, 77, 83, 89–95] Beyond the object–choice task, domestic dogs show other behaviours suggestive of a high-level ability to understand human social behaviour. Dogs know when a human is attentive or inattentive by judging observable cues such as eye contact and orientation of eye, head and body,[78, 86–88] are able to pick up on human play or aggressive signals,[85] engage in functionally referential communication with their owner such as 'showing' behaviour (e.g. informing a human of the location of a hidden object by standing next to it and gazing between object and owner)[96] and rely on information provided by humans when confronted by new problems such as the detour task.[63, 64]

The same cannot be said for wolves, which appear to be less successful than dogs on the object–choice task when presented with the same human social cues. Early studies found that wolves struggle to follow most of the cues, in particular the momentary distal pointing gesture[35, 56, 89] that is considered the hallmark gesture. During this gesture, the pointing hand is far away from the goal and the point is given only briefly. It is considered to be the benchmark because it appears to eliminate low-level or simple learning process interpretations (e.g.

'human hand–food association') that are present in proximal cues such as standing behind, touching or proximal pointing.[56, 90] Virányi and colleagues[90] made the argument that accurate comparison between dogs and wolves could only be made when individuals of both species had been reared in an identical manner; they individually hand-raised neonatal (4–7 days old) dogs and grey wolves in a family home. These hand-raised wolf pups could not attend to the momentary distal pointing gesture. In contrast, even domestic dogs without intensive socialisation could do so at four months of age. After more than seven months of exposure and formal training, the wolves eventually performed as well as dogs on the momentary distal point cue. Monique Udell[97] put a new spin on this research when she found that sanctuary-reared wolves in a US wolf park performed above chance on a modified version of the momentary distal pointing gesture, suggesting that given the right rearing environment (e.g. intensive socialisation with humans during a sensitive developmental window), daily interaction with humans and a testing environment, these skills are evident in wolves without previous experience of the task. Interpretation of results, in terms of comparative analyses, is challenged by methodological variations in these studies (e.g. rearing history, testing environment and methodology).

Some contention exists regarding the factors that have influenced the development of the domestic dogs' exceptional skills of social communication with humans.[98] The process of domestication may provide the most convincing explanation.[35] Through domestication, involving adaptation to humans and to a captive environment, some aspects of the social cognitive abilities of dogs have converged within the phylogenetic constraints of the species – dogs have evolved social communicative skills adapted for social interactions with humans.[35, 71, 72] Researchers Jonathan Cooper[99], Brian Hare and Michael Tomasello[71] suggested that the emergence of this social communicative behaviour in dogs occurred through two selective processes or stages. First, domestication took place, with selection for tameness potentially affecting systems controlling emotional reactivity such as fear and aggression. This was followed by breed diversification, where specific traits were selected for according to specific roles dogs would play in human society. Support for this is found in the ability of domestic foxes, selectively bred only for their tendency to approach humans fearlessly and non-aggressively, to follow human social cues given no prior experience compared with undomesticated captive foxes.[74] Dogs with limited human exposure from the age of six weeks (before the most sensitive period of socialisation with humans, at seven weeks) can instantaneously follow human social cueing.[35, 83, 89, 90, 100] Unfortunately, finding definitive support for the theory of domestication by ruling out the role of learning is essentially impossible. This would require testing dogs without prior experience with the hand delivery of food or other stimuli, or controlling for experiences during development.[98, 101]

It was my aim to test dingoes on the object–choice task using a comprehensive range of human communicative gestures previously tested on canids.[102] I was able to test seven captive dingoes with the help of volunteers at the sanctuary. I had to make sure the person providing the cue was familiar to the dingo (domestic dogs are happy to follow gestures or training commands from any human, familiar or not, whereas dingoes tend to cooperate and trust only those that they know). Each dingo was presented with 11 human-given cues (including controls; Figs 9.13, 9.14). Control trials were performed for each testing block to ensure that only the deliberate cues (and not subtle experimenter movements or odour cues) influenced the behaviour of the dingoes, and that the presence of the food in one of the containers had no effect. Dingoes performed surprisingly well on the task, passing all cues significantly above control, including the benchmark momentary distal pointing, with the exception of gaze only, gaze and point, and pointing from the incorrect location (Fig. 9.15).

Overall, the performance of dingoes appeared to lie somewhere between that of wolves and dogs (better than wolves but not as good as dogs). The results suggest that, at some level, some early domestication events (if they indeed occurred) may have played a role in the dingoes' ability to comprehend human gestures. Through domestication and subsequent co-evolution with humans, dogs may have gained the ability to communicate with humans. The ancestors of dingoes ~5000 years ago may have already had this skill. If so, this would lend support to the dual-stage hypothesis (domestication and selection for tameness + breed diversification) put forward by Hare and Tomasello[71] and supported by Wobber's[103] findings with New Guinea singing dogs, a close relative of the dingo.

Yet something about the domestication hypotheses doesn't sit right with me. It is somewhat unclear to me why dingoes would retain the ability to read human communicative gestures, when as a species they have essentially lived in a wild state isolated from human

Figure 9.13: Experimental setup showing experimenter 1 (left) and experimenter 2 (right) during the proximal pointing-only condition.
Source: Bradley Smith.

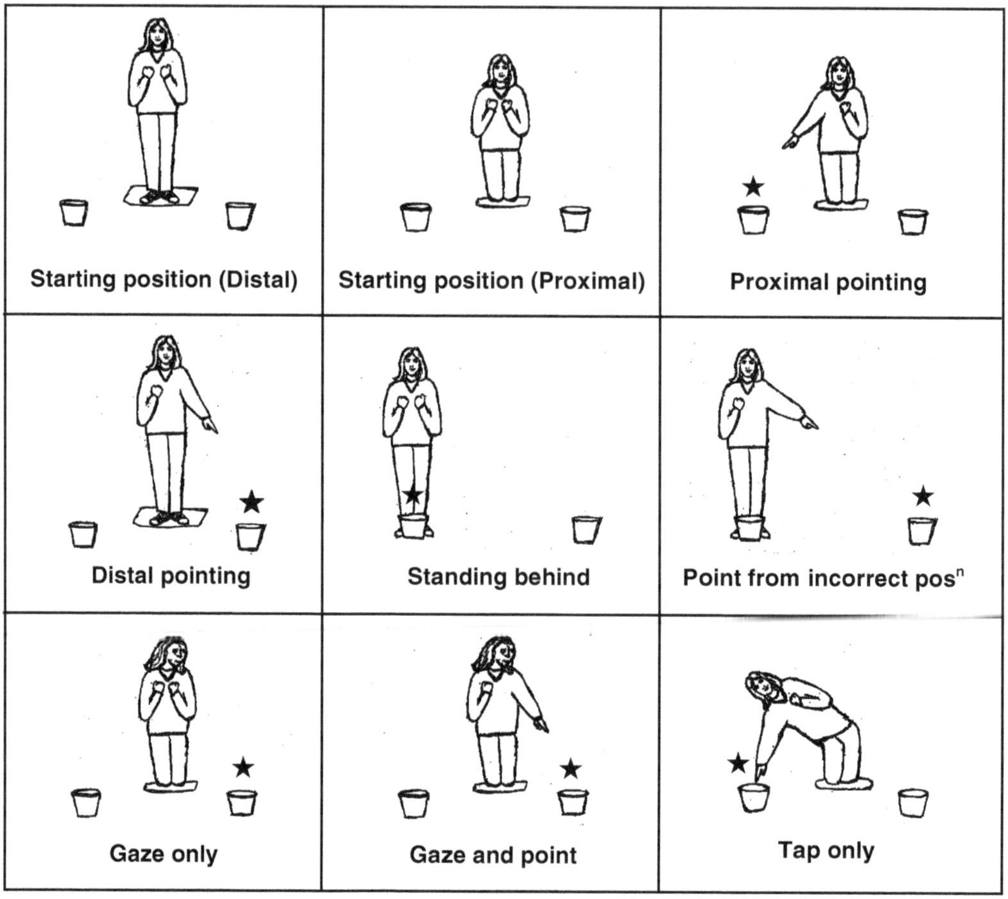

Figure 9.14: Diagrammatic representation of cue gestures, including the starting positions. Star symbol above container = bait location.

Source: Smith and Litchfield.[102]

selection pressures for as long as 5000 years. It is possible that dingoes retained this skill because the selection pressure against it was probably not high, and there was no cost associated with its retention in later generations. But as wolves have been shown to follow many cues under the right circumstances, there are clearly other factors at play. As highly social species, canids are able to quickly develop/learn the ability to read human behaviour. But not all social species can do this, so there must be something special or unique about canids that we don't quite understand yet. To me it suggests that there might be something akin to a basal level of sociality that involves the prominent component of a predisposition to understand cues (which quickly develops in different directions in wolves and dogs). Given the right circumstances (e.g. the right experience and environment), it can be generalised to non-conspecifics. Dingoes, as a proto-dog – an early version of the domestic dog we have today – represent something between a dog and a wolf. Whether dingoes are a result of human domestication events or not, they seem to have the best of both worlds. That is, they can communicate and live with humans quite successfully and more easily than wolves, but remain more 'wild' than dogs

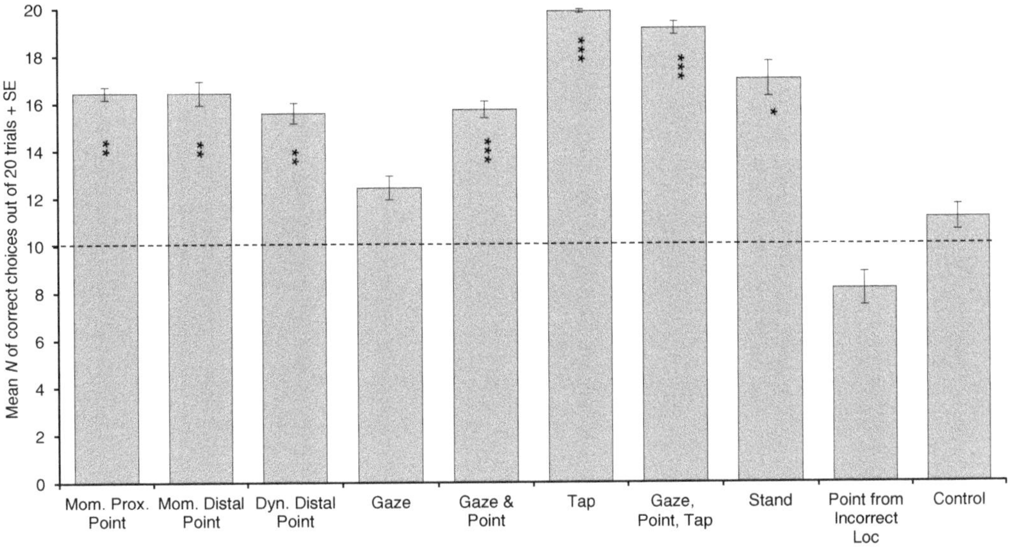

Figure 9.15: Mean number of correct choices (± s.e.m.) by all dingoes in each condition. Dotted line = chance performance level (50%). The asterisks (*p <0.05, **p <0.02, ***p <0.01) indicate that performance for each cue significantly differed from random choice.

Source: Smith and Litchfield.[102]

and able to survive without human assistance. Perhaps the biggest difference between dingoes and wolves is that dingoes can connect more easily with people, while wolves more often cannot.

Conclusion

The observations and studies relating to dingo intelligence discussed in this chapter provide a starting point to uncovering the cognitive and behavioural potential of the dingo. But we have only scratched the surface and are constantly being surprised by new discoveries. You would need to spend your entire life studying dingoes to come even close to understanding them. For now, though, we can say that dingoes are more wolf-like than dog-like in their cognitive ability. That is, dingoes are good at solving problems independently and are nowhere near as proficient as dogs in communicating with humans. Table 9.1 provides a summary of the current state of cognitive research on the dingo, and shows how they compare to wolves and domestic dogs tested on the same experiments.

A fundamental reason why dingoes remain a feature in the Australian landscape, despite centuries of persecution and environmental capriciousness, is their considerable intelligence. This influences many aspects of their life history including their variation and adaptability. Not only has this allowed them to successfully adapt to all environments in Australia (particularly urban areas), it also protects them from constant and various forms of human persecution. The dingo is continually being shaped by nature, and there is simply no place in the wild for a dingo that is too fat, too slow or too dumb. The research discussed in this chapter adds to the mounting evidence suggesting that dingoes should be considered independent from

Table 9.1. Summary of the cognitive studies presented in this chapter, the relevant research question, the performance of wolves, dogs and dingoes, and a conclusion based on the evidence

Concept	Research question	Summary			Conclusion
		Wolf	Dog	Dingo	
High-order behaviour	Are dingoes capable of higher order behaviour?	In the wild can be observed during social/pack hunting.[29] Handful of anecdotal examples of high-order behaviour and observational learning reported in captive wolves[44,4]	Some anecdotes reported[105,106] as well as in the scientific literature.[107] Dogs are much more adept at social cognition, and may use humans as tools during problem-solving [56,72]	Three observations of two captive dingoes exhibiting high-order behaviour were presented. One included an example of tool use.[49] A case where a wild dingo reacted strongly to the death of her pup was also given[58]	Evidence is mounting that dingoes are capable of high-order behaviour. Similar studies have been shown in wolves, but not domestic dogs. Dingoes are the first canid to be documented exhibiting tool use and strong reactions to the death of a conspecific
Spatial problem-solving	How do dingoes compare to dogs and wolves on the detour task?	According to a sole study, wolves have no problem detouring a barrier, and have been shown to perform better than dogs[40,67]	Dogs find this task difficult but can easily perform the detour once shown by a conspecific or human demonstrator[63,64]	Dingoes easily passed this task, and did so quicker and with fewer errors than domestic dogs. Dingoes quickly adapted their problem-solving strategy to variations of a similar task[68]	Results lend support to the idea that wild canids are more adept at non-social (spatial) problem-solving than dogs
Social cognition: problem solving strategies	Do dingoes look back to humans for help during problem-solving?	Wolves do not look back to a nearby human when faced with an unsolvable problem[56]	Dogs make distinctive look back gestures to a nearby human when faced with an unsolvable problem[56,63,90]	Dingo behaviour closely resembled that of wolves. Few dingoes looked back, and looks were fleeting in nature[75]	Results suggest that wild canids (including dingoes and wolves) do not seek human assistance when confronted with a problem which they are having difficulty solving, but dogs do. Dogs have developed a reliance on humans and see them as tools to help solve problems
Social cognition: following human social cues	Can dingoes understand human referential gestures on the object–choice task?	Wolves generally have trouble following human social cues, particularly the momentary distal point, even when reared the same as a dog [56,90]	Dogs are exceptionally good at following many variations of human social cues from a young age and with minimal experience with humans[84]	The dingoes could follow all cues significantly above control, including the benchmark momentary distal pointing, with the exception of gaze only, gaze and point, and pointing from the incorrect location[102]	Dingo performance appears to lie somewhere between wolves and dogs. This suggests that domestication may play a role in this ability, and/or early versions of the domestic dog (proto-dogs) may have had the ability to communicate across the species barrier and start living successfully alongside humans

domestic or feral dogs. Such knowledge could be useful for future attempts to modify the perception of dingoes held by everyday Australians. Human attitudes to dingoes must be challenged and changed if dingo conservation is to be successful. The conservation of whales and dolphins took a turn when people started to realise that they were sentient creatures (able to feel, perceive or be conscious). Surely dingoes, as sentient creatures, deserve the same treatment?

Until now, only wild (e.g. wolves) or domestic canid (e.g. pet dogs) species have been studied, allowing researchers to speculate on how the domestication process may have influenced canid cognition and behaviour. As I hope you will appreciate from reading this chapter, the dingo is a unique canid and able to contribute significantly to this debate. Researchers need to expand the populations to include those such as hybrid canids (wolf–dog and dingo–dog crosses), free-ranging (or feral) domestic dogs and arguably less social species of canids such as foxes, coyotes and jackals. Of particular interest would be whether hybrid or feral dogs retain any of the social skills of their recent domestic dog ancestors. This is important to find out because of the number of feral dogs and dog–dingo hybrids in the Australian bush, about which we simply don't know anything.

Who is the top dog when it comes to intelligence – dogs or dingoes? In many ways they share the same skills and abilities. Both are 'smart', but in their own ways. For these reasons, it is unfair to say one is smarter than the other. They are each unique, and each equipped with the tools they need to survive in their environment. Through the process of domestication, dogs may have lost some skills because they no longer need them. But what they have lost in some areas they have gained in others. The result is the perfect household companion. Adam Miklósi[104] put it best when he said:

> If you assume an animal has to survive without human presence, then wolves are smarter. But if you are thinking that dogs have to survive in a human environment where it's very important to follow the communications of humans, then in this aspect, dogs are smarter.

References

1. Kaleski R (1933) *Australian Barkers and Biters*. Bulletin Newspaper, Sydney.
2. Breckwoldt R (1988) *A Very Elegant Animal: The Dingo*. Angus and Robertson Publishers, Sydney.
3. Hudson L (1974) *Dingoes Don't Bark*. Rigby, Adelaide.
4. Dickman CR, Lunney D (Eds)(2001) *A Symposium on the Dingo*. Royal Zoological Society of New South Wales, Sydney.
5. Parker M (2006) Bringing the dingo home: discursive representations of the dingo by aboriginal, colonial and contemporary Australians. PhD thesis. University of Tasmania
6. Parker M (2007) The cunning dingo. *Society and Animals* **15**, 69–78.
7. Bacon JS (1960) *The Australian Dingo: King of the Bush*. Glasson and Oliver, Sydney.
8. Rolls EC (1993) *From Forest to Sea: Australia's Changing Environment*. University of Queensland Press, Brisbane.
9. Holden P (1991) *Along the Dingo Fence*. Hodder and Stoughton, Sydney.
10. Murray AG (1972) Trapping dingoes: a dying art. *Walkabout* **26**, 32–34.
11. Newsome C (1985) *Dingo Howlers*. Colin J. Newsome, Glen Innes, NSW.

12. Twigg LE, Eldridge SR, Edwards GP, Shakeshaft BJ, de Preu ND, Adams N (2000) The longevity and efficacy of 1080 meat baits used for dingo control in central Australia. *Wildlife Research* **27**, 473–481.
13. Fleming PJS (1996) Ground-placed baits for the control of wild dogs: evalution of a replacement-baiting strategy in north-eastern New South Wales. *Wildlife Research* **23**, 729–740.
14. Fleming PJS, Parker RW (1991) Temporal decline of 1080 within meat baits used for control of wild dogs in New South Wales. *Wildlife Research* **18**, 729–740.
15. Thompson JA, Fleming PJS (1991) The cost of aerial baiting for wild dog management in north-eastern New South Wales. *Rangeland Journal* **13**, 47–56.
16. Thomson PC (1986) The effectiveness of aerial baiting for the control of dingoes in north-western Australia. *Wildlife Research* **13**, 165–176.
17. McIlroy JC, Gifford EJ, Carpenter SM (1988) The effect of rainfall and blowfly larvae on the toxicity of '1080'-treated meat baits used in poisoning campaigns against wild dogs. *Australian Wildlife Research* **15**, 473–483.
18. Glen AS, Dickman CR (2003) Effects of bait-station design on the uptake of baits by non-target animals during control programs for foxes and wild dogs. *Wildlife Research* **30**, 147–149.
19. Glen AS, Gentle MN, Dickman CR (2007) Non-target impacts of poison baiting for predator control in Australia. *Mammal Review* **37**, 191–205.
20. Eldridge SR, Berman DM, Walsh B (2000) Field evaluation of four 1080 baits for dingo control. *Wildlife Research* **27**, 495–500.
21. Corbett LK (2001) *The Dingo in Australia and Asia*. JB Books, Adelaide.
22. Allen L, Gonzalez T (1998) Baiting reduces dingo numbers, changes age structures yet often increases calf losses. In *11th Australian Vertebrate Pest Conference*. Bunbury, Western Australia.
23. EPA (2001) *Fraser Island Dingo Management Strategy*. Environmental Protection Agency/ Queensland Parks and Wildlife Service, Brisbane.
24. O'Neill A (2002) *Living with the Dingo*. Envirobook, Sydney.
25. Peters R, Mech LD (1975) Behavioural and intellectual adaptations of selected mammalian predators to the problem of hunting large animals. In *Socioecology and Psychology of Primates*. (Ed. RH Tuttle) pp. 279–300. Mouton Publishers, The Hague.
26. Frank H (1980) Evolution of canine information processing under conditions of natural and artificial selection. *Zeitschrift für Tierpsychologie* **53**, 389–399.
27. Durbin L, Venkataraman A, Hedges S, Duckworth W (2004) *Dhole cuon alpinus*. In *Wild Canids: Status Survey and Conservation Action Plan*. (Eds C Sillero and DW Macdonald) pp. 210–219. IUCN, Gland.
28. Thomson PC (1992) The behavioural ecology of dingoes in north-western Australia. III. Hunting and feeding behaviour, and diet. *Wildlife Research* **19**, 531–541.
29. Mech LD (2007) Possible use of foresight, understanding, and planning by wolves hunting muskoxen. *Arctic* **60**, 145–149.
30. Packard JM (2003) Wolf behaviour: reproductive, social, and intelligent. In *Wolves: Behaviour, Ecology, and Conservation*. (Eds LD Mech and L Boitani) pp. 35–65. University of Chicago Press, Chicago.
31. Moran G, Fentress JC, Golani I (1981) A description of relational patterns of movement during 'ritualized' fighting in wolves. *Animal Behaviour* **29**, 1146–1165.
32. Schenkel R (1947) Expression studies of the wolf. *Behaviour* **1**, 81–129.
33. Fox MW (1971) *Behaviour of Wolves, Dogs and Related Canids*. Jonathan Cape, London.
34. McLeod PJ, Fentress JC (1997) Developmental changes in the sequential behavior of interacting timber wolf pups. *Behavioural Processes* **39**, 127–136.
35. Hare B, Brown M, Williamson C, Tomasello M (2002) The domestication of social cognition in dogs. *Science* **298**, 1634–1636.
36. Hemmer H (1983/1990) *Domestication: The Decline of Environmental Appreciation*. Cambridge University Press, Cambridge.

37. Hiestand NL (1989) A comparison of problem-solving and spatial orientation in the wolf (*Canis lupus*) and dog (*Canis familiaris*). PhD thesis. University of Connecticut, Storrs.
38. Hiestand L (2011) A comparison of problem-solving and spatial orientation in the wolf (*Canis lupus*) and dog (*Canis familiaris*). *Behavior Genetics* **41**, 840–857.
39. Cheney C (1982) Probability learning in captive wolves. In *Wolves of the World: Perspectives of Behavior, Ecology and Conservation.* (Eds FH Harrington and PC Paquet) pp. 272–281. Noyes Publications, Park Ridge.
40. Frank H, Frank MG (1982) Comparison of problem-solving performance in six-week-old wolves and dogs. *Animal Behaviour* **30**, 95–98.
41. Frank H, Frank MG (1984) Information processing in wolves and dogs. *Acta Zoologica Fennica* **171**, 225–228.
42. Frank H, Frank MG (1985) Comparative manipulation-test performance in ten-week-old wolves (*Canis lupus*) and Alaskan malamutes (*Canis familiaris*): a Piagetian interpretation. *Journal of Comparative Psychology* **99**, 266–274.
43. Frank H, Frank MG, Hasselbach L, Littleton D (1989) Motivation and insight in wolf (*Canis lupus*) and Alaskan malamutes (*Canis familiaris*): visual discrimination learning. *Bulletin of the Psychonomic Society* **27**, 455–458.
44. Fox MW (1971) Possible examples of high-order behavior in wolves. *Journal of Mammalogy* **52**, 640–641.
45. Mech LD (1991) *The Way of the Wolf.* Voyageur Press, Stillwater.
46. Wright SM (1968) *The Way of the Dingo.* Angus and Robertson, Sydney.
47. Macintosh NWG (1975) The origin of the dingo: an enigma. In *The Wild Canids: Their Systematics, Behavioural Ecology and Evolution.* (Ed. MW Fox) pp. 87–106. Van Nostrand Reinhold, New York.
48. Oakman B (2001) The problem with keeping dingoes as pets and conservation. In *A Symposium on the Dingo.* Royal Zoological Society of New South Wales, Sydney.
49. Smith BP, Appleby RG, Litchfield CA (2012) Spontaneous tool-use: an observation of a dingo (*Canis dingo*) using a table to access an out-of-reach food reward. *Behavioural Processes* **89**, 219–224.
50. Morgan CL (1903) *An Introduction to Comparative Psychology.* Walter Scott, London.
51. Epstein R, Kirshnit C, Lanza R, Rubin L (1984) 'Insight' in the pigeon: antecedents and determinants of an intelligent performance. *Nature* **308**, 61–62.
52. Alcock J (1972) The evolution of the use of tools by feeding animals. *Evolution* **26**, 464–473.
53. Beck B (1980) *Animal Tool Behaviour: The Use and Manufacture of Tools by Animals.* Garland STPM Press, New York.
54. Mann J, Sargeant B, Watson-Capps J, Gibson Q, Heithaus M, Connor R, Patterson E (2008) Why do dolphins carry sponges? *PLoS ONE* **3**, e386810.1371/journal.pone.0003868
55. Hunt G (1996) Manufacture and use of hook-tools by New Caledonian crows. *Nature* **379**, 249–251.
56. Miklósi A, Kubinyi E, Topál J, Gácsi M, Virányi Z, Csányi V (2003) A simple reason for a big difference: wolves do not look back at humans, but dogs do. *Current Biology* **13**, 763–766.
57. Oxley J (1820) *Journals of Two Expeditions into the Interior of New South Wales, Undertaken by Order of the British Government, 1817–18.* John Murray, London.
58. Appleby R, Smith B, Jones D (2013) Observations of a free-ranging adult female dingo (*Canis dingo*) and littermates' responses to the death of a pup. *Behavioural Processes* **96**, 42–46.
59. Hartmann G (1935) *Gestalt Psychology: A Survey of Facts and Principles.* Ronald Press, New York.
60. Schiller P (1949) Analysis of detour behaviour. I. Learning of roundabout pathways in fish. *Journal of Comparative and Physiological Psychology* **42**, 463–475.
61. Köhler W (1927) *The Mentality of Apes* (trans. E. Winter). Routledge and Kegan Paul, London.

62. Scott J, Fuller J (1965) *Genetics and the Social Behavior of the Dog*. University of Chicago Press, Chicago.
63. Pongrácz P, Miklósi A, Vida V, Csányi V (2005) The pet dog's ability for learning from a human demonstrator in a detour task is independent from the breed and age. *Applied Animal Behaviour Science* **90**, 309–323.
64. Pongrácz P, Miklósi Á, Kubinyi E, Gurobi K, Topál J, Csányi V (2001) Social learning in dogs: the effect of a human demonstrator on the performance of dogs in a detour task. *Animal Behaviour* **62**, 1109–1117.
65. Pongrácz P, Miklósi A, Kubinyi E, Topál J, Csányi V (2003) Interaction between individual experience and social learning in dogs. *Animal Behaviour* **65**, 595–603.
66. Pongrácz P, Miklósi Á, Timár-Geng K, Csányi V (2003) Preference for copying unambiguous demonstrations in dogs (*Canis familiaris*). *Journal of Comparative Psychology* **117**, 337–343.
67. Frank H, Frank MG (1987) The University of Michigan canine information-processing project (1979–1981). In *Man and Wolf*. (Ed. H Frank) pp. 143–167. Dr W. Junk Publishers, Dordrecht.
68. Smith BP, Litchfield CA (2010) How well do dingoes, *Canis dingo*, perform on the detour task? *Animal Behaviour* **80**, 155–162.
69. Frank H, Frank MG (1982) On the effects of domestication on canine social development and behavior. *Applied Animal Ethology* **8**, 507–525.
70. Topál J, Miklósi A, Csányi V (1997) Dog–human relationship affects problem solving behavior in the dog. *Anthrozoos* **10**, 214–224.
71. Hare B, Tomasello M (2005) Human-like social skills in dogs? *Trends in Cognitive Sciences* **9**, 439–444.
72. Miklósi A, Topál J, Csányi V (2004) Comparative social cognition: what can dogs teach us? *Animal Behaviour* **67**, 995–1004.
73. Miklósi A (2007) *Dog Behaviour, Evolution and Cognition*. Oxford University Press, Oxford.
74. Hare B, Plyusnina I, Ignacio N, Schepina O, Stepika A, Wrangham R, Trut L (2005) Social cognitive evolution in captive foxes is a correlated by-product of experimental domestication. *Current Biology* **15**, 226–230.
75. Smith BP, Litchfield CA (2013) Looking back at 'looking back': operationalising referential gaze for dingoes in an unsolvable task. *Animal Cognition* **16**, 961–971.
76. Frank H, Frank MG (1983) Inhibition training in wolves and dogs. *Behavioural Processes* **8**, 363–377.
77. Hare B, Tomasello M (1999) Domestic dogs (*Canis familiaris*) use human and conspecific social cues to locate hidden food. *Journal of Comparative Psychology* **113**, 173–177.
78. Virányi Z, Topál J, Gácsi M, Miklósi A, Csányi V (2004) Dogs respond appropriately to cues of humans' attentional focus. *Behavioural Processes* **66**, 161–172.
79. Tomasello M, Call J, Hare B (1998) Five primate species follow the visual gaze of conspecifics. *Animal Behaviour* **55**, 1063–1069.
80. Mech LD (1970) *The Wolf: Ecology and Behaviour of an Endangered Species*. Natural History Press, New York.
81. Fox MW (1978) *The Dog: Its Domestication and Behaviour*. Garland STPM Press, New York.
82. Quaranta A, Siniscalchi M, Vallortigara G (2007) Asymmetric tail-wagging responses by dogs to different emotive stimuli. *Current Biology* **17**, R199–R201.
83. Riedel J, Buttelmann D, Call J, Tomasello M (2006) Domestic dogs (*Canis familiaris*) use a physical marker to locate hidden food. *Animal Cognition* **9**, 27–35.
84. Miklósi A, Soproni K (2006) A comparative analysis of animals' understanding of the human pointing gesture. *Animal Cognition* **9**, 81–93.
85. Rooney NJ, Bradshaw JWS, Robinson IH (2001) Do dogs respond to play signals given by humans? *Animal Behaviour* **61**, 715–722.
86. Call J, Bräuer J, Kaminski J, Tomasello M (2003) Domestic dogs (*Canis familiaris*) are sensitive to the attentional state of humans. *Journal of Comparative Psychology* **117**, 257–263.

87. Schwab C, Huber L (2006) Obey or not obey? Dogs (*Canis familiaris*) behave differently in response to attentional states of their owners. *Journal of Comparative Psychology* **120**, 169–175.
88. Gácsi M, Miklósi A, Varga O, Topál J, Csányi V (2004) Are readers of our face readers of our minds? Dogs (*Canis familiaris*) show situation-dependent recognition of humans' attention. *Animal Cognition* **7**, 144–153.
89. Agnetta B, Hare B, Tomasello M (2000) Cues to food location that domestic dogs (*Canis familiaris*) of different ages do and do not use. *Animal Cognition* **3**, 107–112.
90. Virányi Z, Gácsi M, Kubinyi E, Topál J, Belényi B, Ujfalussy D, Miklósi A (2008) Comprehension of human pointing gestures in young human-reared wolves (*Canis lupus*) and dogs (*Canis familiaris*). *Animal Cognition* **11**, 373–387.
91. Miklósi A, Polgárdi R, Topál J, Csányi V (1998) Use of experimenter-given cues in dogs. *Animal Cognition* **1**, 113–121.
92. Miklósi A, Pongrácz P, Lakatos G, Topál J, Csányi V (2005) A comparative study of the use of visual communicative signals in interactions between dogs (*Canis familiaris*) and humans and cats (*Felis catus*) and humans. *Journal of Comparative Psychology* **119**, 179–186.
93. Hare B, Call J, Tomasello M (1998) Communication of food location between human and dog (*Canis familiaris*). *Evolution of Communication* **2**, 137–159.
94. McKinley J, Sambrook TD (2000) Use of human-given cues by domestic dogs (*Canis familiaris*) and horses (*Equus caballus*). *Animal Cognition* **3**, 13–22.
95. Soproni K, Miklósi A, Topál J, Csányi V (2002) Dogs' (*Canis familiaris*) responsiveness to human pointing gestures. *Journal of Comparative Psychology* **116**, 27–34.
96. Miklósi A, Polgárdi R, Topál J, Csányi V (2000) Intentional behaviour in dog–human communication: an experimental analysis of 'showing' behaviour in the dog. *Animal Cognition* **3**, 159–166.
97. Udell MAR, Dorey NR, Wynne CDL (2008) Wolves outperform dogs in following human social cues. *Animal Behaviour* **76**, 1767–1773.
98. Reid P (2009) Adapting to the human world: dog's responsiveness to our social cues. *Behavioural Processes* **80**, 325–333.
99. Cooper JJ, Ashton C, Bishop S, West R, Mills DS, Young RJ (2003) Clever hounds: social cognition in the domestic dog (*Canis familiaris*). *Applied Animal Behaviour Science* **81**, 229–244.
100. Riedel J, Schumann K, Kaminski J, Call J, Tomasello M (2008) The early ontogeny of human–dog communication. *Animal Behaviour* **75**, 1003–1014.
101. Wynne C, Udell M, Lord K (2008) Ontogeny's impacts on human–dog communication. *Animal Behaviour* **76**, e1–e4.
102. Smith BP, Litchfield CA (2010) Dingoes (*Canis dingo*) can use human social cues to locate hidden food. *Animal Cognition* **13**, 367–376.
103. Wobber VE (2005) The evolution of cooperative signal comprehension in the domestic dog (*Canis familiaris*). Hons thesis. Harvard University, Cambridge.
104. Moskowitz C (2010) *Wolves Beat Dogs on Logic Test.* 11May 2009. http://www.livescience.com/animals/090903-wolf-smart.html.
105. Coren S (2006) *The Intelligence of Dogs.* Pocket Books, London.
106. Warden C, Warner L (1928) The sensory capacities and intelligence of dogs, with a report on the ability of the noted dog 'fellow' to respond to verbal stimuli. *Quarterly Review of Biology* **3**, 1–28.
107. Kaminski J, Call J, Fischer J (2004) Word learning in dogs: evidence for 'fast mapping'. *Science* **304**, 1682–1683.

10
The personality, behaviour and suitability of dingoes as companion animals

Bradley Smith

Not content with admiring the dingo from afar, humans throughout history have attempted to keep dingoes as a pet or companion. Many have done so despite the knowledge and the warning that wild canids generally do not make good pets – at least when compared to the more subservient domestic dog. Initially, this chapter will help reveal the reality of living with dingoes by describing some of experiences of people who have raised them as pets. Later it will explore some of the ways by which pet dingoes might differ from dogs in terms of temperament, personality and behaviour using a more scientific approach. The chapter concludes by discussing some of the reasons why dingoes might be an attractive alternative to the domestic dog, despite their seeming unsuitability as a household companion animal.

A history of dingoes as pets

The fascination with dingoes began with Indigenous Australians who, although they did not share the modern phenomenon of pet-keeping, developed an intricate relationship with animals that resembled it in many ways. The management of dingoes was limited by the nature of the dingo as well as by Indigenous people's cultural beliefs and lifestyle factors that are not considered conducive to domestication[1] (see Chapter 4 for a discussion of their relationship with dogs and dingoes).

Following European settlement in Australia, Indigenous people and Europeans seemed to exchange curiosity for their respective canines. The Indigenous communities became fascinated with the European domestic dogs when they saw their benefits as loyal pets and hunting assistants. It quickly became clear that dogs were far easier to control and train than dingoes, and therefore a great benefit to hunting. The 'savage' and mysterious wild dog of the newly settled continent intrigued the Europeans, and many were shipped back to England as trophies and showpieces. In Sydney in 1788, members of the first fleet under the leadership of Captain Arthur Phillip were considered the first to keep dingoes in any type of domestic situation. It is not known exactly under what conditions they were kept, but they are thought to be quite primitive (see Fig. 10.1). They were known to ship dingoes back to England as gifts; the fate of these dingoes remains a mystery. In his diaries published in 1789, Phillip[2] noted the dingo to be 'elegant' but that while it 'has much the manners of the dog, but is of a very savage nature, and not likely to change in this particular' (Chapter XXII). John Hunter, an early explorer and Phillip's successor as Governor of New South Wales, in 1794[3] discussed the challenges involved in keeping dingoes in captivity, arguing that they could never be domesticated:

Figure 10.1: A 1797 drawing depicting a captive wild dingo around the time of the first fleet. The dingo was owned by John White, at the time Surgeon General to the Territory of NSW.
Source: Mitchell Library, State Library of New South Wales, a1764001.

> Of those dogs we have had many which were taken when young, but never could cure them of their natural ferocity; although well fed, they would at all times, but particularly in the dark, fly at young pigs, chickens, or any small animal which they might be able to conquer, and immediately kill, and generally eat them. I had one which was a little puppy when caught, but, notwithstanding I took much pains to correct and cure it of its savageness, I found it took every opportunity, which it met with, to snap off the head of a fowl, or worry a pig, and would do it in defiance of correction. They are a very good natured animal when domesticated, but I believe it to be impossible to cure that savageness, which all I have seen seem to possess (Chapter III).

Among the upper class of Sydney in the 1830s it was considered stylish and fashionable to keep a pet dingo. This fad proved short-lived – owners discovered that dingoes were unsuitable as city dogs once they outgrew puppyhood (see Plate 15). They described the dingoes as intractable, unresponsive to commands and far from well-trained animals.[4, 5] Supporting the realisation that life is good with dingoes until they reach adulthood, Bernice Walters,[6] a breeder of dingoes and founder of the Bargo Dingo Sanctuary (on the outskirts of Sydney), observed that people who own dingoes rarely retain them to old age. She believed that most owners, although well-intentioned, were simply 'ill prepared for a breed as active and independent as the dingo'.

There is no shortage of stories describing the experiences of the few who have lived with dingoes. To learn more about some of the accounts, see writings by scientists Konrad Lorenz,[7] Neil George Macintosh[8] and Eberhard Trumler,[9] dog breeder Robert Kaleski,[10] journalists Lionel Hudson[11] and Sid Wright[12] and dingo breeders Bernice Walters,[6] Gill Ryhorchuk[13, 14] and Barry Oakman.[15] The majority of these experiences demonstrate the difficulties of keeping dingoes and the implication that dingoes are not suited to the urban backyard. Lorenz,[7] widely considered by scientists to be one of the fathers of ethology (the study of animal behaviour) kept a pet dingo in his home in Austria, and detailed this relationship in his book *Man meets Dog*. Although he reported that his dingo 'evidently harbored the warmest feelings' for him, he suggested that 'submission and obedience play no part in these feelings' (p. 22). Trumler,[9] a student of Lorenz, also kept dingoes (the first one, which he called 'Aboriginal', which he obtained from Nuremberg Zoo). He came to a similar conclusion that, like wolves, dingoes do very poorly as domestic animals – even after they have been crossed with domestic dogs. In *Understanding your Dog*[9] he wrote that:

> Dingoes are not suitable as house-dogs in the normal sense of the word; they create too many difficulties. But that is not the end of the story. I have crossed my dingoes with domestic dogs and have gone on to raise several generations of these half-breeds. This too produces dogs intolerable to anyone who sets store by a clean house (p. 17).

Several of the authors mentioned above state that pet dingoes are escape artists, extremely difficult to train, intelligent and very independent and that they retain a powerful prey drive – all traits that make them difficult to manage in urban environments. At one point, the New South Wales Police attempted to train dingoes to conduct police work (e.g. guard dog, sniffer dog), but could produce nothing that resembled obedience.[8] Having kept dingoes himself on the grounds of the University of Sydney (Fig. 10.2), Macintosh[8] wanted to see if others who had domesticated pure dingoes had had a similar experience. He bluntly concluded that the pet dingoes he met were far from 'domesticated' and the owners were either 'self-deluding or anthropomorphising' (accrediting human motivation, characteristics or behaviour to non-human animals, inanimate objects or natural phenomena). The following section provides a detailed account of what it is like living with a pack of dingoes.

Living with a dingo pack: a case study

The dingo-owning experience can differ quite dramatically depending on many factors, including the number of dingoes in the household, how the dingoes are related, whether they are neutered/entire and whether they were born in captivity or rescued from the wild. Living so intimately with dingoes can teach you a lot about the behaviour of the species (and their owners!), so it is worthwhile exploring the experiences of those who have dingoes as pets or companion animals. Most dingo owners typically keep only one or two dingoes, and often report similar experiences to those already outlined in this chapter (e.g. difficult to train, highly intelligent, form a strong bond, fearful in strange situations). But no one has yet described what it is like living with a pack of dingoes.

Figure 10.2: N.G.W. Macintosh in 1953 holding a puppy he bred on the grounds of the University of Sydney. Breeding dingoes allowed him the opportunity to study their behaviour and morphology, which he could compare to wild and fossil dingoes.
Source: Shellshear Museum Collection, University of Sydney.

Darren Griffiths and Leigh Mullan are two dingo enthusiasts from Perth, Western Australia who live with five pure entire dingoes – Loxie, Wylie, Isis, Osirus and Freya. Not only is their story unique and fascinating, they firmly believe the experience has given them a rare perspective on dingo behaviour. Through their respective roles as president and vice-president of the Western Australian Dingo Association (WADA) they have also learned a great deal about keeping dingoes from the stories and experiences of other members.

Darren and Leigh became fascinated with dingoes through breeding blue heelers (Australian cattle dogs), a dog that they believed to have dingo heritage. In 2006, a customer of their pet shop came in with a dingo. It was the first time they had seen a dingo, and they immediately became infatuated with it. They were interested in back-crossing this dingo with a blue heeler to see what influence it would have on the characteristics of the breed. The planned mating never eventuated, but they soon learned of a litter of dingo pups for sale nearby. The day they saw the advertisement they visited the litter of four-week-old dingoes and came home with one they called Loxie. Since 2006, through rescuing a dingo and raising two litters of their own (one planned, one unplanned) the size of their pack has slowly

Figure 10.3: Darren Griffths (left) holding Isis (a dingo pup) with Leigh Mullan (right) with Osirus and the pups' parents Loxie and Wylie.
Source: Mal Fairclough/*The West Australian*.

increased (see Fig. 10.3). All their dingoes live and sleep in the house and are considered integral members of the family. When I spoke with them, the dingoes were strewn across the living room, including on the coffee table. Each had his or her place, and lay there contentedly. The walls of their home were littered with dingo photographs and memorabilia. Their life revolves around their dingoes, so much so that Darren considers himself a 'full-time dingo carer'. The first thing they will tell you is that their dingoes are not pets; they are companions, because 'dingoes deserve more respect than that'.

Darren and Leigh describe living with a pack of dingoes as 'hectic' and like 'having five hyperactive children running around the house. There are times when they are all calm, and then there are times when it's like someone has just given them a can of Coke and a chocolate bar and they're all wound up. On a typical day they spend ~18 hours sleeping, but the rest of the time they go at 100 miles an hour – full throttle'. To make things more difficult, each dingo has a completely different personality. 'There is not one dingo I would say is like any other. Just like people I suppose'.

Given the dingo's known escape abilities, Darren and Leigh have had to make several allowances to adjust to their life with dingoes. At least one of them is with the dingoes 24 hours a day. When they leave the house they take all the dingoes with them in a customised van that serves as a family car, dingo transporter, mobile home and dingo conservation command centre: 'Wherever we go, the dingoes go.' They maintain this practice for two reasons. First, the dingoes 'would trash the place' if they were left home alone and second, 'they wouldn't know how to cope when one of their pack members disappeared'. This dedication has meant that they have made many sacrifices in their lifestyle, including a limited social life and restricted travel opportunities. 'We rarely go out for dinner together, and we can never go on an overseas holiday unless we go separately, because you cannot put them into kennels.' Darren said that when dingoes are placed in a dog kennel they rarely come back the same, and remain highly anxious. This view, shared among dingo owners, is echoed by Bernice Walters[6] who believed that a boarding kennel experience can ruin the temperament of the dingo, and that dingoes seldom recover from a frightening experience or rejection.

Darren and Leigh's house has been modified in that there are multiple doors and barriers that the dingoes would need to get through to escape. 'The hardest part is managing visitors to ensure they don't leave doors open.' Having dingoes also determines the type of furniture in the house. 'There is no point having really expensive or nice things because they will destroy or chew it.' All their furniture has evidence of dingo teeth marks, except for one leather recliner, which is reserved for Darren. Technology such as mobile telephones, remote controls or cameras can't be left lying around because the dingoes will quickly find and destroy them. If clothes or sunglasses are left around, the dingoes will scent-mark them. They occasionally leave their scent on pillows and bedding, so the men are continually washing or cleaning. This doesn't seem to faze them much: they hold that the dingoes see the house as their territory, and scent-marking their territory is a natural behaviour. Another thing that needs to be considered is changing items around the home (furniture, lights, turning on and off ceiling fans etc.) as well as bringing in large new objects (e.g. a large esky/icebox cooler). Dingoes are highly aware of everything in their environment and often 'freak out' when things change, readily inspecting whatever has changed to make sure it is not a threat.

Leigh warned that dingoes are very difficult to control or command. 'The dingoes can quickly tell what is on your mind – for example if you are wanting to catch them, or block them from somewhere. Rattling the treat bowl certainly gets their attention, but you have to be quick because they know what you are up to. Most tricks or methods will only work once or twice if you are lucky, so you soon get good at using reverse psychology. You cannot expect them to obedient and follow commands. They have that higher intelligence – they have their own mind.' When trying to describe the behaviour of a dingo, Darren likes to tell people that the dingo 'is like a cat in a dog suit with a monkey brain'. He validates this description that he coined, by explaining that the dingo has the same demeanour as cats (e.g. aloof and difficult to train), that the dingo looks very dog-like yet is not a dog, and that dingoes are extremely intelligent, like primates.

Owning a dingo is not a decision that should be made lightly. The WADA website warns that owning a dingo needs more thought than entering into a marriage: 'It's important to be 100% sure that you're in it for better or worse. Unlike marriage, you can't divorce your dingo.' The majority of dingoes that need re-homing are euthanased because they don't relocate well. Like WADA, Darren and Leigh do not recommend dingoes as a suitable pet, suggesting that 'if you want a dingo to be a fashion accessory, or for attention, but you want to go off each day to work for 8 hours, then expect your backyard to be trashed. Or for that matter, no dingo when you get home.' Their 'official' position is helping to promote conservation and awareness, not dingo ownership. They strongly believe that 'you don't do the species' reputation any good by allowing or supporting unsuitable people getting a dingo pup'.

Darren and Leigh often find it difficult to communicate the level of commitment involved in owning a dingo, especially when they take their five apparently well-adjusted dingoes out in public. 'People see the bond we have with them and they want that as well, but I try and tell them that you don't want this', warns Darren. There doesn't seem to be a specific type of person that's attracted to getting a dingo, but from Darren's experience they are usually people who are 'a bit aloof, wild of nature, and free spirited'. They need to be easy-going and not likely to hold a grudge or be too domineering. And, most importantly, they need to 'be able to think like an animal, to live on the animal's terms and not treat them like a domestic pet. There needs to be a give and take with rules. If you are looking for an animal that you can control and will do what you want or say, then get a dog – don't get a dingo. A dingo won't obey you.'

Unfortunately, life does not turn out this well for many of the dingoes that are sold as pets to unsuitable owners. According to Leigh, around five out of every 10 dingoes in Western Australia are relinquished within their first year. Worse, only around one in five of those that need re-homing will make it. This is because dingoes seem to bond very early to their environment (people and place) and do not cope well when this is changed in adulthood. Another reason is that the re-homing criteria for many organisations (e.g. the RSPCA, which deals with lost dogs) are specific to dogs, and so a dingo rarely has a chance of meeting them. For example, a dingo reacts to a stranger very differently from how a dog does (e.g. extremely shy, fearful, may fear-bite – one of the tests used to asses temperament and suitability). Another factor, particularly in Western Australia, is that dingo ownership is unregulated. Darren and Leigh want to see dingoes recognised as a native animal, with permits and regulations for ownership in all states and territories. This may reduce the number of backyard breeders and tighten the rules on who can keep dingoes.

Despite all the difficulties, they admit that dingo ownership has its place and can be extremely rewarding for those who are committed to conservation and understand the requirements of owning a dingo. They have learned an incredible amount since living with dingoes that can't be gained from looking at them in a zoo or reading about them in a book. However, the biggest reward is the interspecies interaction. 'Dingoes can be as loyal and loving and affectionate as any human bond you could get. But you will have to earn that trust and respect. It's not something you can tell it to do and it will do (like a domestic animal). It is unique to get that with a *wild* animal. It's not all hugs and kisses, it's a lot of work. But if you have the compassion and the passion to be able to provide the right and stimulating environment, then nothing is more rewarding than when you get that interspecies interaction. I'm sure people get it with their pet dogs and cats, but I don't think they get it to the same degree as with what you can get with a dingo. There is a spiritual connection that I have never experienced that with any other animal.'

Simply owning dingoes isn't enough for this pair. Given their close relationship with their animals, they feel the least they can do is try to help the relatives of their 'urban' companions by working towards the conservation of wild dingoes. Darren believes that if he can change at least one person's opinion about dingoes each day then he has done his job. Alongside WADA members and their pack members, they regularly give presentations to a wide variety of organisations and community groups (see Fig. 10.4). At this stage they concentrate on providing education and awareness, because the organisation can do little else without financial support or funding. Darren and Leigh hope to change the cultural feeling around dingoes, and are working towards the next generation 'understanding the value of dingoes, and that "dingo" is not a dirty word'.

Rules and regulations of dingo ownership

Today, several people keep dingoes as pets in Australian states where ownership is legalised. Unfortunately the official number of pet dingoes in Australia is unknown. The laws in relation to dingo ownership are as broad as they are inconsistent, varying widely between each state and territory. Keeping dingoes is allowed in Western Australia, New South Wales, Northern Territory and Victoria but is illegal in Queensland, Tasmania and South Australia (below the dingo barrier or exclusion fence, which runs through the heart of the state to keep dingoes out of sheep farming country). Each state where ownership is legalised has different requirements

Figure 10.4: Darren Griffths with Leigh Mullan and their dingoes giving a presentation to a community group in Perth. The dingoes have been trained to howl at the end of a presentation as a thank-you for listening.
Source: Barbara York.

regarding registration. For example, in Victoria where dingoes are considered wildlife, a specialist wildlife permit or licence from the state government is required; in New South Wales, dingoes can be registered as a breed of dog with a local council. Mostly, these rules apply only to pure dingoes (according to the DNA test established by Alan Wilton).[16] The same rules do not apply to dingo–dog hybrids, which are treated under the same laws and regulations as a domestic dog.

In Victoria, which has the most defined regulations on pet dingo ownership, the *Wildlife Act 1975* makes it an offence to own a dingo without a permit. A dingo licence authorises the holder to 'possess, keep, breed, buy, sell and dispose of pure-bred dingoes for non-commercial purposes or to possess a dingo at other sites for the purposes of exercising, obedience training, education, or display at shows conducted by a canine association' (Department of Environment and Primary Industries). Laws relating to the ownership of domestic dogs under the *Domestic Animals Act 1994* also apply. These cover the number of dingoes a person can own, the response to dingoes causing a disruption to the neighbourhood, and the requirement to have a microchip implanted under the skin for identification. When dingoes are taken out of the owner's property, they must be controlled by means of a chain, leash or harness, even within designated off-lead areas, and any interactions must be 'done in a manner which minimises the risk of injury to any person or the dingo'. Owning dingoes and displaying them for commercial purposes (e.g. wildlife parks) requires a commercial wildlife licence (see Chapter 11 for details of keeping dingoes in larger commercial facilities).

Dingoes are notorious escape artists[15] (Fig. 10.5) and so most of the recommendations relating to pet ownership concern the size and design of the enclosures. Enclosures must be constructed before obtaining a dingo, and are often inspected by an authorised person before permits are approved and granted. The enclosure standards required for ownership in Victoria provide a good outline of what is generally required or encouraged across Australia for both pet owners and commercial operators. In general, enclosures must:

- be child-proof and of sufficiently secure design and construction to prevent escape of dingoes and unauthorised access to dingoes;
- have a minimum floor area of 30 m^2 for up to two dingoes and an additional 10 m^2 for each additional dingo over nine months;
- have fences of either 3 m in height or a minimum of 2 m in height with an additional 45° inward return at least 1 m in length (or the inward return can be a full secure roof);
- have a 1 m inward-facing mesh return fitted at the base of the fence and fastened to the ground at right angles to the fence (or the fence anchored securely to a cement slab).

It is extremely important for the future of the species that the states and territories develop adequate and consistent regulations for dingo ownership, including a national breeding studbook to manage the existing stock of pure dingoes in captivity. Having dingoes listed or registered as just another breed of dog (e.g. as in New South Wales) leaves little restriction on who can own a dingo, how they house the dingo and/or cross-breeding with domestic dogs. This could lead to events such as contamination of the pure stock of dingoes, resulting in an increase in hybridisation in captive populations (through in-breeding and cross-breeding with

Figure 10.5: Dingoes are excellent escape artists, so it is often difficult keeping them in captivity. This dingo has scaled a 2 m fence and is balancing across the wire.
Source: Lyn Watson.

dogs). Ignorance of the wild characteristics of the dingo could lead to an increase of abuse and/or relinquishment, which increases the pressure on welfare and dingo organisations, or dumping them in the wild or urban areas which may lead to an increase in the wild dog problem. Any legislative changes should utilise the skills and expertise of dingo organisations and owners.

Obtaining dingoes

Dingoes can be obtained from many different sources, including breeders (both responsible breeders registered with state dingo organisations, and irresponsible ones) and larger private dingo sanctuaries that supply dingoes to private owners, zoos and wildlife parks. Selling dingo pups each year is often a significant source of income for not-for-profit organisations, but care is usually taken to sell only to suitable owners. Occasionally a dingo puppy is found or rescued in the wild and given to a dingo organisation, which then finds the pup a new home among its members. It is quite common for people to report having a dingo–dog hybrid obtained from a backyard breeder or from a dog rescue centre. Dingo breeders recommended that dingoes be acquired at around seven to 12 weeks of age to maximise the likelihood that they will successfully bond with the owner and adjust to their new life. This also allows the pups to leave the breeding facility with some socialisation and training. The Victorian dingo licence states that dingoes must be at least seven weeks of age before being sent to their new home.

Perhaps surprisingly, most breeders believe that dingoes are not suitable as pets for most households. They do agree that dingoes can make a close and wonderful companion, but only under the right conditions. Dingo owners stress that dingo ownership is a lifelong commitment: for there to be a successful relationship the owner must understand the requirements of keeping a dingo, and be willing to spend sufficient time training and bonding with it.[6, 15] According to Barry Oakman[15] of the Australian Dingo Conservation Association, 'the very wild nature of the dingo that makes it so fascinating is diametrically opposed to the suburban dictates of backyards, restraint and domesticity' (p. 34). What motivates people to keep dingoes as pets despite warnings from other owners and breeders will be discussed later in the chapter.

Why wild canids don't make suitable pets

Unlike wolves and dingoes, domestic dogs have been living alongside humans for ~10 000 years, making them one of the oldest and most successful of the domestic animals.[17] This is partly because they have been artificially selected for characteristics which humans find desirable and that allow them to conform to the human environment.[18] Today, dogs represent highly variable populations and can be classed genetically into different groups according to the purpose for which they were bred.[19, 20] There are now ~400 varieties or breeds of dogs.[21, 22] Continuous selection of dogs for human environments (independent of morphology or breed history) has led to variation in behaviour and temperament, which remain responsive to the ever-changing desires of dog owners.[23] The following paragraphs detail some of the traits that owners consider desirable and undesirable in a domestic dog. This helps to put into context why wild canids do not fit the same mould.

Desirable traits

Throughout the transformation from wolf to dog, humans selected for certain characteristics in dogs that are important for a successful dog–owner relationship. For example, Bennett and team[24] found that the Australian public's ideal dog is one that is medium-sized, short-haired, neutered, safe with children, fully house-trained, friendly, obedient and healthy. Owners wanted their dog to come when called, not to escape from their property, to enjoy being petted and to display affection. Important behavioural characteristics included calm/compliant, sociable/healthy, energetic/faithful/protective, socially acceptable and non-aggressive.

Undesirable traits

At the same time, humans selected against traits considered undesirable. In domestic dogs, a large survey of Australian dog owners revealed five factors or behaviours that they considered undesirable in a dog. These include disobedience, unfriendliness, nervousness, destructiveness and excitability.[25] A study tracking over 20 000 dogs admitted to dog shelters in Victoria revealed that some dogs are relinquished due to owner issues (e.g. moving house or family circumstances) or lack of owner commitment, but many are surrendered because of behaviour problems that the owners are not able or willing to tolerate, including boisterousness, hyperactivity and aggression.[26]

The keeping of wolves and wolf–dog hybrids appears to be a common practice in the US even though it is prohibited or restricted (e.g. permits) in many states, with an estimated 500 000 wolves and hybrids living as pets in private homes.[27] The experiences of those who have written about keeping wolf–dog hybrids sound remarkably similar to those of dingo owners, and along with government and wolf organisations they generally discourage the practice. Most wolf organisations in the US try to educate people on the difficulty or challenges of keeping wolf hybrids. Tucker and Weide,[28] for example, produced a booklet *Can You Turn a Wolf into a Dog? Commonly Asked Questions about having Wolves and Hybrids for Pets* in an attempt to educate and discourage wolf hybrid ownership. They described how hybrids are poorly adapted as pets, difficult to train, frequently destroy their owners' property, attack people and domestic animals and are generally too wary to be effective guard animals. As pups, most wolves and hybrids behave similarly to dogs and are readily playful and relatively submissive. However, many do not survive their first year because, as they reach sexual maturity, they often become more predatory, wide-ranging, highly territorial and pack-oriented animals.[27, 29, 30] The risk of a wolf or hybrid accidentally injuring the owner during play or when food is involved is quite high.[27]

Several animal behavioural scientists in the 1960s, 1970s and 1980s began observing wolf behaviour in captive settings or even hand-raising wolves to gain a deeper understanding of their social behaviour and development. They included John Fentress,[31] Michael Fox,[32] Harry Frank,[33] Kevin MacDonald,[34] Erich Klinghammer,[29] Jerome Woolpy and Benson Ginsburg[35] and Erik Zimen.[36, 37] Their descriptions of the social, cognitive and behaviour of wolves suggest they do not make good pets for a variety of reasons and are extremely difficult to socialise to humans, even when hand-reared from an early age. However, the practice of hand-raising wolf puppies has been revived to take into account and test new methodologies and theories. Marta Gacsi and team in the Department of Ethology at Eotvos University in Budapest, Hungary[38] hand-raised and socialised 13 wolf puppies in a comparable way to domestic dog puppies. They argue that their results 'demonstrate that already at this early age [three to five weeks] – despite unprecedented intensity of socialisation and the comparable social (human) environment during early development – there are specific behavioural differences between wolves and dogs mostly in regard to their interactions with humans' (p. 111). Compared to wolves, they found that dogs tended to display more communicative signals that could potentially facilitate social interactions, such as distress vocalisation, tail-wagging and gazing at the humans' face. In contrast to dog puppies, wolf pups showed aggressive behaviour towards a familiar experimenter and seemed to be more prone to avoidance.

Like wolves, dingoes have certain innate traits that may make them unsuitable as household pets. For example they have a strong prey drive,[8, 10, 15] are good at escaping and opening locks,[8, 10–12, 15, 39] are hard to train[7, 8, 15] and difficult to socialise to humans, often showing avoidance of humans and a tendency to be easily frightened and difficult to calm.[6, 8, 15] The dingo might be expected to show less of the desirable dog traits that are common among dogs (e.g. trainability, attachment/attention-seeking) and more of the undesirable ones (fearfulness, aggression etc.). It would be interesting to determine whether the dingo (at least in captive environments where such assessments are possible) differs from domestic dogs in behaviour or suitability to domestic environments.

Assessment of pet dingo personality and behaviour

There is anecdotal evidence of dingo temperament and behaviour, but is there any science to back it up? Unfortunately, there isn't. No study has methodically examined the personality and behaviour of wild canids such as wolves and dingoes or their admixtures (e.g. wolf–dog hybrids or dingo–dog hybrids) living in the wild, captivity or human domestic environments. So, using scientific methods previously applied to domestic dogs, I set out to see how dingoes would compare. There are two scientific methods that can be used to distinguish wild and domestic dogs. These include assessments of their personality, and their behaviour, as rated by the owners, both of which will be covered in this section.

The most common method of determining the personality and behaviour of a dog is by conducting behavioural assessments,[40] which seek to code a dog's reaction when presented with various, usually novel, stimuli.[41, 42] This method can be quite intensive and it is difficult to ensure consistency in administering and coding. Another popular approach is the rating method, which is normally a questionnaire utilising a Likert scale for rating individual dogs. A Likert item is simply a statement that the respondent is asked to evaluate according to any kind of subjective or objective criteria – generally the level of agreement or disagreement. The questionnaire is completed by a person who knows the dog well, which in the case of dogs is almost always the owner.[43–45] The human observer is required to interpret the dog's behaviour for predefined behavioural traits. These assessments are considered relatively accurate because dog owners are able to draw on their knowledge of their dog's behaviour in all the situations in which they have observed their dog; this method is therefore potentially more valid than behavioural tests that look for specific behaviours in specific circumstances.[46]

Dingo personality

The study of personality is not restricted to humans. Within the last decade scientists have successfully applied personality assessments to non-human animals, including domestic dogs.[40, 47] An understanding of personality in dogs can provide many applied benefits. For example, it can be used to help identify and predict traits that are beneficial for working animals such as police dogs[48] as well as in selecting appropriate companion animals (e.g. to better match shelter dogs to potential owners[49] or breeding dogs with desired personality traits).[23] Collectively, studies have found dogs to have an array of personality traits or temperaments that can be broadly categorised as reactivity, fearfulness, activity, sociability, responsiveness to training, submissiveness and aggression (see the work by Gosling, a pioneer of the study of animal personality).[40]

Personality studies have been conducted on other carnivores such as hyenas,[50] but the research on canine personality remains almost solely focused on domestic dogs. I wanted to see if there were any personality differences between pet dingoes and pet dogs and whether this could explain some of the behavioural differences and difficulties of trying to keep pet dingoes.[51]

Before I conducted the study, I examined the literature to see how dingo owners describe their dingo. Looking at the words (or adjectives) that people use to describe their pets is useful because it allows us to gain an insight into the general disposition or personality traits of animals that occur through time and/or across different functional contexts.[52, 53] Not surprisingly, the many authors that have shared their experience of living with dingoes describe their behaviour in very similar ways. Commonly used words or descriptions include unpredictable, observant, intelligent, clever, untrainable, smart, nervous and fearful. A more comprehensive list is provided in Table 10.1.

To determine dingo personality I decided to use a rating method via a questionnaire designed for domestic dogs – the Monash Canine Personality Questionnaire, revised edition (MCPQ-R) that was developed and refined by Jacqui Ley from Monash University, Australia.[45, 46] The MCPQ-R is considered a robust test of canine personality, which is one of the most valid[46] and reliable methods of assessing domestic dog personality.[46] The questionnaire involves a 26-item, adjective-based, owner-administered questionnaire, which measures canine personality using five categories: Extraversion (hyperactive, lively, energetic, restless, active, excitable); Training focus (combines the perceived intelligence with trainability, including words

Table 10.1. Adjectives used in the literature to describe captive dingoes, grouped according to the five personality factors used to describe domestic dogs[45]

Extraversion	Self assuredness/ motivation	Training focus	Amicability	Neuroticism
Active[6]	**Independent**[6]	**Intelligent**[6, 8, 10, 11, 15]	**Aggressive**[15]	**Sensitive**[6]
Unpredictable[6, 15]	Individualistic[6]	**Clever**[54]	Free-spirited[6]	**Nervous**[8]
Mischievous[6]	Curious[6]	**Not obedient**[6–8]	Sense of humour[6]	**Cautious**[6]
Quick-witted[6]	Observant[54]	Untrainable[6, 8, 10, 11, 15, 54]	Reserved (with strangers)[6]	**Fearful**[15]
Destructive[6]	Strong-willed[6]	Uncontrollable[6]	Affectionate[6]	Easily frightened[15]
Vocal[15]	Territorial[6]	Defiant[6]	Unaffectionate[6]	Panicked[15]
		Smart[9]	Loyal (to owners)[6]	Shy[6]
		Innovative[6]		
		Pragmatic[6]		

Bolded words indicate adjectives that are also used in the Monash Canine Personality Questionnaire[46] to describe dog personality (with the exception of not obedient and aggressive, where the terms obedient and non- or unaggressive were used).

such as biddable, trainable, reliable, attentive, intelligent, obedient); Motivation (opportunistic, tenacious, persevering, determined, assertive, independent); Amicability (friendly, non-aggressive, easy going, relaxed, sociable); and Neuroticism (nervous, timid, fearful, submissive). Respondents were asked to indicate the extent to which each of these words described their dingo, using a six-point scale ranging from 1('really does not describe my dingo') to 6 ('really describes my dingo').

I recruited dingo owners through an advertisement in the Australian Dingo Foundation's newsletter and through word of mouth. A total of 40 dingoes (22 females; 18 males) that were reportedly kept as companions in areas of Australia where ownership is legalised were included. The dingoes ranged in age from six months to 11 years (average was 3.6 years ± 2.4; seven were aged six months to one year, 14 aged one to three years, and 19 aged over three years), weighed an average of 19.07 kg (± 3.41) and were mostly entire (not de-sexed, spayed or neutered; 62.5%). Animals in the adult study had been de-sexed at an average age of 22 months of age for birth control (56%) and to prevent future behaviour problems (13%). The majority of owners obtained their dingoes from a dingo breeder/association/farm (38%), a wildlife park or sanctuary (28%) or bred them themselves (18%). Dingoes were acquired at the average age of 15.7 weeks (± 25.41), ranging from birth to one year. The majority of dingoes lived in multi-dingo households (72.5%) and were reportedly healthy (95%; one was missing a leg and another had a damaged hip). Most dingoes (78%) did not have any reported behavioural problems, but 20% of owners reported their dingo had minor problems and 3% reported moderate problems. To allow comparison with domestic dogs, the personalities of 455 Australian domestic dogs previously reported by Ley, Bennett and Coleman[46] were used.

As expected, domestic dogs and dingoes living as companion animals differed in various aspects of personality (Table 10.2). Dingoes were rated as significantly more motivated/

Table 10.2. Scores for dogs and dingoes on the MCPQ-R according to the five factors

	Extraversion		Self-assuredness/ motivation*		Training focus*		Amicability		Neuroticism	
	Dog	Dingo	Dog	Dingo	Dog	Dingo	Dog	Dingo	Dog	Dingo
Mean (%)	63.06	66.05	67.09	74.83	77.03	63.41	77.66	80.33	45.79	49.43
s.d.	19.38	15.04	16.78	14.14	14.42	9.34	16.96	15.88	19.94	17.28
Minimum (%)	16.67	30.56	20.00	46.67	30.56	47.22	20.00	50.00	16.67	16.67
Maximum (%)	100.00	94.44	100.00	100.00	100.00	88.89	100.00	100.00	100.00	79.17
Range	83.33	63.89	80.00	53.33	69.44	41.67	80.00	50.00	83.33	62.50
Percentiles										
25	50.0	55.56	56.7	63.33	66.7	58.33	66.7	66.67	29.2	38.02
50	61.1	66.67	66.7	76.67	77.8	61.11	80.0	80.00	41.7	50.00
75	77.8	77.78	80.0	86.67	88.9	69.44	93.3	96.67	62.5	62.50

* indicates differences between dogs and dingoes are significant at $P < 0.01$.
Source: Table replicated from Smith (2014).[51]

self-assured than were domestic dogs. This suggests that 'pet' dingoes are more independent, persevering, dominant, nosy, opportunistic and proud than domestic dogs. This finding concurs with descriptions of dingoes in the literature that dingoes are extremely independent, highly fearful of new things and experiences, and in many situations do not seek or desire human contact.[6, 8, 15] Similarly, studies of dingo problem-solving behaviour indicate that dingoes act independently of humans and persevere longer than dogs do when attempting to problem- solve.[55–57]

Dingoes were rated significantly lower than dogs on training focus, which reflects responsiveness to training, how well the dog works with people and learns in new situations, and how playful it is.[45] This implies dingoes are less attentive, reliable, biddable, trainable and obedient than dogs – confirming the numerous reports on how difficult dingoes are to train and their general lack of obedience.[1, 6–11, 15]

The scores from the training focus subscale in the scale developed by Ley, Bennett and Coleman[45] include the adjectives intelligent and clever, and on first glance the findings might indicate that dingoes are less intelligent than dogs. However, I believe that this is likely to be an inaccurate representation of dingo personality given that most descriptions of the dingo[6, 8, 10, 11, 15] and results from the scientific literature report that dingoes are more proficient at solving problems (at least in terms of non social problem-solving tasks) than domestic dogs[55–57] and are capable of higher-order cognitive processes.[39, 58] It is possible that owners rated the dingoes high for being intelligent and clever but scored them so low on the training-related items that it masked the overall influence of intelligence-related items. Ratings may also vary according to the owner's definition of intelligence, or perhaps owners did not discern between intelligence and trainability.

Dingoes were rated similarly to their domestic counterparts in relation to Extraversion (energetic and extraverted), Amicability (friendly and sociable) and Neuroticism (nervous sensitivity). These findings are somewhat unexpected, particularly for Neuroticism, as the descriptions of dingoes in the literature indicate they are typically shy and withdrawn.

Dingo behaviour

Another popular rating-based questionnaire method of assessing pet dog behaviour is the aptly named C-BARQ (Canine Behavioural Assessment and Research Questionnaire), which was developed by Serpell from the University of Pennsylvania.[44, 59] The C-BARQ is a standardised test for screening dogs for the presence or severity of behavioural problems. It provides information based on the owner's perception of the dog's behaviours, but it has been designed in such a way that that owners are constrained to tell how a dog reacts to very specific stimuli/situations in the recent past (unlike the personality questionnaire, which is based on a list of adjectives that the owner uses to rate their dog). It is considered a good test as it has been checked for validity (question items are actually measuring what they say they are measuring) and reliability (reliably tests what it says it tests) on a large sample of dogs of various breeds. The C-BARQ has been successfully applied to assist in predicting the success of guide dogs.[59] The questionnaire contains 100 items that are grouped into seven sections: training and obedience, aggression, fear and anxiety, separation-related behaviour, excitability, attachment, attention-seeking and miscellaneous. Responses are evaluated in terms of frequency (e.g. 'when off the leash, returns immediately when called'). Responses on a 5-point ordinal scale range from 1 = seldom to 5 = always, or measure the intensity of a particular behaviour (e.g. 'when unfamiliar persons approach you or another member of your family away from your home'). Responses range from 0 = no aggression/no visible signs of aggression to 4 = serious aggression/snaps, bites or attempts to bite. The test has identified 14 distinct behaviour subscales or traits that are common to the majority of dogs (irrespective of breed, age, sex or neuter status).

Like the MCPQ, no comparison has been made between a wild canid and pet dogs on the C-BARQ questionnaire. Serpell has been unable to find enough owners of pet wolves or wolf–dog hybrids to participate, so for now he can only speculate. During a conference presentation (*Canine Science Forum*, 2008, Budapest), Serpell hypothesised that compared to modern breeds of dogs (e.g. golden retrievers) wolves would be neophobic (fearful of new objects or experiences), shy with strangers (potential fear-biters), escape artists, regularly scent-mark their territories/home ranges, have high chase instinct (strong prey drive), be destructive (especially when left alone), difficult to train, howl more often than bark, be poor watch dogs and be unpredictably aggressive when mature. He further predicted that the differences would exist when modern breeds were compared to more primitive breeds of dogs (e.g. basenji, shar-pei, chow chow, saluki, Siberian husky, Alaskan malamute, Afghan hound and dingo).[21] See Table 10.3.

Given the various reports by scientific and non-scientific observers of both captive wolves and dingoes, it was predicted that dingoes would share these characteristics. As the dingo owners previously described, completed the personality assessment of their dingoes, they also completed the C-BARQ. The comparison sample of domestic dogs came from the database of dogs previously collected by Serpell and team (now in excess of 10 000 dogs). The dog scores were based on the average of 89 breeds of domestic dogs that were matched to the dingo sample according to gender and neutered status (60% female, 40% male, 65% entire). Results of the questionnaire and the comparison are reported in Table 10.4.

Table 10.3. Differences in behaviour between phylogenetically ancient (ancestral) and more recent (derived) breeds of dog, according to Serpell

The behaviour of modern versus ancient dogs living as social companions
Primitive breeds are predicted to have: – more aggression, especially towards other dogs – more fear of unfamiliar people and situations – less tolerance of being left alone – more aloofness, less attachment – less trainability – stronger chase motivation – less playfulness – more howling – more destructiveness – more scent marking.

Table 10.4. Item scores (mean rank) on the C-BARQ for domestic dogs of various breeds and dingoes

C-BARQ item	Dog	Dingo	Group that scored higher	Significance level
Escapes	55.13	85.87	Dingo	<0.001
Rolls in faeces	51.86	94.24	Dingo	<0.001
Coprophagia*	60.34	75.38	Dingo	0.034
Chews	56.78	83.30	Dingo	<0.001
Steals food	58.73	78.95	Dingo	0.004
Compulsive staring	57.00	82.80	Dingo	<0.001
Chasing	51.91	94.13	Dingo	<0.001
Stranger-directed fear	56.31	84.34	Dingo	<0.001
Energy	57.07	81.46	Dingo	<0.001
Mounts	80.00	31.63	Dog	<0.001
Urine marking	73.84	45.34	Dog	<0.001
Emotional urination	83.78	23.21	Dog	<0.001
Urination when left alone	72.27	48.83	Dog	0.001
Defecation when left alone	77.00	38.30	Dog	<0.001
Nervous on stairs	74.03	40.50	Dog	<0.001
Snaps at flies	75.00	42.75	Dog	<0.001
Tail chasing	77.07	38.15	Dog	<0.001
Shadow chasing	79.00	33.85	Dog	<0.001
Barks persistently	79.26	33.26	Dog	<0.001
Self-grooming	72.76	45.64	Dog	<0.001
Stranger-directed aggression	76.66	39.05	Dog	<0.001
Owner-directed aggression	78.53	34.90	Dog	<0.001
Trainability	80.11	28.88	Dog	<0.001
Attachment/attention-seeking	69.60	54.78	Dog	0.037
Touch sensitivity	72.79	45.59	Dog	<0.001
Dog-directed fear	72.62	48.05	Dog	0.001
Begs	68.18	57.93	No difference	0.149
Pulls on leash	65.39	62.46	No difference	0.68
Hyperactive	60.92	74.08	No difference	0.064
Allogrooming*	66.22	60.56	No difference	0.426
Other stereotyped behaviour	69.00	56.10	No difference	0.069
Dog-directed aggression	63.43	65.33	No difference	0.79
Dog rivalry	60.02	71.86	No difference	0.097
Excitability	68.16	56.15	No difference	0.092
Non-social fear	60.69	71.75	No difference	0.121
Separation-related problems	69.23	55.59	No difference	0.055

Statistical differences were calculated using the Wilcoxon–Mann–Whitney U test. The group that scored significantly higher ($P < 0.05$) is identified.

* Coprophagia = the consumption of faeces. Allogrooming = social grooming between members of the same species.

It appears that Serpell's predictions about the behavioural differences between pet wild canids and domestic dogs were not only supported by the numerous reports in literature already mentioned in this chapter, but also by multiple items of the C-BARQ questionnaire. For example, dingoes were significantly more likely to attempt to escape, to behave fearfully or warily when approached by strangers, to chase cats, birds and other small animals, and to chew. Dingoes were also less likely to be trainable (willingness to attend to the owner, obey simple commands, learn quickly, fetch/retrieve objects, respond positively to correction, ignore distracting stimuli), to bark persistently and to seek attention from their owners (tendency to maintain close proximity to the owner or other members of household, solicit affection or attention, and become agitated when the owners give attention to third parties). Not surprisingly, there were many behaviours characteristic of canines that were similar among pet dogs and pet dingoes, including pulling on the leash, excitability (tendency to react strongly to potentially exciting or arousing events such as going for walks or car trips, door bells, arrival of visitors or owners arriving home), begging, dog rivalry, allogrooming and separation-related problems (tendency to vocalise and/or escape or engage in destructive behaviour when separated from the owner).

Interpreting the findings

When attempting to quantify personality and behaviour in dingoes there are a few things that we need to consider. One of these is the fact that these studies utilise self-reporting. This opens up the chance for bias; for instance, dingo owners may be inclined not to give an objective evaluation of their animal's personality (either consciously or subconsciously). Owners may be sensitive to the fact that dingoes are not typical household pets and may answer the questionnaire in a way that they believe shows dingoes in a more favourable light. The opposite may also be true; some owners may want to demonstrate how dingoes are unlike dogs and exaggerate some aspects of their behaviour (i.e. to emphasise their exoticism). Dingo owners may have adjusted their perception of their dingo to match their expectations of the breed or animal and, for instance, may exaggerate aspects or specific traits known or suspected to be different. The level of attachment people have towards their dingo may also influence their ratings.[60]

The environment in which a dingo is reared can play a large role in the development of personality and behaviour and so must also be considered. Most of the dingoes surveyed were born and reared in captive facilities, which is different from the home/backyard environments more common to dogs. Rearing histories, including the adoption age (which for these dingoes was 15 weeks, well beyond the suggested seven weeks by organisations such as the Australian Dingo Foundation),[54] are well known to have an effect on future behaviour, personality, temperament and interaction with humans.[61] Further, the effect of neutering (and at what point in their development this occurs) on dog personality could explain some of the differences between dogs and dingoes, particularly because the majority of dogs (at least in Australia) are neutered whereas the majority of wild canids are entire and more likely used for breeding.

The dingoes sampled are already living as successful companion animals or pets and therefore predisposed (or selected) to living in human households. As mentioned earlier, few dingoes kept as pets live to an old age.[6] Walters suggested this is because most owners, although well-intentioned, are ill-prepared. The same can be said of dogs: there is a high rate of domestic dog relinquishment in Australia due to not fitting in with the owner's lifestyle or expectations, or to behavioural problems.[26, 62] It is uncertain whether the differences in dingo personality and behaviour suggests that dingoes do not differ much from dogs beyond a statistical number, or that the dingoes in this sample possess behaviour and temperament within acceptable ranges for pet dogs (i.e. this sample represents the select few that have the required temperament for successful adaptation to domestic environments). That is, dingoes that are considered unsuitable as household pets would most likely have been relinquished and not included in this sample (unfortunately, rates of relinquishment for dingoes are not available).

Certain characteristics in dogs are important for a successful dog–owner relationship (mentioned earlier in this chapter). The question remains as to whether dingoes living in domestic situations meet these personality or behavioural ideals. Only a small number of dingo owners reported any behavioural problems (23%), suggesting that according to their definition of acceptable canid behaviour, dingoes may make suitable pets. It is possible that dingo owners actively refuse to acknowledge any problematic traits or behaviours, as implied by Macintosh.[8] However, it is more realistic to assume that dingo owners have a conception of the ideal dingo pet and have adjusted their expectations to suit. That is, successful dingo owners may be appropriately prepared for ownership. It is presumed that the differences in personality and behaviour between dogs and dingoes found in these studies are not severe enough to cause a great deal of problem for dingo owners, as evidenced by their continued ownership. As Walters[6] and Oakman[15] suggest, if the owner understands the requirements, is willing to spend the time, respects its wild nature and commits to caring for the animal for a lifetime, then a dingo can be a good pet.

Reflective of the rarity of dingo pet-keeping in Australia, only a small number of dingoes and owners could be tested. Further, dingoes were compared with domestic dogs of varying breeds, which is likely to increase the variability of the personality constructs and behaviour differences and thus may have influenced the significance of the findings. Ideally, future studies should be more selective about which breeds dingoes are compared with (e.g. only dogs of similar size). Further, the type of assessment used may need to be reconsidered. MCPQ-R, for example, appears to be specific to domestic dogs and may not be applicable to non-domesticated dogs and wild canids. Either dingo owners have a different perspective on ownership and/or use different adjectives to describe their animals, or the scale itself is not reflective of dingo behaviour. Assessment methods that measure the behaviour of dogs in specific situations might be useful because they are less likely to be affected by the preconceived ideas of owners. Future studies are needed to further our understanding of the dingo–human relationship and will assist in the validation of personality measures for non-domesticated canids. Interview-based studies, in particular, may be a good way of exploring in more detail why people keep dingoes, whether they feel they need to present dingoes in a positive light, and how dingo and owner construct a mutually satisfying relationship.

Why own a pet dingo?

The experiences described in this chapter suggest that successful dingo ownership is possible and that a rewarding relationship can be established, but only in the right circumstances. It requires, among many other factors, a dingo that is acquired at an early age (the earlier the better), early exposure to the people and environment that will be constant throughout its life, and the right temperament (or personality) to adapt to captive human-centred environments. A successful relationship requires owners who are willing to commit to a long-term relationship and who have experience with canids, an adequately sized yard that is secured, the right attitude and patience. They also must have a dingo for the *right* reasons, not as status symbols. A dingo with an unprepared owner will lead to behavioural problems, compromised welfare or the likelihood of death in the case of relinquishment.

Given the number of considerations involved in successfully owning a pet dingo, the choice to keep one is questionable, particularly because domestic dogs can, for less effort and responsibility, offer a valuable alternative. After spending considerable time with both captive and wild dingoes, Roland Breckwoldt,[4] a filmmaker who kept dingoes that he used in his films, summarised this conundrum well: 'Give me a reason for having a dog, and I will suggest any number of suitable domestic breeds before a dingo' (p. 261). Why, then, would anyone choose to have a dingo as a pet?

The keeping of exotic or unique wild animals such as dingoes appears to be an emerging trend, but in general (especially with wild canids, or wild canid–hybrid dogs) it is not recommended, with many important ethical, welfare, safety and conservation implications to consider.[27, 63] It is currently unknown why people choose to own a dingo at all, especially in preference to domestic dogs, when according to their own assessments dingoes may have traits that are not deemed favourable.

The continued practice of dingo pet-keeping indicates that there are obvious benefits to owning a dingo that are perhaps not immediately obvious, or explainable. For instance, Darren and Leigh described a strong and rewarding connection with their dingoes that they had never experienced with any other domestic animal. However, there are several other explanations for choosing a dingo as a pet, including the perception that dingoes are better pets than domestic dogs, as is the case with some owners of wolf–dog hybrids in the US.[27] Owners may perceive dingoes as just a breed of dog, which they prefer to other breeds. Dingoes, as Australia's closest species to a wolf, could be seen as a status symbol, as a way of impressing others, to be different or stand out, as a signal of power or affluence and/or rebellion against society's norms.[64] Dingoes may act as self-objects that evoke or build a sense of self.[65, 66] For instance, people may see themselves in the dingo and thus identify closely with it (e.g. traits such as misunderstood, aloof, wild of nature, free-spirited). Support for this idea is provided by research suggesting that personality characteristics shared between dogs and their owner increases the likelihood of a successful and satisfactory relationship.[67] Lastly, dingo ownership may provide a social benefit. That is, the dingo acts as a social catalyst when in public. Through joining a dingo association or club, owners may gain the feeling of belonging and a sense of worth or achievement from aiding in the conservation of dingoes (primarily through education of the public).

Conclusion

Although dingo puppies are quite endearing at first, as they grow so do their needs, so that even the most experienced dog owner would find them difficult to accommodate. It might be fair to summarise dingoes as cooperative rather than obedient. There are many success stories of people keeping dingoes but there are equal, if not more, failed attempts at ownership. The sad reality is that many captive dingoes end up living a life of stress in a small enclosure, successfully escaping and never being found, shot or euthanased, given away to another dingo owner or organisation, sent back to the breeder or illegally released into the wild. I believe that the high failure rate is generally a result of the owner's expectations. If they expect a dingo to act like a dog, and not make any allowances to their lifestyle or adjust their expectations, they will more than likely relinquish the animal. These aspects appear to be the key to the success of Darren and Leigh's relationship with their five dingoes: they understand what is required and continually make allowances and sacrifices to accommodate those requirements. Few are prepared to go to such extremes. If expectations have a lot to do with it, I wonder what would happen if you reared a dingo that you thought was a dog? Is the behaviour of pet dogs in some ways a product of what we expect to get from them? For example, if someone thinks they have a dingo, they will treat it like they think dingoes need to be treated and thus end up with a different animal than they would have if they believed it to be a dog.

There appear to be some interesting contradictions in the experiences told by dingo breeders and owners like Darren and Leigh. They openly state that the dingo does not make a good domestic pet and should not be treated as one, yet much of the behaviour of the owners and breeders seems to suggest that dingoes can and do make good pets, and in general are treated as though they are domestic pets. Many argue that if dingoes are a wild animal, they should be respected as such and do not belong in any capacity in domestic situations. Both sides make valid arguments. I do not believe we should judge the practice of dingo pet-keeping and, although dingo ownership is not for me, I envy those who have such a close relationship with their dingoes.

Acknowledgements

I wish to thank Darren Griffiths and Leigh Mullan for chatting with me and offering an insight into their fascinating lives with dingoes. Thanks also to the Australian Dingo Foundation and its members for supporting the project and completing the questionnaires. I owe a great deal of gratitude to Bronwyn Chidzey for helping with data entry, and Professor James Serpell for providing access to the C-BARQ and its sample of domestic dogs, as well as for conducting the statistical analysis included in this chapter.

References

1. Smith BP, Litchfield CA (2009) A review of the relationship between indigenous Australians, dingoes (*Canis dingo*) and domestic dogs (*Canis familiaris*). *Anthrozoos* **22**, 111–128.
2. Phillip A (1789) *The Voyage of Governor Phillip to Botany Bay: With an Account of the Establishment of the Colonies of Port Jackson and Norfolk Island/Compiled from the Several Departments*. London.
3. Hunter J (1793) *An Historical Journal of the Transactions at Port Jackson and Norfolk Island with the Discoveries which have been made in New South Wales and in the Southern Ocean, since*

the Publication of Phillip's Voyage, Compiled from the Official Papers; Including the Journals of Governor Phillip and King, and of Lieut. Ball; and the Voyages of the First Sailing of the Sirius in 1787, to the Return of that Ship's Company to England in 1792. John Stockdale, London.

4. Breckwoldt R (1988) *A Very Elegant Animal: The Dingo.* Angus and Robertson, Sydney.
5. Holden P (1991) *Along the Dingo Fence.* Hodder and Stoughton, Sydney.
6. Walters BJ (1995) *The Company of Dingoes: Two Decades with our Native Dog.* Australian Native Dog Conservation Society, Bargo.
7. Lorenz K (1954) *Man meets Dog.* Viking Penguin, New York.
8. Macintosh NWG (1975) The origin of the dingo: an enigma. In *The Wild Canids: Their Systematics, Behavioural Ecology and Evolution.* (Ed. MW Fox) pp. 87–106. Van Nostrand Reinhold, New York.
9. Trumler E (1973) *Understanding your Dog.* Trans. R Barry. Faber and Faber, London.
10. Kaleski R (1933) *Australian Barkers and Biters.* Bulletin Newspaper, Sydney.
11. Hudson L (1974) *Dingoes Don't Bark.* Rigby, Adelaide.
12. Wright SM (1968) *The Way of the Dingo.* Angus and Robertson, Sydney.
13. Ryhorchuk G (1980) *A Dingo in the Family.* Inkata Press, Melbourne.
14. Ryhorchuk G (2013) *Living with Dingoes.* Ex Libris Books.
15. Oakman B (2001) The problem with keeping dingoes as pets and conservation. In *A Symposium on the Dingo.* (Eds CR Dickman and D Lunney). Royal Zoological Society of New South Wales, Sydney.
16. Wilton AN (2001) DNA methods of assessing dingo purity. In *A Symposium on the Dingo.* (Eds CR Dickman and D Lunney) pp. 49–56. Royal Zoological Society of New South Wales, Sydney.
17. Vilà C, Savolainen P, Maldonado J, Amorin I, Rice J, Honeycutt R, Crandall K, Lundeberg J, Wayne R (1997) Multiple and ancient origins of the domestic dog. *Science* **276**, 1687–1689.
18. Clutton-Brock J (1995) Origins of the early dog: domestication and early history. In *The Domestic Dog: Its Evolution, Behaviour and Interactions with People.* (Ed. J Serpell) pp. 7–20. Cambridge University Press, Cambridge.
19. Miklósi A, Topál J, Csányi V (2004) Comparative social cognition: what can dogs teach us? *Animal Behaviour* **67**, 995–1004.
20. Scott J, Fuller J (1965) *Genetics and the Social Behavior of the Dog.* University of Chicago Press, Chicago.
21. Parker HC, Kim LV, Suiter NB, Carlson S, Lorentzen TD, Malek TB, Johnson GS, DeFrance HB, Ostrander EA, Kruglyak L (2004) Genetic structure of the purebred domestic dog. *Science* **304**, 1160–1164.
22. Vilà C, Maldonado JE, Wayne RK (1999) Phylogenetic relationships, evolution, and genetic diversity of the domestic dog. *Journal of Heredity* **90**, 71–77.
23. Svartberg K (2006) Breed-typical behaviour in dogs: historical remnants or recent constructs? *Applied Animal Behaviour Science* **96**, 293–313.
24. King T, Marston L, Bennett P (2009) Describing the ideal Australian companion dog. *Applied Animal Behaviour Science* **120**, 84–93.
25. Bennett PC, Rohlf VI (2007) Owner-companion dog interactions: relationships between demographic variables, potentially problematic behaviours, training engagement and shared activities. *Applied Animal Behaviour Science* **102**, 65–84.
26. Marston L, Bennett P, Coleman G (2004) What happens to shelter dogs? An analysis of data for 1 year from three Australian shelters. *Journal of Applied Animal Welfare Science* **7**, 27–47.
27. Hope J (1994) Wolves and wolf hybrids as pets are big business: bad idea. *Smithsonian* **25**, 34–44.
28. Tucker P, Weide B (1998) *Can You Turn a Wolf into a Dog? Commonly Asked Questions about Wolves and Hybrids in Captivity.* Wild Sentry.
29. Klinghammer E, Goodmann PA (1987) Socialisation and management of wolves in captivity. In *Man and Wolf.* (Ed. H Frank) pp. 31–59. Dr W. Junk Publishers, Dordrecht.

30. Rabb G, Woolpy J, Ginsburg E (1967) Social relationships in a group of captive wolves. *American Zoologist* 7, 305–311.
31. Fentress JC (1967) Observations on the behavioral development of a hand-reared male timber wolf. *American Zoologist* 7, 339–351.
32. Fox MW (1972) Socio-ecological implications of individual differences in wolf litters: a developmental and evolutionary perspective. *Behaviour* **41**, 298–313.
33. Frank H, Frank MG (1987) The University of Michigan canine information-processing project (1979–1981). In *Man and Wolf.* (Ed. H Frank) pp. 143–167. Dr W. Junk Publishers, Dordrecht.
34. MacDonald K (1983) Stability of individual differences in behavior in a litter of wolf cubs (*Canis lupus*). *Journal of Comparative Psychology* **97**, 99–106.
35. Woolpy JH, Ginsburg BE (1967) Wolf socialization: a study of temperament in a wild social species. *American Zoologist* 7, 357–363.
36. Zimen E (1982) A wolf pack sociogram. In *Wolves of the World.* (Eds FH Harrington and PC Paquet) pp. 282–323. Noyes Publications, New Jersey.
37. Zimen E (1987) Ontogeny of approach and flight behavior towards humans in wolves, poodles and wolf–poodle hybrids. In *Man and Wolf.* (Ed. H Frank) pp. 275–292. Dr W. Junk Publishers, Dordrecht.
38. Gácsi M, Győri B, Miklósi A, Virányi Z, Kubinyi E (2005) Species-specific differences and similarities in the behavior of hand-raised dog and wolf pups in social situations with humans. *Developmental Psychobiology* **47**, 111–122.
39. Smith BP, Appleby RG, Litchfield CA (2012) Spontaneous tool-use: an observation of a dingo (*Canis dingo*) using a table to access an out-of-reach food reward. *Behavioural Processes* **89**, 219–224.
40. Jones AC, Gosling SD (2005) Temperament and personality in dogs (*Canis familiaris*): a review and evaluation of past research. *Applied Animal Behaviour Science* **95**, 1–53.
41. Wilsson E, Sundgren P-E (1997) The use of a behaviour test for selection of dogs for service and breeding. II. Heritability for tested parameters and effect of selection based on service dog characteristics. *Applied Animal Behaviour Science* **54**, 235–241.
42. Svartberg K, Forkman B (2002) Personality traits in the domestic dog (*Canis familiaris*). *Applied Animal Behaviour Science* **79**, 133–155.
43. Goodloe LP, Borchelt PL (1998) Companion dog temperament traits. *Journal of Applied Animal Welfare Science* **1**, 303–338.
44. Hsu Y, Serpell JA (2003) Development and validation of a questionnaire for measuring behavior and temperament traits in pet dogs. *Journal of the American Veterinary Medical Association* **223**, 1293–1300.
45. Ley J, Bennett P, Coleman G (2008) Personality dimensions that emerge in companion canines. *Applied Animal Behaviour Science* **110**, 305–317.
46. Ley J, Bennett P, Coleman G (2009) A refinement and validation of the Monash Canine Personality Questionnaire (MCPQ). *Applied Animal Behaviour Science* **116**, 220–227.
47. Gosling SD, Kwan VSY, John OP (2003) A dog's got personality: a cross-species comparative approach to personality judgements in dogs and humans. *Journal of Personality and Social Psychology* **85**, 1161–1169.
48. Svartberg K (2002) Shyness–boldness predicts performance in working dogs. *Applied Animal Behaviour Science* **79**, 157–174.
49. Hennessy M, Voith V, Mazzei S, Buttram J, Miller D, Linden F (2001) Behavior and cortisol levels of dogs in a public animal shelter, and an exploration of the ability of these measures to predict problem behavior after adoption. *Applied Animal Behaviour Science* **73**, 217–233.
50. Gosling SD (1998) Personality dimensions in spotted hyenas (*Crocuta crocuta*). *Journal of Comparative Psychology* **112**, 107–118.
51. Smith B (2014) Living with wild dogs: personality dimensions in captive dingoes (*Canis dingo*) and implications for ownership. *Anthrozoos* **27**, 423–433.

52. Gosling SD (2001) From mice to men: what can we learn about personality from animal research? *Psychological Bulletin* **127**, 45–86.
53. Gosling D (2008) Personality in non-human animals. *Social and Personality Psychology Compass* **2**(2), 985–1001.
54. Australian Dingo Foundation (2012) *Dingo Carer's Handbook*. Australian Dingo Foundation, Melbourne.
55. Smith BP, Litchfield CA (2013) Looking back at 'looking back': operationalising referential gaze for dingoes in an unsolvable task. *Animal Cognition* **16**, 961–971.
56. Smith BP, Litchfield CA (2010) How well do dingoes, *Canis dingo*, perform on the detour task? *Animal Behaviour* **80**, 155–162.
57. Smith BP, Litchfield CA (2010) Dingoes (*Canis dingo*) can use human social cues to locate hidden food. *Animal Cognition* **13**, 367–376.
58. Appleby R, Smith B, Jones D (2013) Observations of a free-ranging adult female dingo (*Canis dingo*) and littermates' responses to the death of a pup. *Behavioural Processes* **96**, 42–46.
59. Duffy D, Serpell J (2012) Predictive validity of a method for evaluating temperament in young guide and service dogs. *Applied Animal Behaviour Science* **138**, 99–109.
60. Serpell J (1996) Evidence for an association between pet behavior and owner attachment levels. *Applied Animal Behaviour Science* **47**, 49–60.
61. Coppinger R, Coppinger L (2001) *Dogs: A New Understanding of Canine Origin, Behavior, and Evolution*. University of Chicago Press, Chicago.
62. Marston L, Bennett P (2003) Re-forging the bond towards successful canine adoption. *Applied Animal Behaviour Science* **83**, 227–245.
63. Schuppli C, Fraser D (2000) A framework for assessing the suitability of different species as companion animals. *Animal Welfare (South Mimms, England)* **9**, 359–372.
64. Serpell J (1986) *In the Company of Animals: A Study of Human–animal Relationships*. Cambridge University Press, New York.
65. Wolf E (1988) *Treating the Self: Elements of Clinical Self Psychology*. Guilford Press, New York.
66. Brown S-E (2007) Companion animals as self objects. *Anthrozoos* **20**, 329–343.
67. Curb L, Abramson C, Grice J, Kennison S (2013) The relationship between personality match and pet satisfaction among dog owners. *Anthrozoos* **26**, 395–404.

11
The role of private sanctuaries in dingo conservation and the management of dingoes in captivity

Bradley Smith and Lyn Watson

Due to hybridisation with domestic dogs and continued targeted control of dingoes across much of the country, the threat to free-ranging pure dingoes is real. According to the IUCN Red List of Threatened Species the dingo is listed as vulnerable and therefore considered to be facing a high risk of extinction in the wild.[1] Unfortunately, there are no tangible conservation strategies in place that will reduce the rate of possible extinction. There has, however, been strong community support to retain the purity of dingoes by keeping and preserving captive populations of pure dingoes (and/or rescuing wild-born dingoes). Such colonies and organisations can be effective tools to educate the community on the plight of the dingo. They can also work collaboratively with governments and scientists to increase our understanding and management of dingoes. Various organisations across Australia are fulfilling these roles, and could represent one of the last lines of defence against the extinction of the dingo. Keeping healthy and content dingoes confined within captive environments, however, creates a variety of challenges, many of which will be outlined in this chapter. To do so, we use a case study of one of Australia's largest and most successful dingo sanctuaries – the Dingo Discovery Sanctuary and Research Centre (or simply the Dingo Discovery Centre), which is supported by the Australian Dingo Foundation, a federally recognised conservation organisation. Managing large groups of dingoes creates distinctly different issues from the management of pet dingoes described in the previous chapter, but the process can reveal a lot about their behaviour and temperament that would be difficult to observe in wild populations.

How can private organisations aid dingo conservation?

There are several private sanctuaries scattered around the country that house colonies of up to 40 dingoes. Besides providing a social and support network for dingo enthusiasts and fulfilling the interest and passion of the sanctuary owners and volunteers, what purpose do they or can they serve in advancing the conservation of dingoes? These organisations play an integral role in dingo conservation. The following outlines some of these positive contributions.

Preservation of pure dingoes

There is a risk that the pure dingo will be lost in coming years as a result of hybridisation with domestic and/or feral dogs diluting the gene pool, potential in-breeding among isolated populations, loss of genetic fitness and exposure to disease (such as those on Fraser Island), and wild dog eradication programs led by state and federal government and the agricultural industry. The extent of the threat is difficult to say with certainty, as estimating the true dingo abundance remains problematic given their geographic range and their external similarity to dingo–dog hybrids (see Chapter 1). Regardless, it is extremely important that the purity of the dingo, as a native Australian mammal, unique canid and integral member of the ecology of Australia, be preserved in the event that extinction does occur.

One method of preserving dingo DNA has been nicknamed the Frozen Zoo (Animal Gene Storage and Resource Centre of Australia), which was established in 1995 by Monash University in Melbourne in conjunction with the Zoological Parks Board of New South Wales. The idea is to collect and freeze sperm and eggs from pure dingoes (and other native Australian species). It will then be possible to use assisted reproduction to either artificially inseminate female dingoes with dingo sperm, or use a domestic dog bitch as a surrogate carrier. Although these methods have been successful with dogs, there has been limited success with dingoes. That is, female dingoes do not take to being artificially inseminated. It is unknown why this has not been successful, but based on our knowledge of breeding dingoes we believe there are two possible explanations. First, the dingo bitch needs to be bonded with a male for some time before conception and second, two matings are often required – the first stimulates ovulation and the second leads to conception. These are both things that are not possible with artificial insemination.

The frozen zoo idea seems a viable option in the event of extreme circumstances as a last resort, but at this stage it is not necessary. At least, not while there are captive facilities and private owners across the country that care for large colonies of pure dingoes from various genetic lineages. Breeding pairs of dingoes can be found outside Australia at Wellington Zoo (New Zealand) and in the US (Fort Wayne Children's Zoo in Indiana and Prospect Park Zoo in New York; Fig. 11.1 and Plate 7), further ensuring preservation of the pure dingo. Although dingoes have been held overseas for many years (e.g. London Zoo bred dingoes during the 1830s,[2] the Bronx Zoo in New York had dingoes around 1933)[3] and some are still scattered throughout Europe, these animals are of unknown ancestry and purity and not seen as adequate stock.

Not knowing the purity of dingoes in captivity has been a constant issue and cause of stress for many dingo sanctuaries until the recent development of a genetic dingo purity test established by Wilton from the University of New South Wales.[4] Before genetic testing was possible, the only scientific way of determining purity (in addition to observable aspects such as an annual breeding cycle and physical appearance) was to measure the skulls of deceased dingoes. This method of purity assessment was developed by CSIRO scientists Corbett and Newsome.[5,6] Basically, the only way to find out if a dingo was pure was to wait for it to die. This meant that dingo breeders could have been raising a dingo for upwards of 15 years without knowing its purity. After its death, owners were required to remove the dingo's head and place it in boiling water to remove the skin before sending the skull to CSIRO for assessment. Such a process would be a horrible experience for any animal owner, not to mention problematic if the animal had been bred from. The method has also been proven to be somewhat unreliable.[7]

Fortunately, the process of determining dingo purity today is far simpler and can be done within a matter of weeks. All that is required is a sample – the owner rubs a cotton bud-ike device along the inside of the dingo's cheek. This gathers biological cells that contain the

Figure 11.1: Two dingo pups being delivered to the staff at the Fort Wayne Children's Zoo from the Dingo Discovery Centre in January 2010.
Source: John Williams.

required DNA. The swab is sent to the School of Biotechnology and Biomolecular Sciences at the University of New South Wales for analysis, which reports the likelihood of purity in a percentage score. This allows sanctuaries to ensure that they breed only from known pure stock and to quickly assess the purity of any wild-born dingoes that are given to them. Those that are considered pure are often added to the breeding programs. Details of the test and how purity is determined is covered in Chapter 1.

Reintroducing dingoes

One of the primary reasons for maintaining populations of dingoes in captivity is the chance that at some point in the future there may be a call to reintroduce dingoes into the wild. Such calls by wildlife ecologists have already begun, particularly in small areas in national parks to restore biodiversity and limit the impact of mesopredators such as foxes and feral cats.[8,9] Reintroductions of this nature have already occurred and been successful, for example the reintroduction of wolves to Yellowstone National Park in the US.[10] At the moment, most dingo sanctuaries employ a hands-on management style with the animals for easier management and for encouraging positive encounters with the public. If reintroductions were to occur, these populations would be unsuitable for a variety of reasons: they would not have skill or at least experience in finding and killing their own food, and they would have little fear of humans and thus would be an easier target for control programs. For reintroductions to be successful, captive dingoes would need to be reared away from human involvement in a large well-resourced sanctuary within a wilderness area. This colony could start with a sanctuary-reared dingo; each successive generation would be more suited to survival in the wild. A breeding pair selected from such populations would be ideal candidates for reintroduction. Our experience with captive dingoes shows that they appear to lose few skills and survival instincts even after successive generations born and raised in captivity.

Political contributions

A group of dedicated people can be quite powerful and influential in getting government laws and policies changed. There are several examples of where dingo organisations have led to change. It might be helpful to provide a brief history of dingo organisations in Australia, as the early events allowed for legal ownership of dingoes and the rise of subsequent dingo organisations and advocates.

History of dingo organisations and obtaining legal ownership

The original dingo organisation, established in the late 1980s, was the National Dingo Association. The National Dingo Association was fashioned around the Dingo Farm located near Castlemaine in Victoria and run by Bruce Jacobs. The Dingo Farm and Association were the first to really attract a fully national band of members who came from all over Australia to attend meetings. Strategically steered by Lyn and Peter Watson, the Association focused on raising awareness and beginning the process of conserving dingoes. It devised a three-part strategy that was implemented during the 1980s and 1990s. First, they achieved recognition of the National Dingo Association by making it an incorporated organisation. Second, they obtained recognition of the dingo as Australia's native dog and developed a breed standard (see Chapter 1) with the Australian National Kennel Council. Peter Watson, who was the Victorian delegate to the National Kennel Council, led the submission. Third, Lyn and Peter, along with the Hon. Max Turner, member for Bendigo, on behalf of the Association convinced the Agriculture Minister of Victoria the Hon. Bill McGrath to successfully seek parliamentary approval for legal ownership of dingoes. Without this, approval of dingo ownership and subsequent dingo organisations and advocates would have not been possible. In fact, the history of all subsequent dingo associations and advocates can be traced to their association with and membership in the now defunct National Dingo Association.

Watchdog, lobbying government and public awareness

Various dingo groups have been involved in developing petitions to present to government, such as a request to ban 1080 poisoning and to get the dingo listed as a protected species. Part of the process of obtaining signatures is educating the community on the importance of such issues. Apart from lobbying, on occasion governments require expertise relating to dingoes when setting policy; it can obtain this from registered dingo organisations. For example, members from the Dingo Care Network were involved in the Dingo Permit Review working group for the Victorian Department of Primary Industries (resulting policies were outlined in

Figure 11.2: Popular Australian television personality and veterinarian Dr Chris Brown filming for a television show at the Dingo Discovery Centre.
Source: Lyn Watson.

Chapter 10), and the Australian Dingo Conservation Association has been involved in Wild Dog Management Plans in accordance with Pest Animal Management committees across Australia. Unfortunately, there is little communication or cooperation between many of the organisations, so the impact they can make individually will never be as great as if they all worked together. Last, private dingo sanctuaries can help achieve positive dingo press by providing opportunities for print and visual media organisations to film and photograph dingoes for news and educational purposes (Fig. 11.2).

Conservation through education

Education of the public is a key aim of dingo organisations, because the general public's lack of knowledge and awareness of dingoes is one of the biggest challenges to conservation. Public education remains one of the primary objectives of many dingo organisations, private sanctuaries and public zoos around Australia. Sanctuaries housing large colonies of dingoes, such as the Dingo Discovery Centre, can directly achieve public dingo education by opening their doors to visitors (particularly during puppy season, see Fig. 11.3) as well as by visiting schools, shopping centres, festivals and other events by request, so that the public can learn about dingoes (and dispel myths) while interacting with them. Being able to interact with animals is one of the best ways to learn about a species and conservation issues, and to encourage

Figure 11.3: Every winter, visitors to the Dingo Discovery Centre get to interact with dingo puppies. This helps to socialise the animals, raises money for the centre and offers a rewarding experience for the public.
Source: Joanne McKay.

people to change their behaviour or attitudes. The many positive outcomes of visiting a zoo or wildlife park and interacting with animals are visible both immediately and months after the visit or encounter. People believe they experience a stronger connection to nature as a result of their visit/encounter, are prompted to reconsider their role in environmental problems and conservation action and begin to see themselves as part of the solution. Visitors can be given educational materials and can purchase merchandise (e.g. photos, toys, figurines etc.) that not only helps spread the message but raises money for further conservation initiatives (Fig. 11.4). Visits reinforce the values and attitudes people already held regarding ecological concepts.[11] In order to achieve these successful encounters, however, there need to be individual dingoes that are able to play the role of dingo ambassadors. They must have the right temperament and have been adequately socialised as a pups.

Dingo organisations that do not have a flagship captive colony *per se* often include members with pet dingoes, such as the Western Australian Dingo Association in Perth. These organisations also work hard to provide a positive dingo experience at public events. Last, there are a few dingo-specific organisations that attract dingo enthusiasts (but not necessarily owners) that aim to spread conservation awareness and/or protect local populations; for example, there are various groups that aim to protect the Fraser Island dingo. It is critical,

Figure 11.4: Volunteers at the Dingo Discovery Centre in front of a stall at a community event.
Source: Lyn Watson.

however, that dingo advocates and groups spread messages that are based on scientific evidence and not on hearsay or speculation, as this can cause more harm than good in many cases and may place management scientists and government offside.

Supply of dingoes to zoos and wildlife parks

The majority of the zoos and parks around Australia that house dingoes do not actively participate in breeding programs and generally neuter or chemically castrate their dingoes. This is because dingoes are not critically endangered, are easy to breed in captivity and are readily available. If they were allowed to breed, it could be difficult to place the dingo pups each year as most zoos and parks cannot accommodate large numbers and do not sell direct to the public. All breeding programs are left to private sanctuaries that supply zoos and fauna parks (both nationally and internationally) with dingo puppies on an order-type basis. A few sanctuaries breed only for this purpose and to maintain breeding lines, while others such as the Dingo Discovery Centre also sell to suitable members of the public (potential owners are screened for suitability as best as possible). The sale of dingo pups is a good source of income for dingo sanctuaries that are completely self-funded. It is also healthy to keep dingoes of breeding age reproducing to avoid potential medical complications from sustained periods of birth control and to ensure certain social and parental behaviours are maintained. Owners of pet dingoes often do not neuter their animals, so that they can be used in a breeding program if necessary.

Facilitation of scientific research

Captive sanctuaries provide a valuable and practical alternative or supplement to research of wild populations. In fact, much of our understanding of the biology and behaviour of canids has come from captive studies. For example, our knowledge of the social behaviour of wolves was initially conducted on captive wolves by Rudolph Schenkel[12] and supported by later studies of wild populations (e.g. by wildlife biologist David Mech).[13] Like wolves, dingoes are notoriously difficult to study in the wild because they are highly mobile, occupy large territories, are difficult to see in many environments and are generally reclusive.[14] Added to this, identification of individuals is often difficult and the individual life histories are incomplete. A more complete understanding of their behaviour can be achieved when field observations and the more passive forms of study mentioned in this chapter are supplemented by observations conducted in captivity. More details about some of the methods used to study wild dingoes are presented in Chapter 7.

Scientists have greater power to manipulate animals in captivity and to take medical, physiological or observational samples. For example, using captive samples scientists from the Sir Arthur Rylah Institute for the Victorian Department of Primary Industries tested dingo urine for its effectiveness as a deterrent for pests such as foxes and feral cats (a bio-boundary), and scientist Michael Parsons conducted research assessing anti-predator reactions to dingoes.[15] The collection of urine for such purposes would be impossible in the wild, and a controlled setting for determining its effect on dingo behaviour in controlled conditions would be difficult. Most organisations housing dingoes have been heavily involved in research, allowing skulls to be measured in the case of Corbett's skull morphology method of assessment and providing genetic samples for the development of Alan Wilton's genetic test. For many years the CSIRO kept colonies of dingoes to conduct research that would complement its fieldwork (Fig. 11.5). For example, CSIRO scientist Catling bred dingoes and deliberately crossed them with domestic dogs to understand more about growth rates and breeding seasonality and behaviour.[16] The effectiveness of tools or strategies to manage dingoes has also been trialled on captive dingoes. Jones and colleagues from Griffith University, for instance, tested the effectiveness of an ultrasonic device as a deterrent to dingoes in captivity before the method was taken to the wild on a larger and more expensive scale.[17] It is important, when studying the behaviour of wild animals in captivity, that they be kept in a way that maximises their opportunities for the expression of normal behaviour.

Many of the scientific studies just mentioned are quite passive and do not require direct management of the animal. Some questions regarding the ethology and psychology of an animal, however, require well-socialised animals that can only be found in captive situations. Not only does interacting with animals provide insights not otherwise possible in the wild, but failure to socialise captive wolves and dingoes may increase their aberrant behaviour because the presence of humans, particularly in close proximity, seriously disrupts the animal's behaviour.[14]

Figure 11.5: CSIRO experimental officer with captive dingoes held at the research centre in the Northern Territory.
Source: National Archives of Australia, A1200, L64195.

Using captive dingoes for scientific research is not entirely straightforward. Many of the cognitive experiments described in Chapter 9 were conducted in captivity because they required dingoes that were socialised to humans and could be handled. In some of the experiments we have conducted it has been difficult selecting dingoes to take part in the study, mostly because we considered them too timid and cautious. Such dingoes are unsuitable because animals exhibiting sensitivity to novelty (neophobia) can lead to false interpretations of cognitive ability.[18] Thus, we require dingoes to reach a certain level of tameness before they participate in an experiment; we select dingoes we consider can be comfortably handled and placed on a leash and which show no obvious fear response to the experimenter, equipment or testing scenario. We then determine, either during the familiarisation period or when testing has commenced, whether a dingo should be excluded from the study. As a general rule, with cognitive studies where a high level of control is required it is not possible to work with adult dingoes. Age appears to be an important factor when testing dingoes and may also contribute to differences in findings within studies. Regardless of history of socialisation (household or sanctuary), as dingoes reach sexual maturity at the age of ~12 months many become increasingly cautious of unfamiliar people, places and scenarios and are therefore difficult to test. These behaviours are similar to those displayed in wolves, where the level of fear of humans and unfamiliar situations increases with age.[19]

How 'wild' are captive dingoes?

There are some concerns from dingo scientist Dr Laurie Corbett[6] and author Roland Breckwoldt[20] that keeping dingoes in captivity does little to assist in conserving them. Corbett goes as far as to suggest that the act of dingo preservation societies breeding dingoes is a major threat to pure dingoes.[6] Corbett believes that a captive dingo ceases to be a dingo and is 'just another dog' because selection pressure for certain traits can easily take place, both deliberately and inadvertently. This could override the natural selection of wild characteristics. He also feels that keeping dingoes as pets is too dangerous, that many of the dingoes kept in captivity are bred from hybrids or untested dingo stock and that people will ultimately breed their dingo with domestic dogs and thus 'effectively increase the hybridisation process'.[6] We believe these claims to be completely unfounded. Although we agree that dingoes can be dangerous in captive settings (as any domestic dog breed can be) and that people should remain vigilant around them, there have been no reported attacks by pet or captive dingoes on an adult or child. The keeping of captive dingoes in fact aids the conservation of dingoes in many ways, outlined in this chapter.

The statements by Corbett and Breckwoldt regarding the purity of captive dingoes were made before the establishment of the genetic purity test. Today, the dingoes that are in captivity and part of breeding programs are all considered pure. As far as we understand, no organisation or private breeder actively breeds dingoes with domestic dogs. On occasion hybrid dog–dingoes are kept but they are usually excluded from breeding programs. The genetic pool of captive dingoes is getting larger, with dingoes from the wild constantly being added to breeding programs. There is a risk that the only dingoes that are bred are those suited to captivity but in our experience this type of selection would happen only slowly – there still remains a large variety of individual differences and temperaments among captive dingo populations. Two parents that are extremely outgoing and friendly towards people do not guarantee a litter with the same temperament. As suggested earlier, in the event of reintroductions, if the dingoes had been reared away from human intervention we would be confident that they would have the tools to survive in the wild.

Managing dingoes in captivity

Private collectors and wildlife parks that keep dingoes can be found across Australia. Whatever the facility, maintaining a large population of dingoes presents a lot of challenges. As a general rule, the more dingoes, the more problems are likely to occur – particularly in the breeding season (refer to case study for further information).

Guidelines for dingo management

For almost all captive species on display, the Zoo and Aquarium Association, which is the peak body representing the zoo and aquarium community throughout Australia, takes responsibility for coordinating breeding programs and sets the level of professional standards and practice for its members. Because many of the zoos and parks around Australia that house dingoes do not actively participate in breeding programs, all dingo breeding is left to private sanctuaries. Unfortunately, there is no official breeding dingo management system or program run by an independent organisation, and the breeding of dingoes goes largely undocumented and without regulation. In 2002 when dingo genetic testing became readily accessible, the Australian Dingo Conservation Association under the governance of Barry Oakman was given responsibility by the Zoo and Aquarium Association (then known as the Australian Regional Association of Zoological Parks and Aquaria) for setting up a studbook of this nature. However, due to political problems among members and the exclusion of many other dingo breeders and private owners by the organisation, a thorough and independent system has never been developed.

There are no universal codes of practice or associated guidelines for the management of dingoes on display. Each private dingo organisation appears to have its own set of rules and guidelines (or code of ethics) regarding membership, housing, purity, breeding and disposal of dingoes. Although existing organisational guidelines are similar, ideally there should be a universal set of guidelines to which all organisations that breed or house captive dingoes should adhere. This would ensure, for example, that genetic lines are preserved and pedigrees known, that dingo housing and treatment is ethical and adequate, that the physical appearance, characteristics and behaviours of pure dingoes can be documented and the numbers and locations of all dingoes known.

In many ways, the care of dingoes is not dissimilar to that of dogs, particularly in relation to medical management (e.g. vaccinations, worming), feeding and manual handling. Stephen Jackson, a zoo-keeper turned wildlife author, provided a good overview of many of the day-to-day issues relating to the management and husbandry of dingoes in zoo environments.[21] But keeping dingoes healthy is only half the battle; some of the most significant management challenges are due to the dingo's nature (see Chapter 10 relating to pet dingoes) and keeping large numbers of dingoes in an unnatural setting. The following case study goes beyond the health needs of the captive dingo, and explains what is really involved in setting up a colony and managing it successfully.

A case study: Dingo Discovery Sanctuary and Research Centre

The keeping of large colonies of dingoes by private organisations is essential to many of the conservation aims outlined at the beginning of this chapter. Keeping dingoes, however, is a difficult and expensive endeavour that requires expertise in wild canid management and a team of dedicated people. This section provides an overview of some of the main issues and requirements that are essential to the effective management of dingoes in captivity. The Dingo Discovery Sanctuary and Research Centre is located in Toolern Vale, Victoria, and is owned and operated by Lyn and Peter Watson. Although every captive dingo facility faces the same challenges, each chooses to manage its animals in different ways. These depend on the manager's individual philosophy, experience and circumstance. There is no right or wrong way, and management methods usually evolve with time, experience and money. This case study is merely one example of how it can be done successfully.

Overview of the sanctuary

The Dingo Discovery Centre is a privately owned and operated facility. Its mission is to 'create a safe captive environment where a collection of remnant DNA-tested pure and entire dingoes could be housed in order to preserve a viable genetic pool, and where close observations could be carried out to further the understanding of, and to complement wild studies on natural social behaviours and biological functioning of the dingo'. From its humble beginnings at sites that included the backyards of Lyn and Peter's various homes, the Centre has developed into an impressive custom-built sanctuary located on a 40 acre property that is surrounded on two sides by state forest. The Centre receives no government assistance; it relies completely on volunteers, fundraising and public donations for all maintenance, dingo care and facility upgrades. The Centre houses ~30 adult dingoes originating from various lineages from across Australia, in addition to their progeny in any given year. Although it can house more, it considers this number as capacity – any more would significantly increase running costs and the animals would not get the individual care considered necessary. The Centre has been actively breeding dingoes since 1990, producing between one and nine litters each season. It is a major supplier for zoos and fauna parks around Australia and the world, and sells to selected private households in states where it is legal to own a dingo as a companion. The facility includes the housing kennels, a kitchen, a small veterinary hospital and eight large exercise yards or paddocks.

The Dingo Discovery Centre is the organisation that maintains the colony of different lineages, but Lyn and Peter have also established the Australian Dingo Foundation that supports the sanctuary and conducts a host of activities that support dingo conservation initiatives. For example, they regularly support a growing number of research studies from many Australian and international universities and the Australian government, as well as providing work experience hours to veterinary and animal science students. Volunteers regularly attend community groups, festivals and many schools and colleges in the surrounding area to give educational presentations and live dingo interactions in an effort to reach as many Australians as possible. The Australian Dingo Foundation is a federally registered conservation organisation, which means it has approved charity status allowing it to undertake various fundraising activities. For example, each weekend during puppy season (around June–September) the Centre runs open educational sessions for visitors; these are a major source of funding. The Foundation's long-term goal is to establish a dingo tourism/interpretive centre, which can become self-sustaining and carry on the work of the organisation and sanctuary in perpetuity.

History and motivations of the founders

The founders Lyn and Peter have been involved with dingoes since the late 1980s when they began visiting Bruce Jacobs at the Dingo Farm. Their passionate fascination with Australia's dingo extended from their love and knowledge of domestic dogs (both having been successful breeders and judges). Their life with dogs eventually led them to dingoes, where they quickly developed a burning desire to conserve a species that they saw as being 'persecuted to extinction'. They began by gathering a collection for preservation in captivity, in the hope the dingoes might one day be returned to wild environments. After seeing that there were few establishments actively breeding dingoes, they decided to develop a dingo facility that drew upon their many years of experience with dogs.

In his book *Bush Aussies*, Alan Nixon[22] covered the story of Lyn and Peter Watson. In it, Lyn exclaimed, 'Until you own a dingo, you know nothing about dogs' (p. 18). Experience with both species suggested to the Watsons that dingoes do not behave like dogs, and people who fall into the common trap of thinking so are destined for failure. Lyn and Peter found the existing guidelines and recommendations offered for captive zoo animals (e.g. relating to behaviour, metabolism, basic needs, diet, coat care, housing, exercise) were of little practical use when there were more than three or four animals, and in some instances the guidelines were seriously misleading. For example, many exhibition parks attempt to keep dingoes in threes rather than as pairs; this can only lead to fights and aggression between dingoes, even neutered animals, unless there is sufficient space for a low-ranked animal to flee or hide from more dominant individuals. The Watsons claim that they had to literally rewrite the book on dingo management, a process that took a lot of time and experience. It has taken about 15 years, including numerous reviews and modifications, before they could confidently claim anything resembling success. The challenge, according to Lyn, was trying to maintain the dingoes' natural behaviours while trying to tame them sufficiently that they can be contained and handled, and their behaviour around people predictable and safe. At the same time, the dingoes need to remain healthy and content.

Considerations when managing captive dingoes

Location of facility

It is important that the colony is not located near disruptive or noisy environments that may cause the dingoes distress. Equally, the dingoes should not be situated so that they cause disruption to neighbours. For example, although dingoes do not create nuisance barking, at certain hours they will howl in groups. While melodic to the ears of some, howling may not be appreciated by everyone. The Dingo Discovery Centre has been positioned in a valley adjacent to forest that is secluded from neighbouring properties, sheltered from the elements and permits constant observation from Lyn and Peter's house, which overlooks the enclosures.

Age of dingoes

An important consideration for anyone thinking of keeping dingoes is their extended lifespan. Captive dingoes are notoriously long-lived compared to their wild counterparts which, due to control and natural hazards, are likely to survive only to around two years of age, five years if they are fortunate. When housed in captivity the dingo can live up to 20 years – a dingo at the Centre died just shy of her 20th year. Such factors have led to a ceiling on the number of dingoes housed at the Centre because it has implications for the long-term future and viability of the operation.

Population size and density

In the wild there are clear social rules and places to hide in or space to run away during disputes. For example, the breeding pair is the head of the family and offspring are able to disperse when they are ready start their own family in their own territory. Captivity represents an unnatural situation and, if individuals have nowhere to hide from aggressive individuals or cannot disperse, the result is often death. This situation occurred for dingo ecologist Laurie Corbett at his CSIRO research station. He had a small enclosure (45 m × 25 m) housing a maximum of 16 dingoes at one stage.[23] This number was below the potential number of dingoes, but it remained low due to a behaviour he described as dominant female infanticide. Normally in the wild only the dominant female will produce pups each year, and her mate prevents the other males from getting close to his female. However, in Corbett's study group, in the captive situation the dominant bitch (alpha, in the traditional sense) let the other females breed but killed the pups once they were born. This behaviour is unlikely to occur in the wild, for two reasons. First, dominant female infanticide has not been observed in wild dingoes and has been rarely seen in other canids. In fact, there are cases where two females from the same territory on Fraser Island raised a litter each (Rob Appleby, pers. comm.). Second, the behaviour is most likely a product of captive situations that present different problems from those in the wild. In this case the enclosure size was far too small to house such a large population of dingoes. It would not be natural for multiple generations of litters to remain in the territory.

To prevent such things occurring in captivity, when a pack extends beyond a breeding pair and one generation of its offspring it is best to separate them into breeding pairs. This is what the Dingo Discovery Centre has chosen to do. Keeping them in pairs allows them to retain the ability to behave socially like dingoes would in the wild. All pups can learn the 'dingo rules' from the adults and each other. This causes far less problems than housing in family groups, which is likely to lead to distress and aggression.

Seasonal changes in behaviour

Lyn has observed over multiple generations that dingoes appear to be heavily influenced by circadian changes in their hormones, resulting in changes in behaviour and temperament as the seasons pass. This means that at particular times of year managing the dingoes can be problematic. Breeding season (from March to the end of June), for example, is the most difficult time as problem behaviours and aggression increase. That is, dingoes that have formed long relationships (pairs) protect their mates from contact with interested parties. A stark shift in behaviour occurs at this time, yet the same animals have the most nurturing and gentle mannerisms towards each other and the newborn young once the peak has passed.

Perimeter fencing and prevention of escape

For many obvious reasons there are strict guidelines/regulations regarding enclosure construction (most are government regulated and similar to those listed in Chapter 9 for pet dingoes). At the Dingo Discovery Centre perimeter fencing consists of 2.4 m high mesh with inward returns all around to preclude escape. The perimeter fencing and yard dividers are constructed of chainlink fences ~2.5 m high. All fencing has a buried extension of ~2 m underground to eliminate the possibility of dingoes digging under the fence. Internal dividing fences for galloping paddocks and area dividers are of similar construction. Galloping areas or exercise paddocks range in area from 400m^2 to 700 m^2 and incorporate hilly landscapes (Fig. 11.9).

In general, these fences manage to keep dingoes in without too much problem, but no enclosure is completely dingo-proof. If a dingo really wants to get out it often can (unless the roof is completely covered in mesh, but this is rarely possible given the expense of covering large open spaces). Almost every facility we have visited tells tales of a dingo escape. Remarkably, when a dingo escapes it rarely runs away; if it does, it returns within a few days. The reason dingoes rarely choose to escape is because the enclosure is their home. It is their territory, familiar to them and a place where they feel safe, where they have enough resources (food) and their friends and family. Most of the time they are content, and it takes a really good reason to warrant an escape. It is more common for dingoes to move within the confines of the sanctuary. For example, a particular dingo at the Dingo Discovery Centre had become frustrated because he had been separated from a dingo that he was usually housed with. So he climbed out of his pen, walked carefully across the top of the wire fence and into the enclosure of the female with which he had an established bond. This is not dissimilar to the case mentioned in Chapter 9 of a dingo opening a latched gate. Such skills in escaping are not uncommon in other species of canids. For example, Raymond Coppinger[24] was studying a captive population of coyotes that were thought to be getting out of their enclosure during the night, although each morning all coyotes were present and accounted for. Curious, Coppinger spread flour around the outside of the enclosure before he left one day and, sure enough, the next morning there were coyote prints all over the place.

Dingo housing facilities and daily routines

The Dingo Discovery Centre features two long shelter buildings where the individual enclosures are open-ended with a roof over the middle to provide cover (Fig. 11.6). They incorporate the food preparation area and an on-site hospital area as well as a nursery area for weaned pups. Constant airflow throughout the undercover area, which totals 170 m^2, is achieved via open mesh walls to east and west. Solid metal dividing walls provide shelter from prevailing hot or cold winds. South- and north-facing walls are full height and solid. This feature prevents the excessive stress that can occur when dingoes have constant eye contact with others, especially during the breeding period. During winter months clear bistro blinds are utilised to prevent cold southerly winds and rain from entering the den areas while still permitting natural light into the space. Each night den is equipped with a raised, enclosed sleeping box constructed of pine, large enough for two dingoes or a mother and litter. The flat tops of these sleeping boxes are also utilised by the dingoes.

As discussed previously, dingoes at the Centre are generally housed in breeding pairs. Most of the enclosures are 2 m × 15 m, including 12 m of flooring made of quarry rubble (gravel) and a 3 m section of concrete slab where the kennel is placed. A roof covers 6 m of the enclosure, with each enclosure separated by 1 m high sheets of corrugated iron to act as a windbreak and prevent dingoes fighting. As the puppies get older they are kept in an enclosure that is ~5 m long by 2.5 m wide. Enrichment objects (e.g. toys, bones, carcasses) are provided on an *ad hoc* basis. Adult dingoes are fed once per day – a mixture of dry dog food and fresh chicken (carcass or ground carcass mince). Juveniles and lactating females are fed a similar diet, but twice per day (Fig. 11.7).

Figure 11.6: The custom-built dingo housing at the Dingo Discovery Sanctuary and Research Centre.
Source: Bradley Smith.

When present, puppies are placed in the main exercise yard each day and allowed to socialise with different combinations of adults. The number of dingoes that are able to interact in the exercise yards varies according to each dingo, rank and season (i.e. between January and September, which covers breeding season, dingoes are housed in breeding pairs or female–female groups to avoid confrontation).

Government housing regulations detail only the required housing conditions and structures of dingoes in captivity (e.g. enclosure sizes and heights, number of dingoes per square meter). They rarely focus on the welfare or safe management of the dingoes, which are just as important. There are a few features that make captive environments more dingo-friendly and facilitate easier management.

Being social creatures, it is important they are housed with other dingoes. They are also exceptionally curious, and are more content if they can scrutinise the activity around their habitat. Several dingoes very quickly figured out how to operate bolts and simple gate latch mechanisms, so modification and more complicated gate closers had to be installed. All gates are completely child-proof via high-placed three-way catches inside the gates. All gates fit tightly so as not to allow protruding dingo parts to be nipped by other inmates passing in walkways or in adjoining pens. This also led to the use of 30 mm welded mesh rather than the previously recommended cyclone mesh, which had been demolished and destroyed by persistent dingo activity and which contributed to very serious interactive injuries. Elevated basking

Figure 11.7: It takes a lot of food and effort to keep 30 dingoes fed. Volunteer Lyn Whitworth is preparing breakfast under the watchful eye of a dingo. The dingoes are fed a combination of fresh chicken carcass mince and high-quality dry dog biscuits. The food must be meat and not starch-based.
Source: Bradley Smith.

benches (Fig. 11.8) afford a view of the dingoes in adjacent enclosures as well as into the distance. These provide positive enrichment and are popular resting places. The secure dens provide a safe and familiar environment for the dingoes, which is an important welfare consideration. Only familiar carers are permitted to enter.

To avoid a floor system of featureless concrete in the enclosures, a system of well-drained deep sand topped by crushed rock was laid down in between a mesh layer (Fig. 11.8). A single strip of concrete runs the length of the housing buildings to keep den boxes level. The porous subfloor remains clean and odour-free, helped by the animals' high-quality diet which results in small stool output. The flooring material can be easily and economically replaced for constant freshness and to fill holes caused by digging.

Exercise/activity

Every dingo is given daily opportunities to exercise and explore one of the eight large galloping areas (access to these areas is rotated) (Fig. 11.9). The grassed exercise yards included natural dens, large hollow logs, stumps, benches, trees, plastic kennels, grasses, rocky areas, water buckets and play items (e.g. bones and enrichment toys). This provides the opportunity to explore, run, dig dens and interact with other dingoes and the constantly varied

Figure 11.8: Close-up of the front section of an enclosure. Note the substrate, raised platform, barriers separating the enclosures and specially designed den boxes. At the half-way mark, the enclosure can be divided into two if animals need to be separated.
Source: Bradley Smith.

enrichment activities and objects. Perhaps the most important activity during this time is olfactory stimulation. Other dingoes, humans and the many birds, reptiles, insects and nocturnal animals that frequent the areas leave various scents that are comprehensively explored. The natural native poa tussocks (*Poa labillardierei*) on the property act as valuable landmarks in the galloping paddocks, as shelter, hiding places, jumping obstacles and scent-marking posts. Large eucalyptus trees offer ample shade and recreational activities. Large hollow tree trunks allow jumping, climbing and hiding activities.

Interestingly, despite daily access to these spacious roaming areas, the dingoes rarely choose to remain outdoors for more than one hour, and mostly they seek a return to undercover resting areas after around 20 minutes. In extreme weather conditions they often refuse to take their free-running opportunities at all.

Figure 11.9: One of eight running paddocks at the Dingo Discovery Centre.
Source: Bradley Smith.

Socialisation with humans

For dingoes raised at the Dingo Discovery Centre, socialisation with humans begins at one to three days of age, with pups handled gently for brief periods by human volunteers in the presence of their mother(s). At four to five weeks, different litters of pups spend their days together in puppy pens. From five weeks of age pups are no longer housed with their mothers at night, and interact with humans and other conspecifics (pups and adults) during the daytime. When given the chance to explore a larger yard, pups are able to interact with one set of parents. Most dingoes receive basic obedience training as pups, such as waving, dropping, crawling, begging, spinning around, running through an obstacle course, and lead training.

All dingoes receive daily human contact from a variety of volunteers (e.g. during grooming, and interactions during feeding and cleaning of pens and yard). The Centre is open to the public by request, so dingoes that either tolerate or seek human contact are able to interact with a diverse range of people on a regular basis. When there are puppies and juvenile dingoes, the Centre runs open days where small groups can interact with them (Fig. 11.3).

Breeding programs

All the captive dingoes at the Dingo Discovery Centre are DNA-tested to ensure genetic purity. Although the director controls all breeding, active selection of specific traits other than the ability to live in a captive environment is avoided as much as possible. Where possible, dingoes of different genetic lineages are bred. Breeding is conducted on rotational bases, so bitches are given the chance to breed every second or third year (Fig. 11.10).

The popular sire model (when an animal of desirable attributes is bred from repeatedly) that is used in most livestock breeding establishments is purposely avoided at the Centre. Conducting intensive dingo conservation breeding, unlike the normal methods used with domestic canines, requires an equal number of males and females to allow for contentment in natural social bonding and the greatest possible utilisation of the available genetic pool. But this population genetics model is not always possible. At the Centre dingoes are known to bond for life and will not cohabit or breed with other partners no matter how much a certain combination of lines may be desired. Attempts to bring about such unions have repeatedly failed. Thus, artificially intervening in the mechanics of mating and whelping is difficult; mating should where possible remain within the desires of the dingoes themselves.

A lot of sanctuaries and facilities may house dingoes, but they do not actively breed. They are simply kept as display animals, and thus management are forced to temporarily (chemically) or permanently neuter the animals. When the desires surrounding breeding are removed the management of dingoes is far less problematic. We are certain that many would like to breed dingoes (puppies can be a great public drawcard), but there simply is nowhere for the

Figure 11.10: A female dingo, Pepper, and her litter in a whelping box. Because the dingoes are well socialised, director Lyn Watson is able to interact with the pups and mother without causing stress or rejection of the litter.
Source: Lyn Watson.

pups to go after they mature. Unfortunately, one of the biggest drawbacks of de-sexing dingoes is that they often do not look or behave the same as dingoes that are entire. For example, they often become overweight and lethargic. The true influence and long-term health impacts are unknown. There may also be problems with temporarily chemically castrating dingoes. For instance, at the Dingo Discovery Centre one dingo was unable to sire a litter even after the chemicals wore off.

Stress in captive environments

One of the most difficult problems that affects dingoes in zoos and wildlife parks is stress. There are many things that can be a source of stress and/or fear; for example, nearby construction work, introduction of an unfamiliar dingo, a change to the enclosure or a new enclosure, loud noises and medical treatment. One of the most familiar causes is a change of keepers or management staff. A bond with the keepers must be established within the dingoes' first year (a period when they are still experiencing and interacting with the world, and forming attachments to the enclosure and the humans involved in their care), because there is little chance of a trusting relationship developing with a different keeper after this time.

When visiting facilities across the country we met several captive dingoes that were living in a constant state of fear and could not be handled by their keepers. At one sanctuary, the keepers and owners who had recently acquired it could enter an enclosure that housed a family of three dingoes, but could not handle them. This becomes a big problem. Not only is it unhealthy for dingoes to live in a constant state of fear, often obvious to the viewing public, it is very problematic when the animals need medical attention or routine care. In such cases, the keepers are forced to sedate the animal, often via a gun or a blow dart – a modern device not dissimilar to the ancient South American method of blowing a poisonous dart through a long pipe. This stresses the dingo even further, and the cycle continues.

Medical management

In terms of physiology and medical management, for the most part dingoes can be treated the same as dogs.[21] However, there are a few important things that can be missed. For example, two DNA-tested pure dingoes were sent to a zoo to go on display and be integrated into a national breeding program. Veterinarians performed a medical check on one dingo after it had injured its hip during a fall. An X-ray revealed evidence of hip dysplasia (an abnormal formation of the hip socket that can lead to lameness and painful arthritis of the joints), which is common issue in larger purebred dogs. The dingo had what they determined 'loose' hips, and the zoo chose the drastic measure of de-sexing it because it considered the dingo unsuitable for use as breeding stock. In the absence of a typical hip score for dingoes, zoo staff used the criteria developed for domestic dogs. We argue that it provides an advantage for dingoes to have looser hips than domestic dogs, whose physiology has been greatly influenced by selective breeding. Thus, spaying the animal was a serious action and certainly warranted further investigation. Other differences between dingoes and domestic dogs may also exist.

Conclusion

Private dingo sanctuaries and organisations are essential to the future of the dingo in Australia. They fulfil many important roles that are critical for promoting dingo conservation, that various wildlife organisations, public zoos and small wildlife parks are unable to achieve. They have the potential to be a powerful voice for the dingo, but only if done in the right way. That is, their actions and information must be based on sound evidence (not opinion), and the delivery or actions must be constructive (not personal or abusive). In order to achieve this, dingo organisations must work closely with scientists and governments and, perhaps most importantly, find a way to work together to achieve maximum impact. It is also evident that keeping a breeding colony of dingoes in captivity presents more challenges than keeping domestic dogs. We hope that by describing some of the methods used to effectively manage dingoes, others may learn from our experience. Keeping dingoes in captivity is not ideal but at the moment it is necessary to conserve the pure dingo gene pool and for various other conservation initiatives. Conservation efforts and dingo management is an expensive, time-consuming and often thankless task, but we remain hopeful that at some stage these organisations will receive assistance or support from the government as they might be called upon if/when the dingo becomes critically endangered.

Acknowledgements

The authors wish to acknowledge and thank the significant amount of work and dedication of those who volunteer their time to either promote dingo conservation and/or care for dingoes housed in captivity at facilities such as the Dingo Discovery Centre.

References

1. Corbett LK (2004) Australia and Oceania (Australasian): dingo (*Canis lupus dingo*). In *Canids: Foxes, Wolves, Jackals and Dogs – 2004 Status Survey and Conservation Action Plan.* (Eds C Sillero-Zubiri, M Hoffmann and DW Macdonald) pp. 223–230. IUCN/SSC Canid Specialist Group, Gland/Cambridge.
2. Zuckerman S (1953) The breeding seasons of mammals in captivity. *Proceedings of the Zoological Society of London* **122**, 827–950.
3. Crandall L (1964) *The Management of Wild Mammals in Captivity.* University of Chicago Press, Chicago.
4. Wilton AN (2001) DNA methods of assessing dingo purity. In *A Symposium on the Dingo.* (Eds CR Dickman and D Lunney) pp. 49–56. Royal Zoological Society of New South Wales, Sydney.
5. Newsome AE, Corbett LK, Carpenter SM (1980) The identity of the dingo. I. Morphological discriminants of dingo and dog skulls. *Australian Journal of Zoology* **28**, 615–625.
6. Corbett LK (2001) *The Dingo in Australia and Asia.* JB Books, Adelaide.
7. Elledge AE, Allen LR, Carlsson BL, Wilton AN, Leung LK (2008) An evaluation of genetic analyses, skull morphology and visual appearance for assessing dingo purity: implications for dingo conservation. *Wildlife Research* **35**, 812–820.
8. Letnic M, Ritchie EG, Dickman CR (2011) Top predators as biodiversity regulators: the dingo *Canis lupus dingo* as a case study. *Biological Reviews of the Cambridge Philosophical Society* **87**, 390–413.

9. Dickman CR, Glen AS, Letnic M (2009) Reintroducing the dingo: can Australia's conservation wastelands be restored? In *Reintroduction of Top Order Predators*. (Eds MW Hayward and MJ Somers) pp. 238–269. Blackwell Publishing, Oxford.
10. Ripple W, Beschta R (2003) Wolf reintroduction, predation risk, and cottonwood recovery in Yellowstone National Park. *Forest Ecology and Management* **184**, 299–313.
11. Falk JH, Reinhard EM, Vernon CL, Bronnenkant K, Deans NL, Heimlich JE (2007) *Why Zoos and Aquariums Matter: Assessing the Impact of a Visit*. Association of Zoos and Aquariums, Silver Spring, MD.
12. Schenkel R (1947) Expression studies of the wolf. *Behaviour* **1**, 81–129.
13. Mech LD (1970) *The Wolf: Ecology and Behaviour of an Endangered Species*. Natural History Press, New York.
14. Klinghammer E, Goodmann PA (1987) Socialisation and management of wolves in captivity. In *Man and Wolf*. (Ed. H Frank) pp. 31–59. Dr W. Junk Publishers, Dordrecht.
15. Parsons MH, Blumstein DT (2010) Familiarity breeds contempt: kangaroos persistently avoid areas with experimentally deployed dingo scents. *PLoS ONE* **5**, e10403.
16. Catling PC (1979) Seasonal variation in plasma testosterone and the testis in captive male dingoes, *Canis familiaris dingo*. *Australian Journal of Zoology* **27**, 939–944.
17. Edgar JP, Appleby RG, Jones DN (2007) Efficacy of an ultrasonic device as a deterrent to dingoes (*Canis lupus dingo*): a preliminary investigation. *Journal of Ethology* **25**, 209–213.
18. Miklósi A, Topál J, Csányi V (2004) Comparative social cognition: what can dogs teach us? *Animal Behaviour* **67**, 995–1004.
19. Woolpy JH, Ginsburg BE (1967) Wolf socialization: a study of temperament in a wild social species. *American Zoologist* **7**, 357–363.
20. Breckwoldt R (1988) *A Very Elegant Animal: The Dingo*. Angus and Robertson, Sydney.
21. Jackson S (2006) Dingoes. In *Australian Mammals: Biology and Captive Management*. (Ed. S Jackson) pp. 381–407. CSIRO Publishing, Melbourne.
22. Nixon A (2008) *Bush Aussies: True Stories of Australian Country Characters*. Penguin Australia, Melbourne.
23. Corbett LK (1988) Social dynamics of a captive dingo pack: population regulation by dominant female infanticide. *Ethology* **78**, 177–198.
24. Coppinger R, Coppinger L (2001) *Dogs: A New Understanding of Canine Origin, Behavior, and Evolution*. University of Chicago Press, Chicago.

12
Forging a new future for the Australian dingo

Bradley Smith and Rob Appleby

Almost every major aspect of discussion surrounding the dingo is steeped in controversy. These debates range from its origin and species status, its identity as a native or introduced animal, its effects on the livestock industry, and its ecological role. In the midst of all this debate, perhaps the overarching dichotomy, and the one that may ultimately serve as a signpost for the dingo's future, is whether dingoes should be killed or conserved. Throughout this book there has been a clear sentiment in favour of dingo conservation and, we believe, with very good reason. But no matter which side of a debate you prefer, there is little doubt that the resultant turmoil has left the dingo to face an uncertain future. The purpose of this book has been to summarise the most recent knowledge we have on the dingo, drawn from a wide array of fields of research, and to aid in cultivating informed opinions based on measured discussion of the many controversial and hotly debated topics. In this final chapter we have chosen to emphasise these topics and issues, as they are likely to ultimately decide the fate of the Australian dingo. We also aim to provide a culmination of these discussions and explore what, if anything, might be done to forge a new and sustainable future for the Australian dingo.

The role of discourse in dingo conservation

Alan Newsome, the former lead CSIRO researcher on dingoes and arguably one of Australia's foremost authorities on dingoes, foretold many of the problems that we are now facing in terms of progressing towards dingo conservation. During a brief 1978 CSIRO video production, *Yellow-dog Dingo*, he had this to say:

> The question we would all like answered of course is whether dingoes are goodies or baddies. Looking at these lot [holding dingo pups] it's a bit hard to imagine that they're baddies. But I have seen one sheep man in this mountain country, driven out by the effects of one dingo on his flock. On the other hand we have all seen, those of us who have worked on dingoes, have seen dingoes very abundant in parts of inland Australia and doing no perceptible damage to stock at all. It seems to depend very much on circumstance: how much native game there is, whether it's a drought or whether it's good times. The difficulty is that there is no real, no one answer for dingo management in the scientific sense for any place in Australia and no one answer therefore, for all of Australia. Trying to predict whether dingoes are going to be a nuisance is a bit like predicting lightning strike. The difficulty is also trying to prevent it. Management appears to be, will have to be, a rather more flexible affair than we imagined, because there is

just no simple answer. And management will presumably have to change from place to place, and from time to time.

The age-old battle lines between dingoes and the livestock industry are well-known and still very much in existence. Newsome spoke in the video before the disappearance of Azaria Chamberlain and the negative human–dingo interactions that have since occurred in places such as Fraser Island, and long before debates on the dingo's ecological role. But the sentiment of his words is equally relevant in these more recent forms of conflict. Hytten[1] conducted a discourse analysis in relation to the dingo (discourse is defined as the words, phrases, metaphors, images, ideas, thoughts and exchanges around a particular subject). Often, by analysing and exploring discourse, for example those surrounding wildlife, we are able to gain some insight into how people construe nature and how this may influence human–wildlife interactions and perceptions. Hytten highlighted some of the key opposing or contradictory views (which she labelled 'dualisms') on existing perceptions about dingoes. That is, she found that no matter what topic, there are almost always two diametrically opposed ways of looking at them. Newsome captured this idea perfectly with his reference to dingoes as 'goodies or baddies'. Such opposing viewpoints also often represent a line in the sand when it comes to dingo conservation, and the discourse that surrounds them serves as a worthy way to examine the various positions taken and what this means for dingo conservation. The next section will explore some of these important dualisms.

The quandary of pedigree: pure or hybrid?

According to Breckwoldt,[2] one of the dingo's worst features is that it looks too much like a dog. This makes conservation difficult for several reasons. For instance, dingoes may not seem important to conserve because they are 'just' dogs and therefore do not belong in the Australian environment. Their likeness to dogs also makes it difficult to discern between pure and hybrid dingoes, and raises questions about what exactly it is that we should be conserving. Yet, as we hope many chapters in this book point out, the dingo is unlike any other native Australian species, or canid in the world for that matter. It is a truly unique animal, whose value can be measured in myriad ways including historical, cultural, ecological, scientific, economic (e.g. as a tourism drawcard or suppressor of agricultural pests) and intrinsic contexts.

In a sense, the dingo's status as a native animal deserving of protection hinges upon its genetic purity. That is, dingoes are often seen as contaminated through interbreeding with domestic dogs, which is now considered a major threat to the future of the pure dingo. Despite Breckwoldt's suggestion that the dingo is remarkably similar to dogs, it is important to emphasise that, although often subtle, there are many differences between a pure dingo (i.e. one that has no modern dog genes) and a dog or hybrid, as well as that most archetypal of wild canids, the wolf. For example, there are many physiological (Chapter 1), biological (Chapters 1 and 2), ecological (Chapter 8), and behavioural differences (Chapters 1, 8, 9 and 10) between dingoes, dogs and wolves. Neither should the subtlety of some of the differences be underestimated, as there are virtually no agreed-upon methods for determining the purity of a dingo. Even genetic testing often relies upon assumptions about a baseline for comparison, that are open to questioning. Yet we can say with some assurance that each difference, be it subtle or obvious, is likely to play an important role in a dingo's survival in the wild. For almost 5000 years in Australia, however you define the dingo, it has primarily been shaped by natural selection. In this sense, the Australian dingo could well be considered an ecotype or ecospecies – one that has adapted to Australia's specific environmental conditions. Note that this differs from the more traditional biological definition of a species as one that cannot produce viable offspring when mated with another species. As wolves, domestic dogs and dingoes are capable of interbreeding and producing viable offspring, they are not usually considered as separate species. Yet few would argue that a poodle is the same thing as a dingo or a wolf. We take the argument further and state that the latter two are also obviously different, having been shaped and moulded by the different natural environs in which they have lived over the millennia that they have been separated. And if not for European settlement, it's fair to say this major process that often leads to speciation, namely geographic isolation, would have persisted between these canids as well. Wolves and dingoes are unlikely to meet anytime soon, so wolf–dingo hybridisation seems highly unlikely, but there is no doubt that since Europeans arrived there have been, and continue to be, plenty of opportunities for dogs and dingoes to meet and produce hybrid offspring. An important question for conservation then becomes exactly 'what is it that we want to conserve?'. If we use natural selection as a conservation yardstick, then we'll probably try to limit hybridisation wherever possible and attempt to preserve something akin to what we imagine the dingo was like before European arrival. In some cases, it may simply be too difficult to do this. We may have to concede that the animal that has come to replace the dingo, if it fits a similar role or niche, may be analogous enough to be worth conserving. No doubt we should still aim to limit the release of domestic dogs into, or the prevalence of feral domestic dogs among, populations of dingoes to the greatest extent possible, but this is likely to be contingent upon finding suitable and agreeable methods for distinguishing between them. If major dingo reintroductions to parts of their former range were to become a reality, it would appear to make the most sense that founders be drawn from populations that most closely resemble the pre-European equivalents for a given part of the range.

We use the notion of purity to illustrate what we believe is a major potential impediment to dingo conservation: our often quarrelsome perceptions about dingoes. In particular, our perceptions about what dingoes are and what they do influence how we think they should be managed. The naming or grouping of dingoes, hybrids and feral dogs poses another interesting scenario. Dingo hybrids (to any degree of dingo ancestry) are almost always grouped with wild domestic dogs and attributed no value. According to Hytten's[1] analysis, in this sense dogs are situated within culture (and have no place or value in the bush) and dingoes are situated within nature and as a consequence do have value in the bush. Hybrids occupy the overlapping space between culture and nature and are therefore not considered to have a legitimate place in either. Indeed, with only a little contrivance, hybridisation can be used as grounds for ongoing persecution of dingoes under the auspices of conservation, because it leverages the fact that getting rid of hybrids in favour of dingoes is probably preferable to most people despite our inability to immediately tell them apart, leading to an obvious predicament.

Perhaps people are often too wrapped up in labels, to the detriment of other, more important aspects such as ecological function. Some might even argue that the boundary between dingoes and dingo hybrids is academic and not significant outside science, as it is defined by as yet unverified criteria. Indeed, it is possible that hybridisation is of little consequence to the resultant individuals, as hybrids may survive and reproduce equally in the wild, and may fulfil the ecosystem functions that dingoes perform equally as well. Perhaps conservationists and scientists are simply trying to preserve their own pet categories and constructs? At the same time, the few pockets where interbreeding has been kept to a minimum represent important arks where dingoes are seen to more closely resemble what dingoes must have been like before European settlement. In the case of purity, it seems clear that if we can't decide what a dingo is and, in turn, isn't, then we are likely to run into problems when it comes to deciding what it is we are trying to conserve and indeed whether it should be conserved at all. As we'll see in the following sections, notions of purity are just one of several dilemmas that must be wrestled with in progressing towards dingo conservation.

Dingoes as pests or protected species

Since the arrival of European settlers, the dingo has been treated and managed as a major pest (see Chapter 5), evident in the various Acts and pieces of legislation condemning it as such. Yet across the country legislation surrounding dingoes has become deeply conflicting. In some states they are still legally classed as pests and in other cases they are legally protected. As Table12.1 illustrates, nowhere are dualisms more prominent than in legislation. Current legislation in Australia relating to dingoes remains disparate, contradictory and confusing, with each state and territory responsible for developing its own system of management.

Internationally, according to the International Union for Conservation of Nature (IUCN) Red List of Threatened Species, the dingo is listed as vulnerable to extinction in the wild.[3] This reflects the fact that although populations of dingoes exist in the wild, the proportion of pure dingoes is believed to be declining as a result of hybridisation with their feral or domestic cousins. This classification currently has no impact on the management of dingoes in Australia, making the designation something of a 'paper tiger'.

Nationally, the dingo is regarded as a regulated native species under the *Environmental Protection and Biodiversity Conservation Act 1999*. This means that the dingo is legally protected in all Commonwealth national parks and reserves, as well as World Heritage and other protected areas. However, this Act also provides for dingoes to be controlled in areas where they are having a demonstrated impact on native ecological communities and, in some cases, where there are negative impacts on agriculture. This conservation-with-a-catch approach almost always sees the dingo lose to competing interests, marginalising their status as a protected species.

As Table 12.1 demonstrates, at a state and territory level, certain key pieces of legislation appear to directly clash with one another (as well as with the EPBC Act), undoubtedly leading to ambiguities in relation to management. For example, in several states and territories, dingoes are a protected species in national parks and nature reserves, in line with the EPBC

Table 12.1. The current legal status of dingoes across Australia, with the exception of Tasmania where there are no dingoes[4]

State/territory	Classification	Relevant legislation	Protection status
Australian Capital Territory	• Protected species	• *Nature Conservation Act 1980*	• Permit required for any control activity
New South Wales	• Threatened species • Pest species	• *National Parks and Wildlife Act 1974; Threatened Species Conservation Act 1995* • *Rural Lands Protection Act 1998*	• Protected in national parks • Government responsible for management on public lands
Northern Territory	• Protected (regarded as indigenous and legally protected)	• *Parks and Wildlife Conservation Act 1993*	• Dingo control occurs, but only if part of a managed and approved program • Landholders not obliged to cull
Queensland	• Pest species	• *Rural Lands Protection Act 1985* • *Nature Conservation Act 1992*	• Landholders obligated to control on their land • Legal protection as a natural resource in protected areas such as national parks
South Australia	• Pest species (south of the barrier fence) • Wildlife species (north of the barrier fence)	• *Animal and Plant Control Board Act 1986* • *Animal and Plant Control Commission 1933; Dog Fence Act 1946*	• Landholders obligated to control on their land • Unprotected, but some protection e.g. control restricted to beyond 35 km of dingo fence
Victoria	• Protected • Pest species	• *National Parks Act 1975* • *Catchment and Land Protection Act 1994*	• Managed through the Wild Dog Policy • Landholders obligated to control on their land
Western Australia	• Declared animals • Unprotected fauna	• *Agriculture and Related Resources Protection Act 1976* • *Western Australian Wildlife Conservation Act 1950*	• Landholders obligated to control on their land • Generally not subject to control in fauna reserves or parks. Control confined to livestock areas
Tasmania	• Prohibited	• *National Parks and Wildlife Act 1970*	• Dingoes have never colonised Tasmania, and the import of dingoes is prohibited • Feral dogs are managed under the *Dog Control Act 1987*

Act, yet in states such as New South Wales and Queensland once dingoes set foot (or paw) outside one of the protected areas they are immediately considered a pest and are no longer afforded any protection. In fact, landholders are obliged to kill dingoes found on their land. This is not limited to sheep growing areas. Anywhere outside of a national park or similar reserve, technically speaking, virtually all land owners are required to lethally control dingoes, even when and where no conflict is occurring. Important dingo populations inside reserves are therefore effectively imprisoned within unseen and all-too easily breached park boundaries. It leaves them to exist as isolated subpopulations with limited prospects for dispersal to other reserve areas and, as a result, increases their vulnerability to extinction processes. Any dingoes that dare to leave such areas face myriad control measures designed to stop them in their tracks.

Along with the often invisible (at least to dingoes) park and reserve boundaries that effectively lock populations into a few scattered geographical islands, there are other spatial boundaries that feature in dingo discourse. For example, dingoes might be considered acceptable north of the dingo barrier fence, whereas to the south they are vilified. Neither does the dingo appear to belong on private land, as people have certain expectations of the landscape and their rights in relation to it; in this sense, the only rightful place for dingoes is in national parks or other public reserves (short of private sanctuaries, of course). Yet, as mentioned above, even in national parks, management policies carry notable caveats that include lethal control options for dingoes that interfere with livestock in adjacent properties, or interact with people in such a way that they are deemed to represent a risk to human safety (e.g. Fraser Island; Chapter 6).

Native or introduced?

There has been a long-running debate on whether the dingo is native to Australia.[5] Conserving native species is one of the central aims of conservation and this is acutely recognised in Australia, given our poor record of extinctions to date.[6] Thus, the question of whether dingoes are native is an important one, and has come under close scrutiny. There are a few ways to define a native animal. One of the more common and accepted definitions includes any animal that was here before European settlement. Chapter 3 covers the dingo's origin and ancestry; as we saw, evidence suggests that dingoes arrived in Australia with seafaring people long before European settlement. It is clear that dingoes would be considered native under this definition. However, the fact that dingoes were 'introduced' by humans at all seems to remain a salient point for some. According to Hytten,[1] there are often conflicts between people's classifications of nature and culture. In this case, the fact that humans transported the dingo, regardless of when, how or in what state, makes them cultural artefacts, potentially reducing their authenticity as part of nature and thus making them less valuable.

The use of the term 'feral' in many references to dingoes is interesting as it implies, following the definition, that they have escaped domestication and become wild. This may be true in some regard but, as Chapter 3 outlined, we do not know the extent to which dingoes were domesticated before they arrived in Australia, and they certainly didn't appear to be further domesticated once they were here. Also of note is that the dingo has been living wild in Australia for just about as long as it has supposedly been living as a domesticated canid in Asia. Its status as 'feral' is therefore contingent upon assumptions about unascertained levels of domestication that occurred millennia ago. The fact that dingoes have been in Australia for at least 4000 years, as opposed to foxes, cats or domestic dogs for that matter, that have been in Australia for only 200 or so years, is of seemingly no significance to some in this particular discourse.

One of the implications of classifying dingoes as feral is that, under legislation, they must be controlled. The term feral is also used to support the position that dingoes should be exterminated, not only to protect livestock but also in some cases to protect Australian wildlife. For instance, the feral discourse is used to question the protected status of the dingo in national parks: if feral, dingoes are not naturally occurring and thus have no place there (see arguments by Coman and Jones).[7] For many ecologists, and undoubtedly many Australians, the suggestion that the dingo is something other than native is absurd. Dingoes have occupied a significant ecological niche as apex predators for several millennia and have established important functional relationships with a large number of species in a wide range of habitats (see Chapter 8). While the feral discourse is probably less commonly espoused than certain other discourses, it persists enough to be a thorn in the side when it comes to dialogue about dingo conservation.

Functional value versus all other values

Since Hytten's work, a new dualism has emerged: one that tends to see dingoes as the saviour of other native species because they limit the impacts of introduced mesopredators (cats and foxes), or as of little value and/or equally threatening in this regard. This functional discourse sees dingoes expressed as somewhat of a biological control tool, where their value is linked to this role. This notion of dingoes as an ecological or biological control mechanism can be extended beyond the usual suspects in this regard: cats and foxes. A poignant example occurred on Townshend Island, off the coast of Queensland. Twenty dingoes were introduced to the island in an effort to reduce a population of ~2000–3000 feral goats.[4, 8, 9] The dingoes were successful in removing 99% of goats within a year and a half, and all goats within three and a half years. The dingoes served admirably in their function as a goat eradication tool, but once the goats were gone the dingoes ceased to have value and eradication efforts turned in their direction. The most recent surveys suggest that few if any dingoes now survive on the island. Such an example leaves us to ponder: if areas overrun with foxes and cats were successfully brought under control with the help of dingoes, would a similar fate await the dingoes following their service?

Hinging conservation on some utilitarian view of dingoes seems prone to problems because value is directly associated only with their function. This is because the absence of any perceived function can also be viewed as an absence of any value, as was the case on Townshend Island. While there is evidence that dingoes suppress cats and foxes in some circumstances, in areas where such suppression is deemed more limited and if this is the only value attributed to dingoes, there will be no impetus to conserve them. In fact, the likely argument will be to affect control practices that either ignore the presence of dingoes (e.g. baiting that indiscriminately targets any predator) or actively seek to remove them. We argue that the value of the dingo in suppressing mesopredators or other species such as rabbits, when and where this occurs, is an additional value to its other important ecological, cultural and intrinsic values and is certainly not the only positive consequence of its persistence in Australian environs. Some more specific elements to this particular debate are covered in greater detail a little later in the chapter.

Confirming the dingo's ecological role

While we've argued above that there is much more to value about the dingo than its potential to suppress invasive species such as cats and foxes, its role in this respect remains an important consideration in overall native animal conservation strategies and may be an additional incentive in terms of its own conservation. Chapter 7 showed that, despite ongoing debate and some concerns about methodology,[10] there is mounting support for the positive ecological role that dingoes play in relation to other native species in the critical weight range (species with average bodyweight in the range 35–5500 g) through, for example, the suppression of introduced mesopredators such as cats and foxes, as well as large herbivores such as kangaroos. Although such support continues to grow it is far from unequivocal, and debate rages. The logical next step in resolving the debate is to use more robust methods in experimentally manipulative studies, a point that would probably find agreement across the board. Here, 'manipulative' often infers removal (i.e. dingoes are removed through lethal control from certain areas and associated effects are compared with areas without removal), but experiments need not be centred on removal. In fact, given that a good portion of the debate revolves around the reintroduction of dingoes, it seems the most appropriate manipulation is to reintroduce dingoes to experimental areas and determine, with a high degree of accuracy, what happens afterwards. This could mirror the work of Moseby and colleagues[11] and use telemetry devices in what might be one of the better methods for exploring relationships between the species of interest: known-fate models. This largely avoids the detection probability issues flagged by Allen and colleagues[12] and Hayward and Marlow[10] because, as the name implies, the fate of every marked (using telemetry devices) animal is known exactly. Of principal interest in known-fate modelling is the survival rate of marked individuals, so it is particularly apt in terms of a manipulative experiment involving reintroduction of dingoes and a marked sample of species including foxes and cats, but also critical weight range animals (which are at the centre of the debate about mesopredator release theory in this context) as well as herbivores such as kangaroos. Telemetric data can also be used to explore spatiotemporal relationships in movement patterns, which would help further elucidate behavioural responses of other species to dingo reintroductions. Consideration would need to be given to adequate replication, the establishment of suitable control sites and other important aspects of experimental design, as for any experiment of this kind.

Hayward and Marlow's[10] arguments regarding the methodological flaws in the reliance on sand-plot or related indices were sound and they concluded that: 'it would be a brave conservation practitioner who leaps at the recommendations proffered' (p. 3). However, there are two interpretations of this point. One is that it would, at present, be foolhardy to reintroduce dingoes to parts of their former range. Granted, it might be difficult to argue for wide-scale reintroductions solely on the basis of existing data, although undoubtedly such an argument could be made. However, the status quo of pursuing dingo eradication or lethal control, which clearly isn't working as intended, hardly constitutes an ideal solution either. Perhaps this is another way to interpret Hayward and Marlow's[10] point above, and is an interpretation we would tend to be more supportive of. Reintroduction experiments could be a key element in resolving the issues surrounding the ecological role of dingoes, particularly in arid environments where critical weight range mammals are under considerable pressure and there is already some evidence that supports the suppressive role of dingoes in relation to mesopredators.

We raise these ideas here because we feel that despite compelling information about the positive role of dingoes in relation to mesopredator release theory (see Chapter 8), ongoing contention erodes confidence in management decisions that might otherwise see dingoes better-protected in certain areas. As Hayward and Marlow[10] and others have pointed out the stakes are high; we argue that they are high enough to warrant the considerable investment of time and money necessary for such research endeavours to proceed.

Scientific research

In order to help governments and management agencies make informed and balanced decisions, it is imperative that defensible, scientific evidence is made available to them. Fortunately, the availability of scientifically credible information is growing due to a significant rise in research studies on the dingo conducted over the past decade (Fig. 12.1). From the 1970s to the 1990s, most research studies on the dingo were primarily funded by livestock industry interests such as Meat and Livestock Australia (an organisation largely representing farmers) under the auspices of the Australian government's primary research organisation, the CSIRO. Reflecting the interests of the funding sources, such research was heavily directed at destruction and eradication (e.g. how effective are certain dingo control measures and how can knowledge of dingo ecology and behaviour improve the effectiveness of such control measures?). The emergence of more diverse funding opportunities, the broader research agendas of funding bodies, the growing number of independent scientists conducting dingo-specific research, our burgeoning understanding of dingo ecology and behaviour and improvements in survey methods and technology have seen a shift in research culture towards understanding the ecological role of dingoes and, increasingly, towards their conservation. Where once it may have been difficult for a scientist to speak up in support of dingoes (e.g. their important role in maintaining biodiversity) or get funding for such research, it is becoming more common to see research reports extolling the virtues of their presence in the Australian environment, backed up by empirical support. This was most apparent in 2013, when a team of ecologists led by Professor Chris Johnson won the prestigious Australian Museum's Eureka Prize for their research involving dingoes. The dingo has never been in a better position in terms of scientific focus, particularly with a conservation bent. Ensuring that funding for such research continues to be available, and getting people to actually listen to the findings and resultant recommendations, is the next challenge.

In terms of dingo management, current practices tend to underestimate dingo cognition and behaviour and often instigate (usually expensive) management strategies with limited engagement with scientists. For example, a lot of money has been spent on fencing infrastructure on Fraser Island to keep dingoes out of townships and resorts. As Chapters 5 and 6 stated, fencing can play an important role in reducing human–dingo conflict, including on Fraser Island. However, Ecosure[13] concluded that fencing has done little to reduce negative human–dingo interactions overall. This may be because most interactions occur outside fenced areas, but an interesting aside is that even where fencing exists dingoes have used their considerable

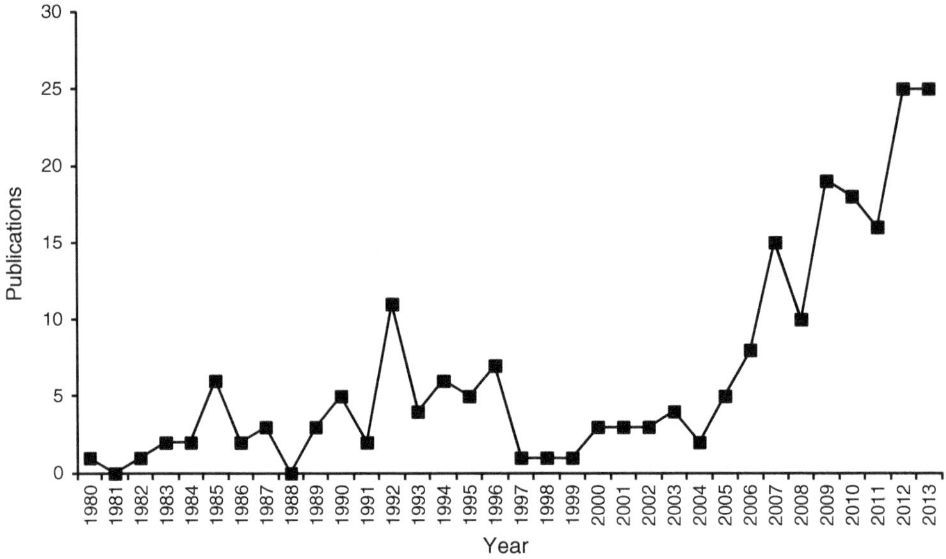

Figure 12.1: The number of scientific publications per year relating to the dingo, 1980–2013. Data sourced from Scopus, the largest abstract and citation database of peer-reviewed scientific literature (keyword search = 'dingo').

Note: This graph does not represent all scientific research that was conducted during this time.

problem-solving skills (Chapter 9) and learned, in some cases, to circumvent the barrier effect of fencing. For example, the modified cattle grids at vehicle entrance points to townships proved no match for some dingoes, who adeptly negotiated grids, forcing managers to further discourage breaches through electrification. People leaving access gates open haven't helped, as they've readily been exploited by ever-opportunistic dingoes. Note that we are not saying that fencing is ineffective at keeping dingoes and people separated in certain key places, but it is only one element in the overall approach to dealing with incident dynamics on Fraser Island and it may ultimately be a limited one. Where funding for research or management is restricted, the associated cost of fencing might need to be weighed against other important aspects of research or management on the island, at least from this point forward. As we come to better understand elements of dingo behaviour and cognition (e.g. their problem-solving prowess), management agencies will benefit from incorporating such knowledge into their decision-making processes and consequently adapting their management initiatives.

What's the reality of dingo conservation in the future?

Is there a realistic future for the dingo in Australia? There has been a lot of talk about dingo conservation but as yet not a lot has been done in this regard. Even on Fraser Island, where the dingo could not be afforded greater legal protection, a number are killed each year. This shows that even in a World Heritage National Park protection is at best conditional, despite it arguably being one of the most important dingo populations in existence. In fact, to date there have been no research or management initiatives directed specifically at dingo conservation on the island. This includes, for example, monitoring the genetic diversity and integrity of the island's dingo population generally (or related pursuits such as population viability analysis), the impacts of humane destructions in this regard or the comprehensive monitoring of diseases, which represent a considerable threat to the long-term persistence of the population. In other national parks dingoes might be afforded some legal protection, but they are essentially left to their own devices and nothing is actively done to ensure their long-term future. If the resident populations are small and isolated (such as on Fraser Island), they are extremely vulnerable. It may ultimately be fence-sitting, in relation to implementing conservation strategies, that threatens such populations, which surely represent some of the last of exemplar collections of dingoes in the wild.

Moving forward ...

We propose a new perspective on dingo management. Over 200 years of dingo control has not fixed the conflict between dingoes and livestock (see Chapter 5), so it seems erroneous to continue on the same path. Killing dingoes, especially indiscriminately, doesn't necessarily equate to resolving problems such as depredation, nor will it necessarily reduce risk to humans in tourism areas. Put another way, lethal control may not be an effective management strategy in at least some circumstances. Rather, it may be what happens when other management strategies fail or cannot be implemented. This might particularly be the case in human–dingo interactions in national parks, but it also rings true in some senses when it comes to livestock conflict. Collectively we have the opportunity to change, and to be more innovative in our approach to dingo management. For example, we believe that if graziers shift away from killing dingoes wherever possible and focus instead on saving livestock, then a far greater suite of options is available (see Chapter 5). A shift from the polarising dualisms described earlier can, at least in part, be achieved with level-headed discussions underpinned by sound scientific evidence. This might help establish some compromises in how, when and where we manage and conserve dingoes, duly acknowledging that they do cause considerable problems for some stakeholders such as livestock producers, but at the same time are a vital and iconic top predator deserving of protection. In addition to the actions already mentioned throughout this chapter (and indeed the book), we propose several key elements that we consider vital in forwarding the cause of dingo conservation.

Recognise the dingo as a species in decline

The first step in ensuring the longer-term survival of dingoes in the wild is acknowledging at both state and federal level that the dingo is a species in decline, with formal nomination to reflect this. Currently, Victoria is the only state that recognises the dingo as a species vulnerable to extinction, in line with the IUCN designation. Numerous attempts have been made to have the dingo federally nominated in one of the categories of protection under the EPBC Act, the two most likely being as a 'vulnerable' or 'conservation dependent' species. As yet, none have proven successful. To achieve greater protection for the dingo, there needs to be national consistency relating to the treatment, status (scientific name, proper recognition as a native species and protection) and accordingly the management of dingoes, both in the wild and in captivity.

Establish a national recovery plan for the dingo

Building upon the research of people such as Stephens,[14] who examined thousands of wild dog genotypes from around Australia, viable conservation, recovery and reintroduction sites should be established along with consummate protection and monitoring mechanisms. Such sites would form the basis of the national recovery plan. Where initial indications of purity are strongly supported, protection measures should include procedures for limiting exposure of the resident dingo populations to domestic dogs. An example is policing (contingent upon the establishment of regulatory powers) national parks and reserves against visitors who ignore bans on domestic dogs. For example, this continues to be a problem on Fraser Island, where trail cameras consistently detect irresponsible visitors walking with domestic dogs. On islands popular with visitors, spot checks at barge crossing points, coupled with cooperation from barge companies, could help to prevent visitors with dogs even getting to the island. National parks in, say, the Tanami Desert, are much more difficult to police to a similar end. Education programs and interpretive strategies targeting visitors (e.g. online at park websites, at park headquarters or at park entrances) could prove useful in encouraging responsible behaviour within national parks, as might opportunistic infringement procedures. In an age of social media, people's reachability is constantly increasing; social media could prove a useful means of spreading such messages. Echoing Corbett,[15] landowners who border national parks and reserves could be approached about de-sexing/neutering domestic dogs and locking them up at night (with exceptions such as guard dogs, which could still be neutered). Within the parks, random genetic sampling (e.g. of scats) to monitor introgression should be undertaken to check that the genetic integrity of populations is being maintained. Where results indicate hybridisation, management strategies should be implemented to limit or halt further introgression.

In wild dog populations where purity is deemed to be limited, decisions about what to do are likely to relate to whether there are realistic prospects for recovering a pure population, and/or whether the resident wild dogs are ecologically equivalent to pure dingoes. Limiting exposure to domestic dogs is likely to be of value even where hybridisation is well established. Reasons include preventing the introduction of disease and preventing levels of hybridisation leading to physiologically or behaviourally divergent wild dogs that are no longer functionally or ecologically equivalent to dingoes.

We still need to be realistic and recognise that dingo control is going to continue in areas of livestock production where there are high levels of conflict but, where lethal control measures are unavoidable, priority must be given to developing and utilising methods that ensure the greatest possible level of animal welfare. Because sentient animals have the ability to suffer (in the same way as humans or domestic dogs), this has implications for the way in which we treat them. Only the most humane methods of lethal control should be utilised, along with procedures for monitoring the appropriateness of techniques and those who implement them. Wherever the opportunity exists, continued research and evaluations of non-lethal control measures are required (e.g. livestock-herding animals, aversive conditioning, strategic land use, e.g. not farming highly susceptible livestock next to national parks) and incentives should be established for landholders to introduce such practices or modify their livestock production. Financial incentives (e.g. grants, tax exemptions, reimbursements) could encourage farmers to reduce or cease dingo control and/or initiate non-lethal control measures (e.g. the use of livestock-guarding animals). This may include schemes to reimburse farmers for stock losses caused by wild dogs, in a similar vein to the US defenders of wildlife.[16] By mirroring such endeavours, Australian conservation groups interested in promoting and supporting dingo conservation could find themselves bridging a divide and encouraging a new dialogue with what might be considered traditional adversaries.

Promote public awareness and education, and change dingo discourse

Corbett[15] closed his comprehensive book on the dingo by discussing the future he envisioned for it: 'in the end, their chances of continued survival in the wild will rest solely on the efforts of an informed public to stop contact between dingoes and domestic dogs, and to take pride in dingoes as native species'. Corbett constructed the dingo a certain way: a native species that is not the same as a hybrid. In a later article, he and a colleague questioned whether it was possible to stop hybridisation, or even if it was all that important provided ecological roles were maintained.[17] It appears that the veteran dingo researcher was himself susceptible to a dualism, or perhaps he shifted his viewpoint to a position of accepting that introgression was an unstoppable force.

It seems that when it comes to wildlife management in Australia, species appear to fall into only two categories: they are either a pest/overabundant species or are a species under threat of extinction. In between, we appear somewhat adrift. In regard to the dingo, it is important that we become aware of our underlying contradictions by critically deconstructing our discourse, particularly when it informs our actions in relation to conservation. We need to reconsider how we conceptualise nature (including the dingo) and avoid being overly human-centric in the way we approach the world. Ultimately, this means we must acknowledge that society is a part of nature, not apart from nature.[1, 18] Humans probably had a lot to do with the initial shaping of the dingo, and are assumed to have had a hand in bringing it to Australia so many centuries ago. It is a certainty that we are crucial in shaping the dingo's future. It is important that dingo and wildlife conservation organisations, scientists, media and governments actively engage in appropriate and accurate education and discussion about dingoes.

A national consortium on the dingo

In 1999, the Royal Zoological Society of New South Wales convened a symposium on the dingo to bring various stakeholders together to discuss whether the dingo qualified as a threatened species under the New South Wales *Threatened Species Conservation Act 1995*.[19] The conference was successful in bringing together those with a vested interest in dingoes and in beginning discussions relating to dingo management and conservation. It's disappointing that in the 15 years since, not much in the way of a follow-up conference has taken place. It is imperative that those with an interest in dingo management and conservation, such as state and federal governments, farmers, conservation organisations, environmental lawyers and scientists (geneticists, ecologists, ethologists) meet in a similar spirit to the 1999 symposium.

This would allow a diverse array of stakeholders to be heard and to discuss innovative approaches to dealing with conflict and resolving long-held discord in relation to the various points of discourse outlined above. Such a meeting would also allow interested parties to discuss and develop either an outright nomination for federal protection for the dingo, or assemble supporting information for such a nomination.

Conclusion

As shown throughout this book, the dingo is highly adaptable, intelligent, resourceful and hardy. Despite unrelenting extermination efforts since the arrival of Europeans, it has managed to survive in some form throughout much of Australia. In a sense, dingoes are victims of their own success and the reasons that some have come to so admire them are the very same reasons that have led others to revile them. Ideally, if we can learn to co-exist with the dingo, we can all benefit. The dingo's future ultimately depends on how we – governments, landholders, the media, scientists, conservationists, the public – choose to see them and how this influences our views on how to manage them.

We believe that the arguments presented in this book are grounds for a strong reconsideration of wild dog and dingo management in Australia, away from the demonstrably unsuccessful and ecologically and economically unsustainable pursuit of dingo eradication, and more towards sustainable coexistence models that favour non-lethal approaches to conflict resolution. While this paradigm shift is conservation-centric, it also represents an opportunity to explore innovative approaches to protecting livestock, rather than what often just amounts to killing wild dogs. Although some have argued that the status quo of wide-scale indiscriminate baiting should continue until there is unequivocal evidence in support of the beneficial role of dingoes (almost specifically in terms of limiting foxes and cats or other invasive species), this business-as-usual argument doesn't sufficiently acknowledge the extent and gravity of the problem, let alone resolve it. A new discourse is needed that, while balancing the need for caution in relation to dingo reintroduction and conservation programs, also recognises the urgency of the situation.

In 1988, Breckwoldt[2] concluded his much-admired book on dingoes with the following poignant statement: 'If we do not make peace there is a danger of losing the dingo by failing to accept that not everything in Australia was designed only for human benefit' (p. 264). This sobering statement is both a warning of what might happen if we are complacent, and an opportunity to plot a new course away from conflict and towards coexistence. It is fair to say that, despite Breckwoldt's caution a good deal more than a decade ago, it is not a course that has yet been well plotted. Neither, though, is it unchangeable. It therefore remains to be seen whether in the next decade the dingo survives or even thrives, or instead is destined to head towards extinction like so many Australian native mammals before it.

References

1. Hytten KF (2009) Dingo dualisms: exploring the ambiguous identity of Australian dingoes. *Australian Zoologist* **35**, 18–27.
2. Breckwoldt R (1988) *A Very Elegant Animal: The Dingo*. Angus and Robertson, Sydney.

3. Corbett LK (2008) *Canis lupus* ssp. *Dingo*. In *IUCN Red List of Threatened Species. Version 2014.1.* IUCN, Switzerland.
4. Fleming P, Corbett L, Harden R, Thomson P (2001) *Managing the Impacts of Dingoes and other Wild Dogs*. Bureau of Rural Sciences, Canberra.
5. Barker BCW, Macintosh A (1979) The dingo: a review. *Archaeology and Physical Anthropology in Oceania* **14**, 27–53.
6. Johnson CN (2006) *Australia's Mammal Extinctions: A 50,000 Year History*. Cambridge University Press, New York.
7. Coman BJ, Jones E (2007) The loaded dog. *Quadrant* **51**, 10–15.
8. Allen LR, Lee J, Edwards J (1996) Managing feral goats and their impact on Townshend Island in Shoalwater Bay training area. In *Environmentally Responsible Defence*. (Eds P Crabb, J Kesby and L Olive) pp. 79–86. Australian Defence Studies Centre, University of New South Wales, Sydney.
9. Allen LR, Lee J, Gonzalez A (1998) *The Management and Eradication of Feral Goats from Townshend Island*. Department of Defence/Department of Natural Resources, Toowoomba.
10. Hayward M, Marlow N (2014) Will dingoes really conserve wildlife and can our methods tell? *Journal of Applied Ecology* **51**, 835–838.
11. Moseby KE, Neilly H, Read JL, Crisp HA (2012) Interactions between a top order predator and exotic mesopredators in the Australian rangelands. *International Journal of Ecology* **2012**, 250352.
12. Allen BL, Fleming PJS, Allen LR, Engeman RM, Ballard G, Leung LKP (2013) As clear as mud: a critical review of evidence for the ecological roles of Australian dingoes. *Biological Conservation* **159**, 158–174.
13. Ecosure (2012) Fraser island dingo management strategy review. Report to Department of Environment and Heritage Protection, Brisbane.
14. Stephens D (2011) The molecular ecology of Australian wild dogs: hybridisation, gene flow and genetic structure at multiple geographic scales. PhD thesis. University of Western Australia, Perth.
15. Corbett LK (2001) *The Dingo in Australia and Asia*. JB Books, Adelaide.
16. Wagner KK, Schmidt RH, Conover MR (1997) Compensation programs for wildlife damage in North America. *Wildlife Society Bulletin* **25**, 312–319.
17. Daniels MJ, Corbett L (2003) Redefining introgressed protected mammals: when is a wildcat a wild cat and a dingo a wild dog? *Wildlife Research* **30**, 213–218.
18. Kleese D (2002) Contested natures:wolves in late modernity. *Society and Natural Resources* **15**, 313–326.
19. Dickman CR, Lunney D (Eds) (2001) *A Symposium on the Dingo*. Royal Zoological Society of New South Wales, Sydney.

Index

www.ingramcontent.com/pod-product-compliance
Ingram Content Group UK Ltd.
Pitfield, Milton Keynes, MK11 3LW, UK
UKHW061957041026
15591UKWH00016B/1047

9 781486 300297